Bioenergetics 3

Bioenergetics 3

DAVID G. NICHOLLS

Buck Institute for Age Research,
Novato,
California, USA

STUART J. FERGUSON

Department of Biochemistry,
University of Oxford,
W.R. Miller Fellow of St Edmund Hall,
Oxford, UK

ACADEMIC PRESS

An imprint of Elsevier Science

Amsterdam Boston London New York Oxford Paris
San Diego San Francisco Singapore Sydney Tokyo

This book is printed on acid-free paper.

First edition published 1982
Second edition published 1992
Second printing 1997
Third printing 1999
Fourth printing 2001

Academic Press
An imprint of Elsevier Science
84 Theobald's Road, London WC1X 8RR, UK
http://www.academicpress.com

Academic Press
An imprint of Elsevier Science
525 B Street, Suite 1900, San Diego, California 92101-4495, USA
http://www.academicpress.com

ISBN 0-12-518121-3
Library of Congress Catalog Number: 2002101624

A catalogue record for this book is available from the British Library

Typeset by Charon Tec Pvt. Ltd, Chennai, India
Printed and bound in Spain by Grafos SA Arte Sobre Papel, Barcelona
02 03 04 05 06 07 GF 9 8 7 6 5 4 3 2 1

CONTENTS

Colour plates of selected molecular structures are located between pages 126 and 127.

PREFACE

The context for the first edition of this book in 1982 was that Mitchell's chemiosmotic theory of energy transduction had been widely accepted, as acknowledged by the award of the Nobel Prize in 1978, yet the underpinning principles of this theory were widely misunderstood and its full scope was not appreciated. The second edition in 1992 was written against the background that on the one hand many general textbooks still gave too superficial a treatment to chemiosmotic mechanisms, whilst on the other hand the high resolution structure of a bacterial photosynthetic reaction centre that operates according to Mitchell's ideas had recently been reported and recognized with the Nobel Prize in 1988.

Nobel Prizes seem to serve as triggers because this third edition follows the 1997 Nobel prize to Paul Boyer and John Walker for their work on the ATP synthase enzyme. In fact, it is not only the acquisition of structural information for this enzyme that has made the last ten years in Bioenergetics so exciting but also the remarkable developments concerning high resolution structures for many components of respiratory chains, ion translocators and, in 2001, the two photosystems of plant photosynthesis. It is indeed striking that a majority of the presently known membrane protein crystal structures are for 'bioenergetic' proteins.

These developments in themselves warranted preparation of a new version of *Bioenergetics* but an unexpected turn of events has provided an equal stimulus for producing *Bioenergetics 3*. This is the realization that, particularly in the mammalian cell, mitochondria are involved with an increasing number of cell processes beyond just the provision of ATP. A spectacular example is the role of mitochondria in programmed cell death, apoptosis. Such developments have meant that it is increasingly important to be able to study the bioenergetics of mitochondria in the cell rather than only *in vitro*. This has brought a new constituency of scientists to the field of bioenergetics for whom the details of the chemiosmotic mechanism are as relatively unknown as they were to many researchers in bioenergetics around 1970. Thus a number of high profile papers in the complex field of cellular bioenergetics draw erroneous conclusions stemming from this lack of understanding.

A primary purpose of this book, therefore, continues to be the description of the principles of chemiosmotic aspects of membrane bioenergetics. We explain, for example, why the mitochondrial membrane potential cannot vary arbitrarily between values of, say,

200 and 400 mV depending on the state of the cell and why its magnitude must be related to the size of the pH gradient across the membrane. We hope that we have avoided the pitfalls of explaining things too superficially, something that general textbooks, at least one of which still explains uncoupling incorrectly, find hard to avoid given their space limitations. The extension of the bioenergetics of mitochondria into the context of the eukaryotic cell has necessitated the addition of an extra chapter in which both the measurements of bioenergetic parameters for mitochondria *in situ* and the functioning of mitochondria have been included.

The other chapters are completely updated counterparts of those in *Bioenergetics 2*. As noted above, there has been a great increase in molecular information about proteins involved in energy transduction, both at the structural and functional levels. The advent of large quantities of structural information presents new challenges to the authors of a book such as this. One could elect to include very large numbers of multi-colour 3D structures of every relevant protein for which such information is available, but frequently these structures are so complex that the reader, having closed the book, would be able to retain scarcely any manageable information about how structure relates to function. Thus we have sought to produce sketches, in a uniform style, of protein systems for which structures are known; sketches that are designed to be memorable and convey the key functional/mechanistic information. We have, of course, included a selection of 3D structures, but we have chosen only those, or truncated versions of them, that provide direct insights into the function. By this route we hope that readers will grasp the bioenergetic essentials; if they wish to see the full structures, including in some cases subunits that appear to have no bioenergetic function, then these are easily accessible via an appendix.

The advent of genome sequencing has meant that the electron transport systems of many prokaryotes, for instance the disease-causing *Helicobacter pylori*, have been revealed indirectly. Thus the distribution of many different types of oxidases in terminal respiration has become apparent, as has the presence of enzymes that use other electron acceptors, for example the trimethylamine-N-oxide reductase which generates the trimethylamine smell of bad fish. Now that these enzymes are seen to be widespread, and to underpin the physiology of many bacteria, their molecular features have attracted more interest than hitherto; we believe that our expanded chapter on electron transport systems will allow the reader to make at least an initial acquaintance with these systems that are unfamiliar to those who study only mitochondria or chloroplasts.

The world wide web poses both problems and opportunities for authors. The opportunities include the facility to issue corrections and supply supplementary material. The problem is to update a web site sufficiently regularly to make it useful. The web site associated with this book (for details see note to reader) will contain corrections, of which we hope there will not be too many, and periodic, rather than regular, updates on major developments. Readers who would like to notify us of errors are encouraged to do so via the web site. We are also including on the site some material from the second edition that has been omitted from the third edition because of space constraints. For example, we have omitted a section from Chapter 1 that dealt with the historical background, but a modified version of this is included on the web site for those who wish to learn how the subject developed. One of us (SJF) knows from many hours of small group teaching in Oxford how surprisingly difficult many undergraduate students find some aspects of bioenergetics, for example the mode of action of uncouplers. It is intended that if new generations of students find parts of the present text indigestible then the web site will provide further clarifying exposition.

Writing a book such as this is becoming more difficult and not just because of the information explosion. Within the universities writing books is not as well regarded in the sciences as it once was because it can be seen as distracting from research work. Thus there are pressures not to allow preparation of a book to attenuate research effort. Such pressures have meant that we have had to use our spare time (slight as it is!) to write this book and thus we particularly thank our families for their patience and support. But we have inevitably been distracted from time to time during normal working hours and we thank members of our laboratories for their understanding. We are particularly grateful to Dr Vilmos Fülöp of the University of Warwick who kindly prepared several of the pictures of 3D structures for us. Choosing the best angle of view and the colour scheme is almost a new art form and takes longer than most readers will realize.

We have been brave and for most part backed our own judgement of what to include and how to explain it. Doubtless there are places where we have made an erroneous interpretation or omitted something that many others would have expected to have seen included. We alone are responsible for any such disappointments and can only apologize to those affected. As in previous editions, we have not provided extensive references throughout the text. Whilst their inclusion would have permitted the reader immediate access to our source for a point, they would have broken up the text. Consequently, we have mainly restricted ourselves to listing recent reviews at section heads, but from time to time have also listed a specific paper where that seemed warranted. The new Chapter 9 provides more specific references to the papers than the others; this is a consequence of the pace of change in this subject area.

David G. Nicholls
Stuart J. Ferguson
Novato and Oxford
November 2001

NOTE TO THE READER

Two points of nomenclature deserve special attention. First, we have used the symbol Δp for protonmotive force in units of millivolts. In the first edition, as is frequently done elsewhere, we used $\Delta \tilde{\mu}_{H^+}$, but strictly speaking the latter has units of $kJ\,mol^{-1}$ and so we have adopted Δp in this edition. Second, we have defined throughout the side of a membrane to which protons are pumped as the P (positive) side and the side from which they are pumped as the N (negative) side. This allows a uniform nomenclature and overcomes the confusion that can arise when describing the matrix side of the inner mitochondrial membrane as being on the inside in mitochondria but on the outside in inverted submitochondrial particles. We realize that P is also used by electron microscopists to define the protoplasmic side of a membrane, e.g. the interior surface of a bacterial cytoplasmic membrane and that this is the N-side in our convention, but we believe the advantages and increasing use of the P and N nomenclature outweigh any slight chance of the two conventions being confused.

A dedicated website for this book containing appendices and updates can be accessed at http://www.academicpress.com/bioenergetics/.

GLOSSARY

Ac	Acetate
AcAc	Acetoacetate
Acetate	Ethanoate
Acetic acid	Ethanoic acid
ADP/O	The number of molecules of ADP phosphorylated to ATP when two electrons are transferred from a substrate through a respiratory chain to reduce one 'O' ($\frac{1}{2}O_2$) (dimensionless)
ADP/2e$^-$	As ADP/O, except more general, as the final electron acceptor can be other than oxygen (dimensionless)
ANT	Adenine nucleotide translocator
Bchl	Bacteriochlorophyll
bR	Bacteriorhodopsin
Bpheo	Bacteriopheophytin
BQ	Benzoquinone
BQH$_2$	Benzoquinol
C	Flux control coefficient
$[Ca^{2+}]_c$	Cytoplasmic free calcium ion concentration
$[Ca^{2+}]_m$	Mitochondrial matrix free calcium ion concentration
Chl	Chlorophyll
C_MH^+	The effective proton conductance of a membrane or a membrane component (dimensions: nmol H$^+$ min^{-1} mg protein^{-1} mV protonmotive force^{-1})
Cyt	Cytochrome (a haemoprotein in which one or more haems is alternately oxidized and reduced in electron transfer processes). A letter (a, b, etc.) denotes the type of haem, a three-digit subscript indicates an absorbance maximum for the reduced form.
Cyt aa_3	Another name for complex IV (cytochrome c oxidase or cytochrome oxidase)
Cyt bc_1 complex	Another name for complex III (ubiquinol–cytochrome c oxidoreductase)
dO/dt	Respiratory rate (dimensions: nmol O min^{-1} mg protein^{-1})

DAD	Diaminodurene (or 2,3,5,6 tetramethyl-p-phenylenediamine)
DBMIB	2,5-Dibromo-3-methyl-6-isopropylbenzoquinone
DCCD	N, N'-dicyclohexylcarbodiimide (inhibitor of the F_o sector of ATP synthase)
DCMU	3-(3,4-Dichlorophenyl)-1,1-dimethylurea
DCPIP	2,6-Dichlorophenolindophenol
E	Redox potential at any specified set of component concentrations (dimensions: mV)
E^o	Standard redox potential (all components) in their standard states, i.e. 1 M solutions and 1 atm gases (dimensions: mV)
$E^{o\prime}$	Standard redox potential except that pH specified, usually pH = 7 (all other components in their standard states, i.e. 1 M solutions and 1 atm gases) (dimensions: mV)
E_m	Mid-point potential at a defined pH (equivalent to E^o at pH = 0 because concentrations of oxidized and reduced species cancel, and thus their ratio is unity, as in E^o definition) (dimensions: mV)
$E_{m,7}$	Mid-point potential at pH = 7 (also equivalent to $E^{o\prime}$, because concentrations of oxidized and reduced species cancel, and thus their ratio is unity as in $E^{o\prime}$ definition) (dimensions: mV)
E_h	Actual redox potential at a defined pH (dimensions: mV)
$E_{h,7}$	Actual redox potential at pH = 7 (dimensions: mV)
EP(S)R	Electron paramagnetic (spin) resonance
ER	Endoplasmic reticulum
ETF	Electron-transferring flavoprotein
F	Faraday constant ($= 0.0965\,\mathrm{kJ\,mol^{-1}\,mV^{-1}}$); to convert from mV to $\mathrm{kJ\,mol^{-1}}$, multiply by F
$F_1.F_o$	The proton translocating ATP synthase/ATPase (meaning fraction one/fraction oligomycin)
FCCP	Carbonyl cyanide p-trifluoromethoxyphenylhydrazone
Fd	Ferredoxin
Fe/S	Iron–sulfur centre
Ferricyanide	Hexacyanoferrate (III)
Ferrocyanide	Hexacyanoferrate (II)
FTIR	Fourier transform infrared spectroscopy
G	Gibbs (free) energy
GSH	Reduced monomeric glutathione
GSSG	Oxidized dimeric glutathione with a disulfide bond between the monomers
H	Enthalpy
H^+/ATP	The number of protons translocated through the ATP synthase for the synthesis of one molecule of ATP (dimensionless) (usually refers to ATP synthase alone but, in the case of mitochondria, may also subsume H^+ movement associated with adenine nucleotide and phosphate translocation)
H^+/O	The number of protons translocated by a respiratory chain during the passage of two electrons from substrate to oxygen (dimensionless)

$H^+/2e^-$	As H^+/O, except more general, as final electron acceptor need not be oxygen (dimensionless)
hv	The energy in a photon (dimensions: kJ)
J_{H^+}	Proton current (dimensions: nmol H^+ min^{-1} mg protein^{-1})
K	Absolute equilibrium constant
K'	Apparent equilibrium constant
LH 1, LH 2	Bacterial light-harvesting complexes 1 and 2
LHC II	A major thylakoid light-harvesting complex
MCA	Metabolic control analysis
MGD	Molybdopterin guanine dinucleotide (a cofactor that chelates Mo at several enzyme active sites)
MPP^+	1-Methyl-4-phenyl-pyridinium ion
MPT	Mitochondrial permeability transition
MQH_2/MQ	Menaquinol/menaquinone
mV	Millivolt
MV^+	Reduced methyl viologen
MV^{2+}	Oxidized methyl viologen
N-side/N-phase	Negative side of a membrane from which protons are pumped
Nbf-Cl	4-Chloro-7-nitrobenzofurazan
NMDA	N-Methyl-D-aspartate
NMR	Nuclear magnetic resonance
Nuo	NADH–ubiquinone oxidoreductase
O	$\frac{1}{2}O_2$
OSCP	Oligomycin sensitivity conferral protein
Oxidase	A haem-containing (usually) protein that binds and reduces oxygen, generally to water; oxygen reductase is the function
P/O	As ADP/O
$P/2e^-$	As ADP/$2e^-$
P-side/P-phase	Postive side of a membrane to which protons are pumped
P_{870}, etc.	The primary photochemically active component in a reaction centre
P_i	Phosphate anion (ionization state not specified)
PC	Plastocyanin
Pheo	Pheophytin
pmf	Protonmotive force (dimension: mV)
PMS	Phenazinemethosulphate
PQ	Plastoquinone
PQH_2	Plastoquinol
PSI, PSII	Photosystem I, II
PTP	Permeability transition pore
q^+/O	The number of charges translocated across a membrane when two electrons are transferred from a substrate to oxygen via an electron transport system
$q^+/2e^-$	As q^+/O, except more general, so as to specify the movement of electrons through any segment of an electron transport system
Q_p, Q_n	Ubiquinone/ubiquinol binding sites towards the P- or N-sides, respectively, of complex III (cytochrome bc_1 complex)

R	The gas constant ($8.3\,\mathrm{J\,mol^{-1}K^{-1}}$)
S	Entropy
SHAM	Salicylhydroxamic acid
SMP	Submitochondrial particle
TMPD	N,N,N',N'-tetramethyl-p-phenylenediamine; a redox mediator, especially between ascorbate and cytochrome c
$TPMP^+$	Triphenylmethyl phosphonium cation
TPP^+	Tetraphenyl-phosphonium cation
UCP	Uncoupling protein
UQ, UQH_2	Ubiquinone, ubiquinol
$UQ^{\bullet-}$	Ubisemiquinone
VDAC	Voltage-dependent anion channel
Γ	Mass action ratio
Γ'	Apparent mass action ratio
Δp	Protonmotive force (dimensions: mV)
$\Delta\psi$	Membrane potential, i.e. the electrical potential difference between two bulk phases separated by a membrane (dimensions: mV)
ΔpH	The pH difference between two bulk phases on either side of a membrane (dimensionless)
ΔE_h	Difference between two redox couples (dimensions: mV)
ΔG	Gibbs energy change at any specified set of reactant and product concentrations (activities)
ΔG^o	Standard Gibbs energy change when all reactants and products are in their standard states (i.e. 1 M for solutes, pure liquid for solvents and 1 atm for gases (dimensions: $\mathrm{kJ\,mol^{-1}}$)
$\Delta G^{o\prime}$	Standard Gibbs energy change, except that H^+ concentration is 10^{-7} (i.e. pH = 7) rather than 1 (i.e. pH = 0)
ΔG_p	The phosphorylation potential, i.e. the ΔG for ATP synthesis at any given set of ATP, ADP and P_i concentrations (dimensions: $\mathrm{kJ\,mol^{-1}}$)
ΔH	Enthalpy change
ΔS	Entropy change
$\Delta\tilde{\mu}_{x+}$	Ion electrochemical gradient (dimensions: $\mathrm{kJ\,mol^{-1}}$)
$\Delta\tilde{\mu}_{H+}$	Proton electrochemical gradient (dimensions: $\mathrm{kJ\,mol^{-1}}$)
$\Delta\psi_m$, $\Delta\psi_p$	Mitochondrial and plasma membrane potentials (in the context of intact cell bioenergetics)
$\delta O/\delta t$	Respiratory rate (dimensions: typically $\mathrm{nmol\,O\,mg^{-1}\,min^{-1}}$)
ε	Elasticity coefficient
/	Antiporter
:	Symporter

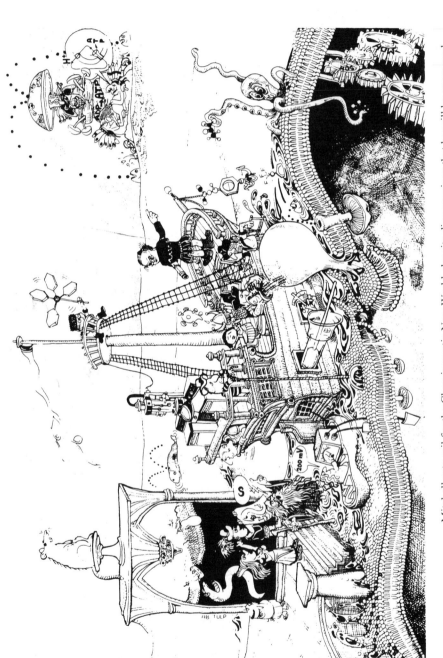

Mitchell sets sail for the Chemiosmotic New World, despite dire warnings that he will be consumed

1 CHEMIOSMOTIC ENERGY TRANSDUCTION

1.1 INTRODUCTION

All biochemical reactions involve energy changes; thus the term 'bioenergetics' could validly be applied to the whole of life sciences. Bioenergetics as a discipline rose to prominence in the 1950s as a highly directed search for the solution to the mechanism by which energy made available by the oxidation of substrates, or the absorption of light, could be coupled to 'uphill' reactions such as the synthesis of ATP from ADP and P_i, or the accumulation of ions across a membrane. Within this narrow definition of bioenergetics the central concept – the chemiosmotic theory, which forms the core of this book – has been firmly established for more than a decade. The second edition of this book, published in 1992, was written in a period when bioenergetics was considered to be a relatively quiescent field, suffering from the very success of research in the period 1965–1980. How dramatically this has changed in the past 10 years is evidenced by the enormous advances that have been made both in the elucidation of the molecular mechanisms of the protein complexes, many of which are now understood at atomic levels of resolution, and at the other extreme by the explosion of the field of 'mitochondrial physiology' – the investigation of the role of mitochondria in the healthy and diseased cell. Indeed the mitochondrion, which was once considered so reliable that it was futile to investigate mitochondrial dysfunction, is now revealed to be an organelle operating close to its design limits and extremely prone to damage, with potentially disastrous consequences for the organelle itself and its host cell. More and more disorders, particularly the chronic neurodegenerative diseases, stroke and heart reperfusion injury, are being associated to a greater or lesser extent with mitochondrial dysfunction.

1.2 THE CHEMIOSMOTIC THEORY: FUNDAMENTALS

Although some ATP synthesis is catalysed by soluble enzyme systems, by far the largest proportion is associated with membrane-bound enzyme complexes that are restricted to a particular class of membrane. These 'energy-transducing' membranes include the plasma membrane of simple prokaryotic cells such as bacteria or blue–green algae, the inner

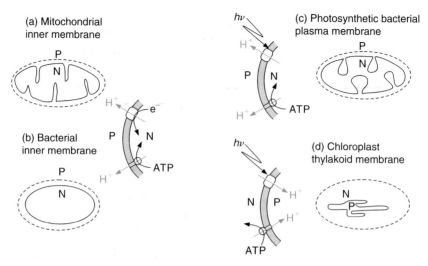

Figure 1.1   Energy-transducing membranes contain pairs of proton pumps with the same orientation.
In each case the primary pump utilizing either electrons (e^-) or photons ($h\nu$) pumps protons from the N (negative) compartment to the P (positive) compartment. Note that the ATP synthase in each case is shown acting in the direction of ATP hydrolysis, when it would also pump protons from the N- to the P-phase.

membrane of mitochondria and the thylakoid membrane of chloroplasts (Fig. 1.1). These membranes have a related evolutionary origin, since chloroplasts and mitochondria are commonly thought to have evolved from a symbiotic relationship between a primitive, non-respiring eukaryotic cell and an invading prokaryote. Thus the mechanism of ATP synthesis and ion transport associated with these diverse membranes is sufficiently related, despite the differing natures of their primary energy sources, to form the core of classical bioenergetics.

Energy-transducing membranes possess a number of distinguishing features. Each membrane has two distinct types of proton pump. The nature of the primary proton pump depends on the energy source used by the membrane. In the case of mitochondria or respiring bacteria, an electron transfer chain catalyses the 'downhill' transfer of electrons from substrates to final acceptors such as O_2 and uses this energy to generate a gradient of protons (details of this will be covered in Chapter 5). Photosynthetic bacteria exploit the energy available from the absorption of quanta of visible light to generate a gradient of protons, while chloroplast thylakoids not only accomplish this but also drive electrons 'uphill' from water to acceptors such as $NADP^+$ (Chapter 6). The topologies of the membranes differ and, to facilitate comparison, it is a useful convention to define the side of the membrane to which protons are pumped as the P or positive side, and the side from which they have originated as the N or negative side (Fig. 1.1).

In contrast to the variety of primary proton pumps, all energy-transducing membranes contain a highly conserved secondary proton pump termed the ATP synthase or the H^+-translocating ATPase (Chapter 7). If this pump was operating in isolation in a membrane, it would hydrolyse ATP to ADP and P_i and pump protons in the same direction as the

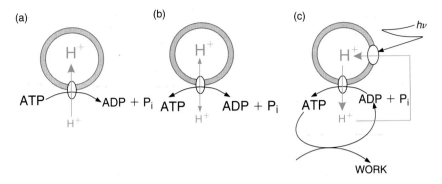

Figure 1.2   A hypothetical 'thylakoid' to demonstrate chemiosmotic coupling.
An ATP synthase complex is incorporated into a phospholipid membrane such
that the ATP binding site is on the outside. (a) ATP is added, the nucleotide starts
to be *hydrolysed* to ADP + P_i and protons are pumped into the vesicle lumen. As
ATP is converted to ADP + P_i the energy available from the hydrolysis steadily
decreases, while the energy required to pump further protons against the gradient which has already been established steadily increases. (b) Eventually an
equilibrium is attained. (c) If this equilibrium is now disturbed, for example, by
removing ATP, the ATP synthase will reverse and attempt to re-establish the equilibrium by *synthesizing* ATP. Net synthesis, however, would be very small as the
gradient of protons would rapidly collapse and a new equilibrium would be
established. For continuous ATP synthesis, a primary proton pump, driven in this
example by photons ($h\nu$), is required to pump protons across the same membrane
and replenish the gradient of protons. *A proton circuit has now been established.*
This is what occurs across energy-conserving membranes: ATP is continuously
removed for cytoplasmic ATP-consuming reactions, while the gradient of protons,
Δp, is continuously replenished by the respiratory or photosynthetic electron-
transfer chains.

primary pump (Fig. 1.1). However, the essence of the chemiosmotic theory is that the primary proton pump generates a sufficient gradient of protons to force the secondary pump
to reverse and *synthesize* ATP from ADP and P_i (Fig. 1.2). It should be noted that metabolism (i.e. electron flow or phosphorylation) within both the primary and secondary pumps
is tightly coupled to proton translocation: the one cannot occur without the other.

What do we mean by a gradient of protons? The quantitative thermodynamic measure is
the proton electrochemical gradient $\Delta\tilde{\mu}_{H^+}$. An ion electrochemical gradient, expressed in
$kJ\,mol^{-1}$, is a thermodynamic measure of the extent to which an ion gradient is removed
from equilibrium (and hence capable of doing work) and will be derived in Chapter 3.
For the present it is sufficient to note that $\Delta\tilde{\mu}_{H^+}$ has two components: one due to the concentration difference of protons across the membrane ΔpH and one due to the difference in
electrical potential between the two aqueous phases separated by the membrane, the mem-
brane potential, $\Delta\psi$. A bioenergetic convention is to convert $\Delta\tilde{\mu}_{H^+}$ into units of electrical
potential, i.e. millivolts, and to refer to this as the *protonmotive force*, or *pmf*, expressed
by the symbol Δp.

In only a few cases, such as the chloroplast, does Δp exist mainly as a pH difference
across the energy-conserving membrane. In this example, the pH gradient, ΔpH, across the
thylakoid membrane can exceed 3 units. Although the thylakoid space is therefore highly

acidic, there are no enzymes in this compartment that might be compromised by the low pH. The more common situation is where $\Delta\psi$ is the dominant component and the pH gradient is small: perhaps only 0.5 pH units. This occurs, for example, in the mitochondrion, allowing enzymes in both the mitochondrial matrix and cell cytoplasm to operate close to neutral pH.

Figure 1.2 constructs a hypothetical ATP-synthesizing organelle from first principles. A central feature is the *proton circuit* linking the primary pump with the ATP synthase. *The most important single concept to grasp is that the proton circuit* (Fig. 1.3) *is closely analogous to an electrical circuit, and the analogy holds even when discussing detailed and complex energy flows* (see Chapter 4). Thus:

(a) Both circuits have generators of potential difference (the battery and the respiratory chain, respectively).
(b) Both potentials (voltage difference and Δp) can be expressed in millivolts.
(c) Both potentials can be used to perform useful work (running the light bulb and ATP synthesis, respectively).
(d) The current flowing in both circuits (amps or proton flux, J_{H^+}) is defined by Ohm's law (i.e. current = voltage/resistance).
(e) The rate of chemical conversion in the battery (or respiratory chain) is tightly linked to the current of electrons (or protons) flowing in the rest of the circuit, which in turn depends on the resistance of the circuit.
(f) Both circuits can be shorted (by, respectively, a piece of wire or a protonophore – an agent which makes membranes permeable to protons, see Chapter 2).
(g) The potentials fall as the currents drawn increase.

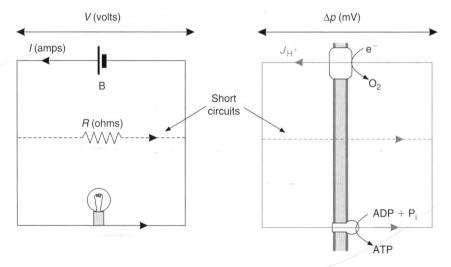

Figure 1.3   Proton circuits and electrical circuits are analogous.
A simple electrical circuit comprising battery and light bulb is analogous to a basic proton circuit. Voltage (Δp equivalent to V), current (J_{H^+} equivalent to I) and conductance $C_M H^+$ (equivalent to electrical conductance – reciprocal ohms) terms can be derived. Short-circuits have similar effects and more complex circuits with parallel batteries can be devised to mimic the multiple proton pumps in the mitochondrion (see Chapter 4).

To avoid short-circuits, it is evident that the membrane must be closed and possess a high resistance to protons. Protonophores, also called *uncouplers*, are synthetic compounds which break the energetic coupling between the primary pump and the ATP synthase. Uncouplers were described long before the chemiosmotic theory was propounded, and one of the most successful predictions of the theory was that they act by increasing the proton conductance of the membrane and inducing just such a short-circuit (Fig. 1.3).

Mitochondrial and bacterial membranes have not only to maintain a proton circuit across their membranes, but must also provide mechanisms for the uptake and excretion of ions and metabolites. It is energetically unfavourable for a negatively charged metabolite to enter the negative interior of a mitochondrion or bacterium (see Chapter 3), and transport systems have evolved in which metabolites are transported together with protons, or by an equivalent exchange with OH^-. Alternatively, components of Δp can be exploited in other ways so as to drive transport in the desired direction (see Chapter 8).

1.3 THE BASIC MORPHOLOGY OF ENERGY-TRANSDUCING MEMBRANES

1.3.1 Mitochondria and submitochondrial particles

Review Frey and Mannella 2000

The classical mitochondrial cross-section (Fig. 1.4) is obtained from thin sections viewed under the electron microscope. Their shape is not fixed but can change continuously in the cell, and the appearance of the cristae can be quite different in mitochondria isolated from

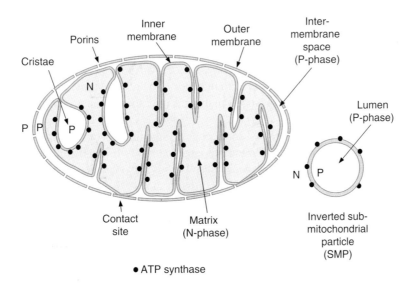

Figure 1.4 Schematic representation of a typical mitochondrion and sub-mitochondrial particle.
P and N refer to the positive and negative compartments. Note that the shape of the cristae is highly variable and that communication between cristae and inter-membrane space may be restricted.

different tissues or even with the same mitochondria suspended in different media. Thus heart mitochondria, for which periods of high respiratory activity are required, tend to have a greater surface area of cristae than liver mitochondria. This view, in which there is a relatively free connection between the cristal space and that between the inner and outer membranes (the intermembrane space) is now being challenged as a result of high-resolution electron microscopy of serial sections and reconstruction of the three-dimensional (3D) structure by computer tomography. The cristae are revealed as tortuous structures with only small tubular contacts with the intermembrane space. It is unclear whether this restriction is sufficient to limit the interchanges of metabolites between the cristal and intermembrane spaces. In some cells, mitochondria appear to be fused into a continuous reticulum, whereas in others, such as neurons, they are discrete filaments which are independently mobile along the axons and dendrites. In many cells, mitochondria appear to be in close contact with elements of endoplasmic reticulum (ER), and this may aid the rapid exchange of Ca^{2+} between the ER and mitochondria (Chapter 9).

The outer mitochondrial membrane possesses proteins, termed porins, which act as non-specific pores for solutes of molecular weight less than 10 kDa, and is therefore freely permeable to ions and most metabolites. The mitochondrial porin is also termed the 'voltage-dependent anion channel' or VDAC, although it should be emphasized that there is no potential gradient across the highly permeable outer membrane and the voltage dependency is only seen in synthetic reconstitution experiments.

The inner membrane is energy transducing. In mitochondrial preparations that have been negatively stained with phosphotungstate, it is possible to see knobs on the matrix face (N-side; Fig. 1.4) of the inner membrane. These are the catalytic components of the ATP synthase where adenine nucleotides and phosphate bind. The enzymes of the citric acid cycle are in the matrix, except for succinate dehydrogenase, which is bound to the N-face of the inner membrane. It must be borne in mind that the concentration of protein in the matrix can approach 500 mg ml^{-1} (*sic*) and there may be a considerable structural organization within this enormously concentrated solution that more closely resembles a glue than an ideal dilute medium. The matrix pools of NAD^+ and $NADP^+$ are separate from those in the cytosol, while matrix ADP and ATP communicate with the cytoplasm through the adenine nucleotide exchanger (Chapter 8). Additionally, specific carrier proteins exist for the transport of many metabolites.

Mitochondria are usually prepared by gentle homogenization of the tissue in isotonic sucrose (for osmotic support and to minimize aggregation) followed by differential centrifugation to separate mitochondria from nuclei, cell debris and microsomes (fragmented ER). Although this method is effective with fragile tissues such as liver, tougher tissues such as heart must either first be incubated with a protease, such as nagarse, or be exposed briefly to a blender to break the muscle fibres. Yeast mitochondria are isolated following digestion of the cell wall with snail-gut enzyme.

Ultrasonic disintegration of mitochondria produces inverted submitochondrial particles (SMPs) (Fig. 1.4). Because these have the substrate binding sites for both the respiratory chain and the ATP synthase on the outside, they have been much exploited for investigations into the mechanism of energy transduction.

Finally, increasingly sophisticated techniques are being developed to investigate mitochondrial function *in situ* within the cell, using mainly fluorescence techniques. These approaches will be discussed in Chapter 9.

1.3.2 Respiratory bacteria and derived preparations

Energy transduction in bacteria is associated with the cytoplasmic membrane (Fig. 1.5). In Gram-negative bacteria (which are typically of similar size to mitochondria) this membrane is separated from a peptidoglycan layer and an outer membrane by the periplasm, which is approximately 100 Å wide. In Gram-positive bacteria the periplasm is absent and the cell wall is closely juxtaposed to the cytoplasmic membrane. Figure 1.5 is an oversimplification because in some organisms with a very high rate of respiration there are substantial infoldings of the cytoplasmic membrane. The archaea are an evolutionary distinct group of bacteria, but which nevertheless catalyse energy transduction processes on their cytoplasmic membranes via a chemiosmotic mechanism.

It is difficult to study energy transduction with intact bacteria because:

(a) Many reagents do not penetrate the outer membrane of Gram-negative organisms;
(b) ADP, ATP, NAD^+ and NADH do not cross the cytoplasmic membrane;
(c) cells are frequently difficult to starve of endogenous substrates and thus there can be ambiguity as to the substrate which is donating electrons to a respiratory chain;
(d) finally, the study of transport can be complicated by subsequent metabolism of the substrate.

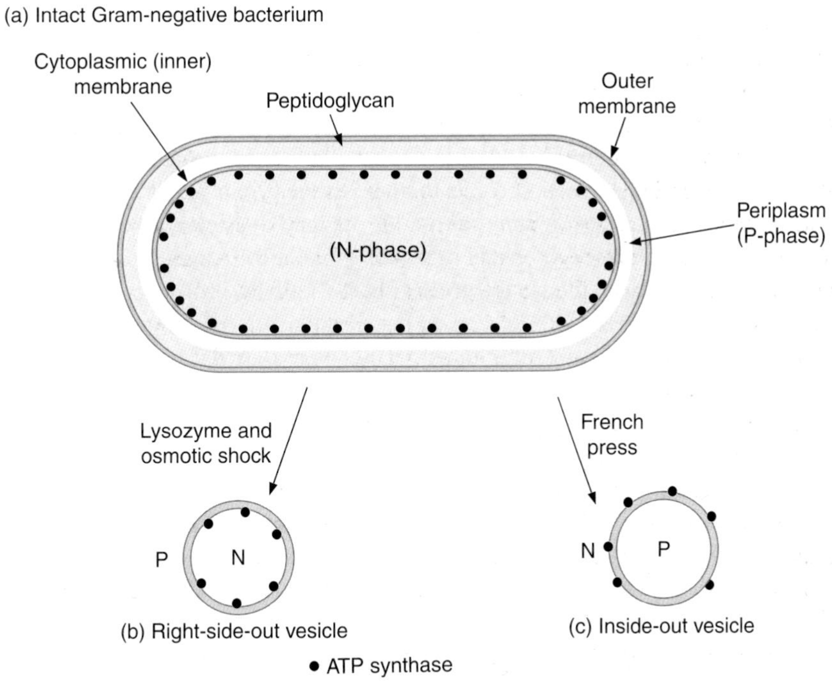

(a) Intact Gram-negative bacterium

Cytoplasmic (inner) membrane

Peptidoglycan

Outer membrane

(N-phase)

Periplasm (P-phase)

Lysozyme and osmotic shock

French press

P N

N P

(b) Right-side-out vesicle

(c) Inside-out vesicle

● ATP synthase

Figure 1.5   Gram-negative bacteria and vesicle preparations.
P and N refer to positive and negative compartments. The periplasm is part of the P-phase, which also includes the bulk external medium, since the outer membrane is freely permeable to ions. Note that Gram-positive bacteria differ by lacking an outer membrane and a periplasm. Nevertheless, similar vesicle preparations can be made from these organisms as is also the case for the archaea.

Cell-free vesicular systems can overcome these problems. For most transport studies *right-side-out vesicles* are required. These can often be obtained by weakening the cell wall, e.g. with lysozyme, and then exposing the resulting spheroplasts or protoplasts to osmotic shock. Vesicles with this orientation can only oxidize substrates that have an external binding site or can permeate the cytoplasmic membrane. They cannot hydrolyse or synthesize ATP, in contrast to *inside-out vesicles*, which can frequently be prepared by extruding cells at very high pressure through an orifice in a French press. These vesicles can oxidize NADH and phosphorylate added ADP. The method of vesicle preparation varies between genera; occasionally, osmotic shock may give inside-out vesicles or a mixture of the two orientations. This last feature need not be a major problem because, for example, in a study of ATP synthesis, the reaction would be confined to the inside-out population (see above). Nevertheless, failure to characterize the orientation of vesicles has caused confusion in the past.

Vesicle preparations have some disadvantages, such as the loss of periplasmic electron-transport or solute-binding proteins; the latter play key roles in many aspects of bacterial energy transduction (Chapters 5 and 8). Also the membrane of a vesicle may be somewhat leaky with the result that the stoichiometry of an energy transduction reaction may be adversely affected.

1.3.3 Chloroplasts and their thylakoids

Chloroplasts are plastids, organelles peculiar to plants (Fig. 1.6); there may be from one to a hundred or more chloroplasts per cell. Chloroplasts are considerably larger than the average mitochondrion, being 4–$10\,\mu$m in diameter and 1–$2\,\mu$m thick and bounded by an envelope of two closely juxtaposed membranes, the matrix within the inner membrane being the stroma (Fig. 1.6). Within stroma are flattened vesicles called thylakoids, the membranes of which have regions that are folded so that the contiguous membrane has a stacked appearance, referred to as the grana (Fig. 1.6). Energy conservation occurs across the thylakoid membranes and light causes the translocation of protons into the internal thylakoid spaces (usually called the lumen). The chloroplast ATP synthase is part of the thylakoid membrane and is orientated with its 'knobs' on the stromal face of the membrane. Thus the lumen space inside the thylakoid is the P-compartment and the stroma the N-compartment. The ATP and NADPH generated by photosynthetic phosphorylation is used by the CO_2-fixing dark reactions of the Calvin cycle located in the stroma.

Although at first sight the structure of chloroplasts appears to be very different from that of mitochondria, the only topological distinction is that the thylakoids, in contrast to the mitochondria cristae, can be thought of as having become separated from the inner membrane, with the result that the thylakoid lumen is a separate compartment, unlike the 'cristal space', which is continuous with the intermembrane space of mitochondria. Note, however, that even in mitochondria there are suggestions that communication between cristal and intermembrane spaces may be restricted (Section 1.3.1).

Chloroplasts are prepared by gentle homogenization in isotonic sucrose or sorbitol of leaves (e.g. from peas, spinach or lettuce), but avoiding material rich in polyphenols or acid. After removal of cell debris, the chloroplasts are sedimented by low-speed centrifugation. A rapid and careful preparation will contain a high proportion of intact chloroplasts capable

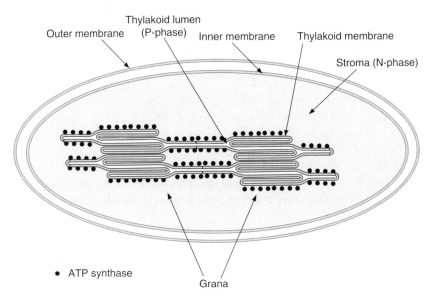

Thylakoid lumen
Outer membrane   (P-phase)   Inner membrane   Thylakoid membrane

Stroma (N-phase)

• ATP synthase

Grana

Figure 1.6   Chloroplasts and their thylakoids.
Note it is probable that there is a single continuous lumen (the internal thylakoid space). The thylakoid membrane is heterogeneous with, for example, the ATP synthase being excluded from the grana (appressed regions) where the membrane is closely stacked (see Chapter 6). Light-driven proton pumping occurs from the N- to the P-phase (note, however, that in steady-state light the membrane potential across a thylakoid membrane is negligible and that the pH gradient dominates – see Chapter 6).

of high rates of CO_2 fixation. Slightly harsher conditions yield 'broken chloroplasts', which have lost the envelope membranes and hence the stroma contents. These broken chloroplasts (thylakoid membrane preparations) do not fix CO_2 but are capable of high rates of reduction of artifical electron acceptors and of photophosphorylation. They are often the choice material for bioenergetic investigations because the chloroplast envelope prevents access of substances such as ADP or $NADP^+$.

1.3.4 Photosynthetic bacteria and chromatophores

Three groups of prokaryotes catalyse photosynthetic electron transfer: the green bacteria, the purple bacteria, and the cyanobacteria (or blue–green algae). The purple bacteria are divided into two groups: the Rhodospirillaceae (or non-sulfur), and the Chromatiaceae (or sulfur). Cyanobacteria carry out non-cyclic electron transfer (Chapter 6), use H_2O as electron donor, and are in this respect similar to chloroplasts. Of the remaining groups, the purple bacteria, and especially the Rhodospirillaceae, have been the more intensively investigated, and several factors make them suitable for bioenergetic studies. Thus mechanical disruption of the cells (e.g. in a French press) enables the characteristic invaginations of the cytoplasmic membrane to bud off and form isolated closed vesicles called chromatophores (Fig. 1.7). Chromatophores retain the capacity for photosynthetic energy transduction and possess the same orientation as the inside-out vesicles discussed in

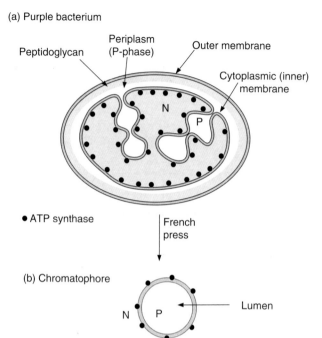

(a) Purple bacterium

Peptidoglycan

Periplasm (P-phase)

Outer membrane

Cytoplasmic (inner) membrane

N

P

● ATP synthase

French press

(b) Chromatophore

N   P

Lumen

Figure 1.7   Photosynthetic bacteria and chromatophores.
The cytoplasmic membrane of photosynthetically grown organisms such as *Rhodobacter sphaeroides* is highly invaginated. When the cells are forced through a narrow orifice at high pressure (the French press), the membranes pinch off as shown to give chromatophores.

Section 1.3.2. Light-driven ATP synthesis can be studied, and they have been important for chemiosmotic studies, especially since they are so small (diameters of the order of 500 Å) that light scattering is negligible and suspensions are optically clear.

A further advantage of the purple bacteria is that the reaction centres (the primary photochemical complexes) can be readily isolated (Chapter 6). Finally, these organisms will grow in the dark, for example, by aerobic respiration, permitting the study of mutants defective in the photosynthetic apparatus.

In addition to these bacteria, members of the archaea called halobacteria carry out a unique light-dependent energy transduction in which a single protein, bacteriorhodopsin, acts as a light-driven proton pump (Chapter 6).

1.3.5 Reconstituted systems

An essential feature of the chemiosmotic theory is that the primary and secondary proton pumps should be functionally and structurally separable. In order to observe proton translocation, the purification of proton-translocating complexes must be followed by their reincorporation into synthetic, closed membranes that have low permeabilities to ions. Historically, such 'reconstitutions' allowed aspects of the chemiosmotic theory to be tested, such as whether each complex was capable of pumping protons as an autonomous unit. Currently,

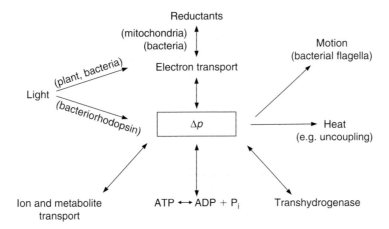

Figure 1.8   Pathways of energy transduction.
The protonmotive force interconnects multiple forms of energy.

reconstitution is an important technique to investigate mechanism, cofactors, etc. for respiratory chain complexes and metabolite transporters.

Membrane proteins are generally purified following solubilization of the membrane with a detergent (usually non-ionic) that disrupts protein–lipid but not protein–protein interactions. Once purified, there are two principal ways in which they can be reconstituted into a membrane structure. The first is to mix with phospholipids the purified protein dispersed in a suitable detergent, preferably one with a high critical micellar concentration (the concentration at which micelles form from monomers in solution), and then to allow the concentration of detergent to fall slowly either by dialysis or gel filtration. Under optimal conditions this can lead to the formation of unilamellar phospholipid vesicles.

The protein can in principle be oriented in either of two ways. If the protein uses a substrate, e.g. ATP, to which the phospholipid bilayer is impermeable, then mixed orientation is not a problem because only those molecules with their catalytic site facing outward will be accessible to the substrate. On the other hand, if the protein is a photosynthetic system, then asymmetry can obviously not be imposed in this way. Fortunately, proteins frequently orient asymmetrically, since the differences in radius of curvature for the two sides of the vesicle may be an important factor. A more demanding type of reconstitution is when the presence of two different proteins (e.g. a primary and secondary pump) is required in the same membrane. The problem here is to ensure not only that at least a majority of vesicles contain both proteins, but also that the relative orientations of the two proteins allow coupling between them via the proton circuit. An example is given in Chapter 4 (Fig. 4.16).

A second procedure for reconstitution is to incorporate the purified protein into a planar bilayer, which can be formed over a tiny orifice that separates two reaction chambers. The insertion of protein is frequently achieved by fusing phospholipid vesicles containing the protein of interest with the planar bilayer. Alternatively, in some cases it has been possible to form the bilayer directly by application of a protein–phospholipid mixture in a suitable volatile solvent to the aperture. The amount of enzyme incorporated into such bilayers is usually so small that biochemical or chemical assays of activity are not possible. However, the crucial advantage of this type of system is that macroscopic electrodes can be inserted into the two chambers and thus direct electric measurements (either current or voltage) of any ion or electron movements driven by the reconstituted protein can be made.

1.4 OVERVIEW

Figure 1.8 summarizes the very wide range of different processes capable of generating or being driven by the protonmotive force. Chapters 2–4 will deal with the ways in which these processes are linked, Chapters 5–8 will cover the molecular mechanisms and Chapter 9 will discuss some of these findings in the context of the intact cell.

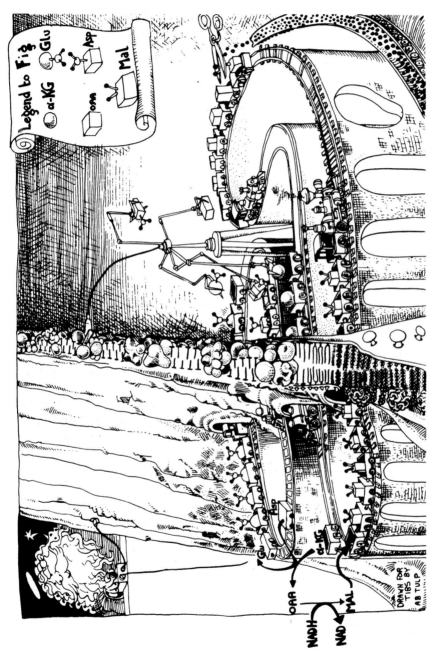

The malate–asparate shuttle enables reducing equivalents from cytosolic NADH to be transported into the mitochondrial matrix (Fig. 8.4)

2 ION TRANSPORT ACROSS ENERGY-CONSERVING MEMBRANES

2.1 INTRODUCTION

The chemiosmotic theory requires that the transport of ions be considered an integral part of bioenergetics. It was this more than any other single factor that helped to remove the artificial distinction, present in the early days of bioenergetics, between events occurring in energy-conserving membranes, which were considered the valid preserve of the bioenergeticist, and closely related transport events occurring in the eukaryotic plasma membrane, endoplasmic reticulum (ER), secretory vesicles, etc., which were assigned to a different field of research.

In this chapter we shall describe the basic permeability properties of membranes and the abilities of ionophores to induce additional pathways of ion permeation, bearing in mind that what is discussed is equally applicable to energy-conserving and non-energy-conserving membranes – after all, the only significant distinction between the two is the presence of the proton pumps in the former.

> For an ion to be transported across a membrane both a driving force and a pathway are required. Driving forces can be metabolic energy (such as ATP hydrolysis), concentration gradients, electrical potentials, or combinations of these. These forces will be discussed in Chapter 3; this chapter will deal with the natural and induced pathways in energy-conserving membranes.

2.2 THE CLASSIFICATION OF ION TRANSPORT

To reduce the complexity of membrane transport events, it is useful to classify any transport process in terms of the following four criteria:

1. Does transport occur across the bilayer or is it protein-mediated (Fig. 2.1a)?

2. Is transport passive or *directly* coupled to metabolism (Fig. 2.1b)?
3. Does the transport process involve a single ion or metabolite, or are fluxes of two or more species *directly* coupled (Fig. 2.1c)?
4. Does the transport process involve net charge transfer across the membrane (Fig. 2.1d)?

2.2.1 Bilayer-mediated versus protein-catalysed transport

A consequence of the fluid-mosaic model of membrane structure is that transport can occur either through lipid bilayer regions of the membrane or be catalysed by integral, membrane-spanning proteins. The distinction between protein-catalysed transport and transport across the bilayer regions of the membrane is fundamental and will be emphasized in this chapter.

While the fluid-mosaic model is usually represented with protein 'icebergs' floating in a sea of lipid, the high proportion of protein in energy-conserving membranes (in the case of the mitochondrial inner membrane 50% of the membrane is integral protein, 25% peripheral protein and 25% lipid) results in a relatively close packing of the proteins. Unlike plasma membrane proteins, no attachment to cytoskeletal elements occurs.

Consistent with the proposal that mitochondria and chloroplasts evolved from respiring or photosynthetic bacteria, energy-conserving membranes tend to have distinctive lipid compositions: 10% of the mitochondrial inner membrane lipid is cardiolipin, while only 16% of the chloroplast thylakoid membrane lipid is phospholipid, the remainder being galactolipids (40%), sulfolipids (4%) and photosynthetic pigments (40%). Despite this heterogeneity of lipid composition, the native and ionophore-induced permeability properties of the bilayer regions of the different membranes are sufficiently similar to justify extrapolations between energy-transducing membranes and artificial bilayer preparations. However, protein-catalysed transport can be unique, not only to a given organelle but also to an individual tissue, depending on the genes expressed in that cell. For example, the inner membranes of rat liver mitochondria possess protein-catalysed transport properties which are absent in mitochondria from rat heart (see Chapter 8).

2.2.2 Transport directly coupled to metabolism versus passive transport

A tight coupling of transport to metabolism occurs in the ion pumps, which are central to chemiosmotic energy transduction and the mechanism of the ATP-hydrolysing Na^+ and Ca^{2+} pumps in non-mitochondrial membranes. Ions can be accumulated without direct metabolic coupling if there is a membrane potential or if transport is coupled to the 'downhill' movement of a second ion. For example, while Ca^{2+} is accumulated into the sarcoplasmic reticulum by an ion pump (the Ca^{2+}-ATPase), the same ion is accumulated across the mitochondrial inner membrane by a uniport mechanism (Fig. 2.1) driven by the membrane potential (see Chapter 8). Only the former is strictly 'active', since mitochondrial Ca^{2+} accumulation occurs down the electrochemical gradient for the ion. In some texts the terms primary and secondary active transport are used to distinguish these examples. However, confusingly, the term 'active transport' is often used for any process in which a concentration

(a) Does transport occur across the bilayer or is it protein-mediated?

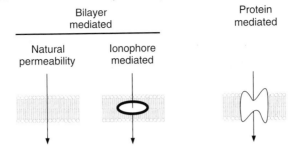

Bilayer mediated

Protein mediated

Natural permeability

Ionophore mediated

(b) Is transport passive or directly coupled to metabolism?

Passive

Directly coupled to metabolism

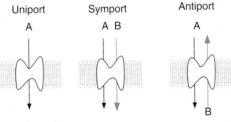

ATP

ADP + P_i

(c) Does a transport process involve a single ion or metabolite, or are fluxes of two or more species directly coupled?

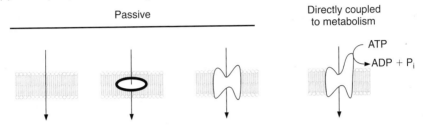

Uniport

A

Symport

A B

Antiport

A

B

(d) Does a transport process involve net charge transfer across the membrane?

Electroneutral

Electrical

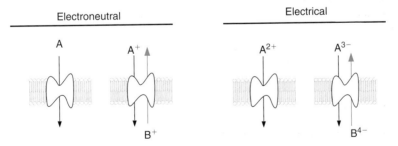

A

A^+

B^+

A^{2+}

A^{3-}

B^{4-}

Figure 2.1   The classification of ion and metabolite transport.
(a) Transport may be bilayer mediated (via either natural permeation across the membrane or an ionophore-induced pathway) or protein catalysed. (b) Transport by any of the three pathways in (a) can be passive (not directly coupled to metabolism) or, in the case of protein-catalysed transport alone, directly coupled to metabolism, e.g. ATP hydrolysis. (c) Transport, by any of the pathways in (b), may occur as a single species, as two or more ions whose transport is tightly coupled together in by symport (or co-transport), or by antiport (or exchange diffusion). (d) Any of the mechanisms in (c) may be electroneutral or electrical (electrogenic, electrophoretic).

gradient of a solute or ion is established across a membrane. Naturally only protein-catalysed transport can be directly coupled to metabolism (Fig. 2.1b).

2.2.3 Uniport, symport and antiport

The molecular mechanism of a transport process can involve a single ion or the tightly coupled transport of two or more species (Fig. 2.1c). A transport process involving a single ion is termed a uniport. Examples of uniports include the uptake pathway for Ca^{2+} across the inner mitochondrial membrane (Section 8.2.2) and the proton permeability induced in bilayers by the addition of proton translocators such as dinitrophenol. A transport process involving the obligatory coupling of two or more ions in parallel is termed symport or co-transport. In this book we shall use the shorthand $A : B$ to denote symport of the species A and B. Examples of proton symport are found at the bacterial membrane where the mechanism is used to drive the uptake of metabolites into the cell (Section 8.7).

The equivalent tightly coupled process where the transport of one ion is linked to the transport of another species in the opposite direction is termed antiport or exchange-diffusion, represented here in the form A/B for the antiport of A against B (Fig. 2.1c). Examples include the Na^+/H^+ antiport activity, which is present in the inner mitochondrial membrane (Section 8.2.1), and the K^+/H^+ antiport catalysed by the ionophore nigericin in bilayers (Section 2.3.4). Note, however, that if one of the ions involved in a nominal symport or antiport mechanism is a proton or hydroxyl ion, it is usually impossible to distinguish between the symport of a species with a H^+ and the antiport of the species with a OH^-. For example, the mitochondrial phosphate carrier (Section 8.5) may be variously represented as a P_i^-/OH^- antiport or a $H^+ : P_i^-$ symport.

Closely related transport pathways exist across non-energy-conserving membranes. At the plasma membrane the Na^+ ion can be involved in uniport (through a voltage-activated channel), symport (e.g. Na^+: glucose co-transport) and antiport (e.g. the $3Na^+/Ca^{2+}$ exchanger), while more complex stoichiometries may occur; for example, some neuronal membranes possess a carrier that catalyses the co-transport of Na^+ and glutamate coupled to the antiport of a third ion, K^+.

2.2.4 Electroneutral versus electrical transport

Electroneutral transport involves no net charge transfer across the membrane. Transport may be electroneutral either because an uncharged species is transported by a uniport, or as the result of the symport of a cation and an anion or the antiport of two ions of equal charge (Fig. 2.1d), an example of the last being the K^+/H^+ antiport catalysed by nigericin. Electrical transport is frequently termed either electrogenic ('creating a potential', proton pumping driven by ATP hydrolysis would be an example) or electrophoretic ('moving in response to a pre-existing potential', Ca^{2+} uniport into mitochondria would be an example). As these terms can refer to the same pathway observed under different conditions, the overall term 'electrical' will be used here.

It is important to distinguish between movement of charge at the molecular level, as discussed here, and the overall electroneutrality of the total ion movements across a given membrane. The latter follows from the impossibility of separating more than minute quantities of positive and negative charge across a membrane without building up a large membrane potential. Thus the separation of 1 nmol of charge across the inner membranes of 1 mg of mitochondria results in the build-up of more than 200 mV of potential. Or, put another way, a single turnover of all the electron transport components in an individual mitochondrion or bacterium will translocate sufficient charge to establish a membrane potential approaching 200 mV. The establishment of such potentials by the movement of so little charge is a consequence of the low electrical capacitance of biological membranes (typically estimated as $1\,\mu F\,cm^{-2}$). However, this property does not preclude the occurrence of steady-state charge movements at the molecular level as long as these compensate each other. In addition, it is necessary to appreciate that the effect on an energy-transducing membrane of a tightly coupled electroneutral antiporter, such as the ionophore nigericin, which catalyses an electroneutral K^+/H^+ antiport, is not the same as that caused by the addition of two electrical uniporters for the same ions (e.g. valinomycin plus a protonophore).

The four criteria discussed above allow a comprehensive description of a transport process; for example, proton pumping by the ATP synthase is an example of a protein-catalysed, metabolism-coupled electrical uniport.

2.3 BILAYER-MEDIATED TRANSPORT

2.3.1 The natural permeability properties of bilayers

Review   Zeuthen 2001

The hydrophobic core possessed by lipid bilayers creates an effective barrier to the passage of charged species. With a few important exceptions (Section 2.2), cations and anions do not permeate bilayers. This impermeability extends to the proton, and this property is vital for energy transduction to avoid short-circuiting the proton circuit. Not only does the bilayer have a high electrical resistance, but it can also withstand very high electrical fields. An energy-conserving membrane with a membrane potential of 200 mV across it has an electrical field in excess of $300\,000\,V\,cm^{-1}$ across its hydrophobic core.

A variety of uncharged species can cross bilayers. O_2 and CO_2 are all highly permeable, as are the uncharged forms of a number of low molecular weight acids and bases, such as ammonia and acetic (ethanoic) acid. These last permeabilities provide a useful tool for the investigation of pH gradients across membranes (Section 3.5). The mystery of how the most polar of compounds, water, crosses membranes has been resolved by the discovery of aquaporins, a large family of water-permeating channels present in membranes and catalysing the direct transport of water.

2.3.2 Ionophore-induced permeability properties of bilayer regions

The high activation energy required to insert an ion into a hydrophobic region accounts for the extremely low ion permeability of bilayer regions. It follows that, if the charge can be delocalized and shielded from the bilayer, the ion permeability might be expected to increase. This is accomplished by a variety of antibiotics synthesized by some micro-organisms, as well as by some synthetic compounds. These are known collectively as ionophores.

These are typically compounds with a molecular weight of 500–2000 possessing a hydrophobic exterior, making them lipid soluble, together with a hydrophilic interior to bind the ion. Ionophores are *not* natural constituents of energy-conserving membranes, but as investigative tools they are invaluable.

Ionophores can function as mobile carriers or as channel formers (Fig. 2.2). Mobile carriers diffuse within the membrane, and can typically catalyse the transport of about 1000 ions s^{-1} across the membrane. They can show an extremely high discrimination between different ions, can work across thick synthetic membranes and are affected by the fluidity of the membrane. In contrast, channel-forming ionophores discriminate poorly between ions but can be very active, transporting up to 10^7 ions per channel per second. Ionophores can also be categorized according to the ion transport that they catalyse.

2.3.3 Carriers of charge but not protons

Valinomycin (Fig. 2.2) is a mobile carrier ionophore that catalyses the electrical uniport of Cs^+, Rb^+, K^+ or NH_4^+. The ability to transport Na^+ is at least 10^4 less than for K^+. Valinomycin is a natural antibiotic from *Streptomyces* and is a depsipeptide, i.e. it consists of alternating hydroxy and amino acids. The ions lose their water of hydration when they bind to the ionophore. Na^+ cannot be transported because the unhydrated Na^+ ion is too small to interact effectively with the inward-facing carbonyls of valinomycin, with the result that the complexation energy does not balance that required for the loss of the water of hydration. Because valinomycin is uncharged and contains no ionizable groups, it acquires the charge of the complexed ion. Both the uncomplexed and complexed forms of valinomycin are able to diffuse across the membrane. Therefore, a catalytic amount of ionophore can induce the bulk transport of cations. It is effective in concentrations as low as 10^{-9} M in mitochondria, chloroplasts, synthetic bilayers and to a more limited extent in bacteria (the outer membrane can exclude it from Gram-negative organisms). Other ionophores catalysing K^+ uniport include the enniatins and the nactins (nonactin, monactin, dinactin, etc., so-called from the number of ethyl groups in the structure). However, these ionophores do not have such a spectacular selectivity for K^+ over Na^+ as valinomycin. Energy-conserving membranes generally lack a native electrical K^+ permeability, and valinomycin can be exploited to induce such a permeability, in order to estimate or clamp membrane potentials, or to investigate anion transport.

Gramicidin is an ionophore that forms transient conducting dimers in the bilayer (Fig. 2.2). Its properties are typical of channel-forming ionophores, with a poor selectivity between protons, monovalent cations and NH_4^+, the ions permeating in their hydrated forms. The capacity to conduct ions is limited only by diffusion, with the result that one channel can conduct up to 10^7 ions s^{-1}.

2.3.4 Carriers of protons but not charge

Nigericin is a linear molecule with heterocyclic oxygen-containing rings together with a hydroxyl group. In the membrane the molecule cyclizes to form a structure similar to that of valinomycin, with the oxygen atoms forming a hydrophobic interior. Unlike valinomycin, nigericin loses a proton when it binds a cation, forming a neutral complex, which can then

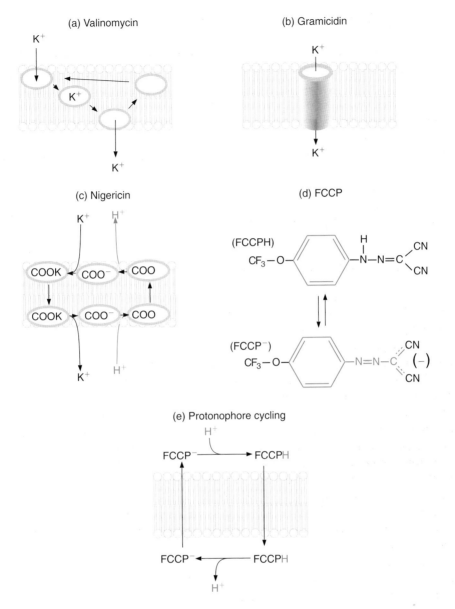

Figure 2.2   Ionophores.
Schematic function of four ionophores. (a) Valinomycin is a mobile carrier ionophore able to cross the lipid bilayer transporting a K^+ ion. Note that the ion's hydration sphere is lost and replaced by the ionophore. (b) Gramicidin is a channel-forming ionophore, with less selectivity than valinomycin but much higher activity. (c) Nigericin is a hydrophobic weak carboxylic acid permeable across lipid bilayer regions as either the protonated acid or the neutral salt. Nigericin has a selectivity $K^+ > Rb^+ > Na^+$. (d) FCCP is the most commonly employed example of a protonophore, although many such compounds exist. The blue bonds represents the extent of the π-orbital system. If a Δp exists across the membrane, the protonophore will cycle catalytically in an attempt to collapse the potential (e). $FCCP^-$ will be driven to the P-face of the membrane by the membrane potential, while FCCPH will be driven towards the alkaline or N-phase due to ΔpH. When sufficient FCCP is present (for most membranes 10^{-9}–10^{-5} M), the cycling can reduce both $\Delta\psi$ and ΔpH to near zero.

diffuse across the membrane as a mobile carrier. Nigericin is also mobile in its protonated non-complexed form, with the result that the ionophore can catalyse the overall electro-neutral exchange of K^+ for H^+ (Fig. 2.2). Other ionophores that catalyse a similar electro-neutral exchange include X-537A, monensin and dianemycin. The latter two show a slight preference for Na^+ over K^+, while X-537A will complex virtually every cation, including organic amines. Nigericin has been employed to study anion transport (Section 2.5) and to modify the pH gradient across energy-conserving membranes. It is often stated that nigericin abolishes ΔpH across a membrane; in fact the ionophore equalizes the K^+ and H^+ concentration gradients, the final ion gradients depending on the experimental conditions.

A23187 and ionomycin are dicarboxylic ionophores with a high specificity for divalent cations. A23187 catalyses the electroneutral exchange of Ca^{2+} or Mg^{2+} for two H^+ without disturbing monovalent ion gradients. Ionomycin has a higher selectivity for Ca^{2+} and has the additional advantage that it is non-fluorescent, allowing its use in experiments using fluorescent indicators.

2.3.5 Carriers of protons and charge

Protonophores, also known as proton translocators or uncouplers, have dissociable protons and permeate bilayers either as protonated acids, with pK_a values not too far below 7, or as the conjugate base, e.g. FCCP (Fig. 2.2). This is possible because these ionophores possess extensive π-orbital systems, which so delocalize the charge of the *anionic* form that lipid solubility is retained. By cycling across the membrane they can catalyse the net electrical uniport of protons and increase the proton conductance of the membrane. In so doing the proton circuit is short-circuited, allowing the process of Δp generation to be uncoupled from ATP synthesis. Uncouplers were described long before the formulation of the chemiosmotic theory. In fact the demonstration that the majority of these compounds act by increasing the proton conductance of synthetic bilayers was important evidence in favour of the theory.

An indirect proton translocation can be induced in membranes by the combination of a uniport for an ion together with an electroneutral antiport of the same ion in exchange for a proton. For example, the combination of valinomycin and nigericin induces a net uniport for H^+, while K^+ cycles around the membrane. The $Ca^{2+}/2H^+$ ionophores discussed above can also uncouple mitochondria in the presence of Ca^{2+}, since a dissipative cycling is set up between the native Ca^{2+} uniport (Section 8.2.3) and the ionophore.

2.3.6 The use of ionophores in intact cells

While ionophores were introduced largely for investigations of isolated mitochondria, they have also been applied to intact cells in attempts to modify *in situ* mitochondrial function. However, since they display no membrane selectivity, one must be aware of the consequences of introducing these ion permeabilities into other membranes. Thus valinomycin will hyperpolarize cells in low K^+ media by clamping the plasma membrane potential close to the K^+-diffusion potential (Section 3.8). Unfortunately, the ionophore will also collapse the mitochondrial $\Delta\psi$, since the K^+ concentrations in the cytoplasm and matrix are both about 100 mM. Nigericin can be added at low concentrations with no deleterious effect except to slightly hyperpolarize the mitochondria. In contrast, protonophores have multiple effects, most seriously depleting the cytoplasm of ATP by allowing the ATP synthase to reverse and hydrolyse glycolytically generated ATP. It is important also to appreciate that ionophores such as valinomycin may fail to act on intact bacteria owing to their absorption to cell walls.

2.3.7 Lipophilic cations and anions

Further reading   Karadjov *et al.* 1986, Wingrove and Gunter 1986, Lombardi *et al.* 1998

The ability of π-orbital systems to shield charge and enhance lipid solubility has been exploited in the synthesis of a number of cations and anions that are capable of being transported across bilayer membranes even though they carry charge. Examples include the tetraphenyl phosphonium cation (TPP^+) and the tetraphenylborate anion TPB^-. These ions are not strictly ionophores, since they do not act catalytically, but are instead accumulated in response to $\Delta\psi$ (Section 4.2.2). Lipophilic cations and anions were of value historically in demonstrations of their energy-dependent accumulation in mitochondria and inverted sub-mitochondrial particles, respectively. These experiments eliminated the possibility of specific cation pumps driven by chemical intermediates. Subsequently the cations have been employed for the estimation of $\Delta\psi$ (Section 4.2.2). Fluorescent lipophilic cations are employed to monitor changes in $\Delta\psi$ both in isolated mitochondria (Chapter 4) and mitochondria *in situ* within intact cells (Chapter 9). TPP^+ may also inhibit mitochondrial Ca^{2+} efflux (Chapter 8).

2.4 PROTEIN-CATALYSED TRANSPORT

The characteristics of protein-catalysed transport across energy-conserving membranes are usually sufficiently distinct from those of bilayer-dependent transport, whether in the absence or presence of ionophores, to make the correct assignment straightforward. The transport proteins of the mitochondrial inner membrane and bacterial cytoplasmic membrane will be discussed in detail in Chapter 8; here we shall merely summarize the distinctions between protein-catalysed and bilayer-mediated transport.

Transport proteins share the features of other enzymes; they can display stereospecificity, can frequently be inhibited specifically and are genetically determined. This last feature means that it is not possible to make the same kinds of generalizations as for bilayer transport. For example, if FCCP induces proton permeability in mitochondria, it can generally be assumed that the effect will be similar in thylakoids, bacteria and synthetic bilayers. In contrast, a transport protein may not only be specific to a given organelle but may be restricted to the organelle from one tissue. Thus the citrate carrier is present in liver mitochondria (Section 8.3.2), where it is involved in the export of intermediates for fatty acid synthesis, but is absent from heart mitochondria. The strongest evidence for the involvement of a protein in a transport process is often the existence of specific inhibitors. For example, whereas pyruvate was for many years considered to permeate into mitochondria through the bilayer, which is feasible as it is a monocarboxylic weak acid, it was later found that cyanohydroxycinnamate (Section 8.3.2) was a specific transport inhibitor. This provided the first firm evidence for a transport protein for this substrate.

Transport proteins have been studied by many approaches, and this has led to a plethora of names, including carriers, permeases, porters and translocases, all of which are synonyms for transport protein. The term 'carrier' is particularly inappropriate because there is no evidence that any protein functions by the carrier type of mechanism exemplified by valino-mycin. Instead, most mitochondrial transporters function as dimers and undergo subtle con-formational changes to expose their binding sites alternately to N- and P-phases (Chapter 8).

2.5 SWELLING AND THE CO-ORDINATE MOVEMENT OF IONS ACROSS MEMBRANES

The driving forces for the movement of ions across membranes will be derived quantita-tively in the next chapter. Here we shall discuss qualitatively how the movement of ions on different carriers within the same membrane may be coupled to each other.

The overriding principle of bulk ion movement across a closed membrane is that there must at no time be no more than a slight charge imbalance across the membrane. We have seen that the electrical capacity of a mitochondrion or bacterium is tiny, and thus that the uncompensated movement of less than 1 nmol of a charged ion per mg protein is sufficient to build up a $\Delta\psi$ of $>200\,mV$. Thus, during the operation of a proton circuit, the charge imbalance would never exceed this amount, even though the proton current might exceed $1000\,nmol\ H^+\,min^{-1}(mg\ protein)^{-1}$.

In order to illustrate the coordinate movement of ions, we shall first discuss a simple but powerful technique, which was much used to establish the pathways and mechanisms of ion transport across the mitochondrial inner membrane: osmotic swelling. Mitochondria will swell and ultimately burst unless they are suspended in a medium that is isotonic with the matrix and impermeant across the inner membrane. Swelling does not mean that the inner membrane stretches like a balloon, but rather that the inner membrane unfolds as the matrix volume increases, ultimately rupturing the outer membrane. Mitochondria will also swell in an isotonic solution in which the principal solute is permeable across the inner membrane. To observe osmotic swelling of mitochondria in ionic media, both the cation and anion of the major osmotic component of the medium must be permeable and the requirement for overall charge balance across the membrane must be respected.

Suspensions of mitochondria are turbid and scatter light, owing to the difference in refractive index between matrix contents and the medium, and any process which decreases this difference will decrease the scattered light. Thus, paradoxically, an increase in matrix volume owing to the influx of a permeable solute results in a decrease in the light scattered as the matrix refractive index approaches that of the medium. This provides a very simple semiquantitative method for the study of solute fluxes across the mitochondrial inner mem-brane by monitoring the decrease in light scattered in the $90°$ geometry of a fluorimeter, or the increase in transmitted light in a normal spectrophotometer, during the swelling process. Swelling can proceed sufficiently to rupture the outer membrane and release adenylate kinase and cytochrome c, which are located in the intermembrane space.

The simplest case to consider is that of mitochondria where both the respiratory chain and the ATP synthase are inhibited. In the example shown in Fig. 2.3, the permeability of rat liver mitochondria for Cl^- and SCN^- is investigated. Mitochondria suspended in isotonic (120 mM) concentrations of either KCl or KSCN undergo little decrease in light scattering. However, this does not necessarily demonstrate that the inner membrane is impermeable to these anions, as cation and anion must both be transported and the K^+ ion is poorly

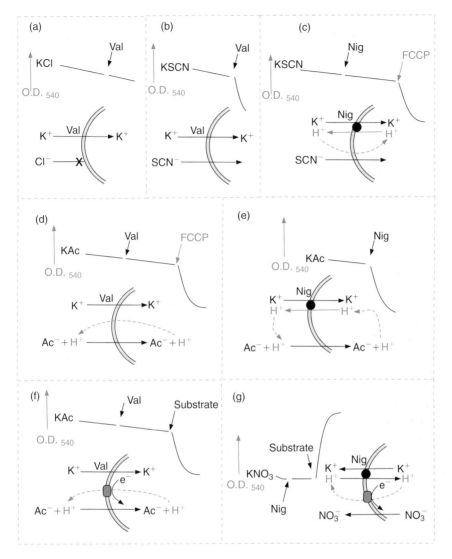

Figure 2.3   Ion permeability requirements for osmotic swelling or contraction of non-respiring and respiring mitochondria.
Mitochondria are suspended in iso-osmotic KCl, KSCN or K-acetate. Electrical (valinomycin, Val) or electroneutral (nigericin, Nig) potassium permeability is induced by ionophore addition and swelling monitored by decreased light scattering (schematic traces: downwards equals decreased light scattering and hence swelling). Protonophore (FCCP) is added where indicated. (a) Valinomycin does not initiate swelling in KCl since Cl$^-$ is impermeant; (b) however, mitochondria swell in KSCN since SCN$^-$ is highly permeant. (c) Mitochondria do not swell in KSCN plus nigericin since charges do not balance, addition of FCCP to allow proton re-entry initiates swelling. (d) Conversely, FCCP is needed for proton efflux in K-acetate after valinomycin addition, but not (e) after nigericin; note that acetic acid is the permeant species. (f) On initiation of respiration, mitochondria swell rapidly in K-acetate in the presence of valinomycin, while (g) mitochondria pre-swollen in KNO$_3$ contract on addition of nigericin plus substrate since nitrate (NO$_3^-$) is a permeant ion. O.D. = optical density.

permeable. To overcome this, an electrical uniport for K^+ can be induced by addition of valinomycin (Section 2.3.3). Rapid swelling is now observed in KSCN, but slow swelling in KCl. It can therefore be concluded that the inner membrane is permeable to SCN^-. However, the failure of KCl and valinomycin to support swelling does not yet prove that Cl^- is impermeable, because the requirement for charge balance has yet to be investigated.

Swelling in potassium acetate in the presence of valinomycin or nigericin illustrates the need for charge balance. Nigericin supports swelling in potassium acetate, but valinomycin is ineffective while allowing rapid swelling to occur in KSCN. The reason for the difference is the need for charge balance. K^+ entry catalysed by valinomycin is electrical, while acetate permeates the bilayer as the neutral protonated acid. Therefore, in the presence of valinomycin and potassium acetate, a membrane potential (positive inside) rapidly builds up preventing further K^+ entry. The permeation of acetic acid also ceases, as dissociation of the co-transported proton within the matrix builds up a pH gradient (acid in the matrix), which opposes further acetic acid entry. These problems are not encountered with potassium acetate in the presence of nigericin, since cation and anion entry are *both* now electroneutral. Also, the proton entering with acetic acid is re-exported by the ionophore in exchange for K^+.

It should now be clear that mitochondria swell in KSCN plus valinomycin, but not in KSCN plus nigericin, because permeation of the protonated species HSCN is negligible. It is therefore possible to use swelling not only to determine if a species is permeable but also to determine the mode of entry. Returning to the case of KCl, since swelling occurs in the presence of neither nigericin nor valinomycin, Cl^- cannot cross the membrane either as the anion or as HCl. The requirement for swelling that both ions enter by the same mode, be that electrical or electroneutral, can be overcome if an electrical proton uniport is induced by the addition of a protonophore (Fig. 2.3). Thus with FCCP present swelling occurs in the presence of KSCN plus nigericin or potassium acetate plus valinomycin. The ion fluxes that lead to accumulation of either KSCN or potassium acetate in the matrix are shown in Fig. 2.3.

Matrix volume changes occurring in respiring mitochondria have to take account of the contribution of the protons pumped across the membrane by the respiratory chain. Respiration-dependent swelling occurs in the presence of an electrically permeant cation (which is accumulated due to the membrane potential) and an electroneutrally permeant weak acid (accumulated due to ΔpH; Section 3.5), as shown in Fig. 2.3. Conversely, a rapid contraction of pre-swollen mitochondria occurs on initiation of respiration when the matrix contains an electroneutrally permeant cation (expelled by ΔpH) and an electrically permeant anion (expelled by $\Delta\psi$) (Fig. 2.3e).

Swelling induced under conditions of Ca^{2+} overload (the mitochondrial permeability transition, Section 8.2.5) is due to an alteration in the conformation of the inner membrane adenine nucleotide transporter, which produces a non-selective pore permeable to cations and anions up to 1.4 kDa.

Consideration of charge balance is also necessary in reconstitution experiments. For example, if a pH indicator is trapped inside a phospholipid vesicle with an acidic lumen and the external pH is subsequently raised into the alkaline range, the indicator trapped inside will continue to indicate an acid pH even if the protonophore FCCP is added. Significant proton efflux is not possible because efflux of a tiny quantity of protons generates a membrane potential, positive outside. However, if external K^+ is available and valinomycin is added, protons can efflux via FCCP because the requirement for charge balance is satisfied by the influx of K^+. The internal indicator then signals an alkaline pH.

AB TULP

3 QUANTITATIVE BIOENERGETICS: THE MEASUREMENT OF DRIVING FORCES

3.1 INTRODUCTION

Thermodynamics provides the quantitative core of bioenergetics, and this chapter is intended to provide an introduction to that part of thermodynamics of specific bioenergetic relevance. We have attempted to de-mythologize some of the more important relationships by deriving them from what we hope are commonsense origins. The reader is strongly advised to follow through the derivations, if only to exorcise the idea, which amazingly still exists in many general biochemistry textbooks, that ATP is a 'high-energy' compound. Thermodynamic ignorance is also responsible for some extraordinary errors found in the current literature, particularly in the field of mitochondrial control of cell survival (see Chapter 9).

3.1.1 Systems

In thermodynamics three types of system are studied. *Isolated* (or adiabatic) systems are completely autonomous, exchanging neither material nor energy with their surroundings (e.g. a closed, insulating vessel). *Closed* systems are materially self-contained, but exchange energy across their boundaries (e.g. a hot water bottle). *Open* systems exchange both energy and material with their environment (e.g. all living organisms). The complexity of the thermodynamic treatment of these systems increases as their isolation decreases. Open systems strictly require a non-equilibrium thermodynamic treatment; classical equilibrium thermodynamics cannot be applied precisely to open systems because the flow of matter across their boundaries precludes the establishment of a true equilibrium.

The most significant contribution of equilibrium thermodynamics to bioenergetics comes from considering individual reactions or groups of reactions as closed systems, asking questions about the nature of the equilibrium state for that reaction, and establishing how far removed the actual reaction is from that equilibrium state. Despite this limitation, equilibrium thermodynamics is immensely powerful in bioenergetics since it can be applied to the following problems:

(a) Calculating the conditions required for equilibrium in an energy transduction, such as the utilization of the protonmotive force to produce ATP, and by extension determining how far such a reaction is displaced from equilibrium under the actual experimental conditions. It is this displacement from equilibrium that defines the capacity of the reaction to perform useful work.

(b) Eliminating thermodynamically 'impossible' reactions. While no thermodynamic treatment can prove the existence of a given mechanism, equilibrium thermodynamics can readily disprove any proposed mechanism that disobeys its laws. For example, if reliable values are available for the protonmotive force and the free energy for ATP synthesis (concepts that will be developed quantitatively below), it is possible to state unambiguously the lowest value of the H^+/ATP stoichiometry (the number of H^+ that must enter through the ATP synthase to make an ATP), which would allow ATP synthesis to occur.

3.1.2 Entropy and Gibbs energy change

The universe is by definition an isolated system, and in an isolated system the driving force for a reaction is an increase in entropy, which may be broadly equated to the degree of disorder of the system. In a closed system a process will occur spontaneously if the entropy of the system *plus its surroundings* increases. Although it is not possible to measure directly the entropy changes in the rest of the universe caused by the energy flow across the boundary of the system, this parameter can be calculated under conditions of constant temperature and pressure from the flow of enthalpy (heat) across the boundaries of the system (Fig. 3.1). The thermodynamic function that takes account of this enthalpy flow is the *Gibbs energy change*, ΔG, which is the quantitative measure of the net driving force (at constant temperature and pressure):

$$\Delta G = \Delta H - T\Delta S \qquad\qquad [3.1]$$

where ΔH is the enthalpy change of the system and ΔS is the entropy change, again of the system. This is the Gibbs–Helmholtz equation. Thus ΔG can be determined using only those parameters which refer to the system itself, while entropy changes in the surroundings need not be determined. A process that results in a decrease in Gibbs energy ($\Delta G < 0$) is one which causes a net increase in the entropy of the system plus surroundings and is therefore able to occur spontaneously *if* a mechanism is available.

The Gibbs energy change (also termed the free energy change) occurs in bioenergetics in four different guises; indeed, the subject might well be defined as the study of the mechanisms by which the different manifestations of Gibbs energy are interconverted.

(1) Gibbs energy changes themselves are used in the description of substrate reactions feeding into the respiratory chain and of the ATP that is ultimately synthesized.

(2) The oxido-reduction reactions occurring in the electron transfer pathways in respiration and photosynthesis are usually quantified not in terms of Gibbs energy changes but in terms of closely derived redox potential changes.

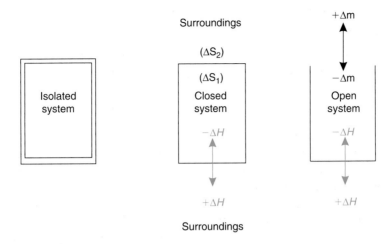

Figure 3.1 Gibbs energy and entropy changes in isolated and closed systems.
Isolated, closed and open systems exchange, respectively, neither enthalpy nor
material, enthalpy alone (ΔH), and enthalpy plus material (Δm) with their
surroundings.

(3) The available energy in a gradient of ions is quantified by a further variant of the Gibbs
energy change, namely the ion electrochemical gradient.

(4) In photosynthetic systems the Gibbs energy available from the absorption of quanta of
light can be compared directly with the other Gibbs energy functions within the reaction centre or the cell.

It should be emphasized that these different conventions merely reflect the diverse
historical background of the topics that are brought together in chemiosmotic energy
transduction.

3.2 GIBBS ENERGY AND DISPLACEMENT FROM EQUILIBRIUM

Since the first edition of *Bioenergetics* we have taken a rather unorthodox
approach to Gibbs energy changes. Whereas the classical physical chemistry
approach emphasizes standard free energies (a hypothetical condition where
all components are present at unit activity), we prefer to discuss reactions
purely in terms of displacement from equilibrium. This leads to considerably
simpler, more intuitive and more symmetrical equations without sacrificing
precision.

Consider a simple reaction A $\rightarrow$ B, occurring in a closed system. By observ-
ing and measuring the concentration of reactant A ($[A]_{obs}$) and product B ($[B]_{obs}$),
we can calculate the *observed mass action ratio* Γ (capital gamma), equal to

$[B]_{obs}/[A]_{obs}$. If the mixture of reactant and product happens to be at equilibrium, the mass action ratio of these equilibrium concentrations $[B]_{equil}/[A]_{equil}$ is termed the *equilibrium constant K*. The absolute value of the Gibbs energy (G) increases the further Γ is displaced from K, i.e. the further the reaction is from equilibrium. When G is plotted as a function of the logarithm of Γ/K (Fig. 3.2), a parabola is obtained. The curve shows the following features:

(a) The Gibbs energy content G is at a minimum when the reaction is at equilibrium. Thus any change in Γ away from the equilibrium ratio requires an increase in the Gibbs energy content of the system and so cannot occur spontaneously.

(b) The slope of the curve is zero at equilibrium. This means that a conversion of A to B that occurs at equilibrium without changing the mass action ratio Γ (e.g. by replacing the reacted A and removing excess B as it is formed) would cause no change in the Gibbs energy content. Another way of saying this is that the slope ΔG (in units of $kJ\,mol^{-1}$) is zero at equilibrium.

(c) When the reaction A $\to$ B has not yet proceeded as far as equilibrium, a conversion of A to B without changing the mass action ratio Γ results in a decrease in G, i.e. the slope ΔG is negative. This implies that such an interconversion can occur spontaneously, *provided that a mechanism exists*.

(d) The slope of the curve decreases as equilibrium is approached. Thus ΔG decreases the closer the reaction is to equilibrium. Note that ΔG does *not* equal the Gibbs energy that would be available if the reaction were allowed to run down to equilibrium, but rather gives the Gibbs energy that would be liberated per mole if the reaction proceeded with no change in substrate and product concentrations. This closely reflects the conditions prevailing *in vivo* where substrates are continuously supplied and products removed.

(e) For the reaction to proceed beyond the equilibrium point would require an input of Gibbs energy, this therefore cannot occur spontaneously.

The discussion may be generalized and placed on a quantitative footing by considering the reaction where a moles of A and b moles of B react to give c moles of C and d moles of D, i.e.

$$aA + bB \rightleftharpoons cC + dD \qquad [3.2]$$

The equilibrium constant K for the reaction is defined as follows:

$$K = \frac{[C]^c_{eq}[D]^d_{eq}}{[A]^a_{eq}[B]^b_{eq}} \; Molar^{(c+d-a-b)} \qquad [3.3]$$

where the equilibrium concentration of each component is inserted into the equation to obtain an equilibrium mass action ratio.

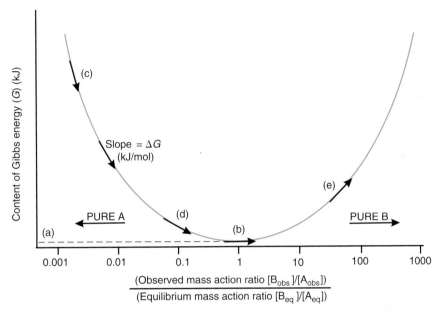

Figure 3.2   Gibbs energy content of a reaction as a function of its displacement from equilibrium.
Consider a closed system containing components A and B at concentrations [A] and [B], which can be interconverted by a reaction [A] ↔ [B]. The reaction is at equilibrium when the mass action ratio [B]/[A] = K, the equilibrium constant. The curve shows qualitatively how the Gibbs energy content (G) of the system varies when the total [A] + [B] is held constant but the mass action ratio is varied away from equilibrium. The slope of the tangential arrows represent schematically the Gibbs energy change, ΔG, for an interconversion of A to B, which occurs at different displacements from equilibrium, without changing the mass action ratio (e.g. by continuously supplying substrate and removing product). For details of (a)–(e), see text.

This complicated-looking equation will not be around for long, however, since we can now define the observed mass action ratio Γ when the reaction is held away from equilibrium by:

$$\Gamma = \frac{[C]^c_{obs}[D]^d_{obs}}{[A]^a_{obs}[B]^b_{obs}} \, \text{Molar}^{(c+d-a-b)} \qquad [3.4]$$

Note the symmetry between equations 3.3 and 3.4.

We shall now state, without deriving it, the key equation which relates the Gibbs energy change, ΔG, for the generalized reaction given in equation 3.2 to its equilibrium constant and observed mass-action ratio given in equations 3.3 and 3.4, respectively:

$$\Delta G = -2.3RT\log_{10} K/\Gamma \qquad [3.5]$$

where the factor 2.3 comes from the conversion from natural logarithms, R is the gas constant, and T is the absolute temperature.

This key equation tells us the following:

- ΔG has a value which is a function of the displacement from equilibrium. The numerical value of the factor $2.3RT$ means that at 25°C a reaction which is maintained one order of magnitude away from equilibrium possesses a ΔG of 5.7 kJ mol^{-1}.
- ΔG is negative if $\Gamma < K$ and positive if $\Gamma > K$.

Note again that ΔG is a differential, i.e. it measures the change in Gibbs energy that would occur if 1 mol of substrate were converted to product without changing the mass action ratio Γ (e.g. by continuously replenishing substrate and removing product). It does not answer the question 'how much energy is available from running down this reaction to equilibrium?'.

3.2.1 ΔG for the ATP hydrolysis reaction

The consequences of equation 3.5 may be illustrated by reference to the hydrolysis of ATP to ADP and P$_i$. At pH 7.0, and in the presence of an approximately physiological 10^{-2}M Mg^{2+}, this reaction has an apparent equilibrium constant K' of about 10^5M. By *apparent* equilibrium constant, we mean that obtained by putting the total chemical concentrations of reactants and products into the equation and ignoring the concentration of water, the pH and the effect of ionization state! That such a surprising oversimplification is possible will be explained later. The equation for the equilibrium of the ATP hydrolysis reaction is:

$$K' = \frac{[\Sigma ADP][\Sigma P_i]}{[\Sigma ATP]} M = 10^5 M \qquad [3.6]$$

where each concentration represents the total sum of the concentrations of the different ionized species of each component, including that complexed to Mg^{2+} (see below).

As equilibrium is attained when the apparent mass action ratio Γ', obtained using exactly the same simplifications as for K', is 10^5M, the equilibrium concentration of ATP in the presence of 10^{-2}M P$_i$ and 10^{-3}M ADP (which are approximate figures for the cytoplasm) would be only 10^{-10}M, or about 1 part per ten million of the total adenine nucleotide pool!

The variations of ΔG with the displacement of the ATP hydrolysis mass action ratio from equilibrium are shown in Table 3.1. Mitochondria are able to maintain a mass action ratio in the incubation medium which is as low as 10^{-5}M, ten orders of magnitude away from equilibrium. Under these conditions the incubation might contain 10^{-2}M P$_i$, 10^{-2}M ATP and only 10^{-5} ADP. To synthesize ATP under these conditions requires an input of Gibbs energy of 57 kJ per mole of ATP produced. The reason for a lower ΔG for the ATP/ADP pool in the matrix will be discussed in Chapter 8.

Note that ΔG for ATP *synthesis* (sometimes referred to as the 'phosphorylation potential', ΔG_p) is obtained from the corresponding value for ATP hydrolysis by simply changing the sign.

Table 3.1   The Gibbs energy change for the hydrolysis of ATP to ADP + P$_i$ as a function of the displacement from equilibrium

For $K' = 10^5$ M, pH 7, 10 mM Mg^{2+}, 10 mM P$_i$

Γ' (M)	K'/Γ'	ΔG (kJ mol^{-1})	[ATP]/[ADP]	Relevant condition
10^5	1	0	0.0000001	Equilibrium
10^3	10^2	-11	0.00001	
1	10^5	-28	0.01	'Standard conditions'
10^{-1}	10^6	-34	0.1	
10^{-3}	10^8	-46	10	Matrix
10^{-5}	10^{10}	-57	1000	Cytoplasm

(Note: [ATP]/[ADP][P$_i$] = 1 under 'standard conditions'.)

3.2.2 The uses and pitfalls of standard Gibbs energy, $\Delta G°$

A special case of the general equation for ΔG (equation 3.5) occurs under the totally hypothetical condition when the concentration of all reactants and products are in their 'standard states', i.e. 1 M for solutes, a pure liquid or a pure gas at 1 atmosphere. These conditions define the standard Gibbs energy change $\Delta G°$.

Considering again our generalized reaction in equation 3.2, under these 'standard' conditions, Γ has a value of 1 M$^{(c+d-a-b)}$ and equation 3.5 reduces to:

$$\Delta G° = -2.3RT \log_{10} K \qquad [3.7]$$

(note that the term $\log_{10} K$ is dimensionless because the units of K are cancelled by those of Γ).

Equation 3.7 is frequently misunderstood. It is important to appreciate that $\Delta G°$ is simply related to the logarithm of the equilibrium constant and as such gives in itself no information whatsoever concerning the Gibbs energy of the reaction in the cell. It is therefore absolutely incorrect to use $\Delta G°$ values to predict whether a reaction can occur spontaneously or to estimate the Gibbs energy available from a reaction.

Equation 3.7 can, however, be used to derive the more commonly used form of the Gibbs energy equation in which the equilibrium constant is substituted by $\Delta G°$. If we take equation 3.5 and divide both K and Γ by the standard state concentrations to make them dimensionless, and then rearrange the equation, we get:

$$\Delta G = -2.3RT \log_{10} K + 2.3RT \log_{10} \Gamma \qquad [3.8]$$

Combining with equation 3.7 and eliminating K gives:

$$\Delta G = \Delta G° + 2.3RT \log_{10} \Gamma$$

or as usually written:

$$\Delta G = \Delta G° + 2.3RT \log_{10} \left(\frac{[C]_{obs}^c [D]_{obs}^d}{[A]_{obs}^a [B]_{obs}^b} \right) \qquad [3.9]$$

Equation 3.9 is the most common form of the Gibbs energy equation and the one found in most textbooks. Just as equation 3.5 has terms for Γ and K, so equation 3.9 has terms for Γ and ΔG°. Note that equation 3.9 reverts to equation 3.7 at equilibrium when $\Delta G = 0$ and, of course, $\Gamma = K$. Both equation 3.5 and 3.9 can be used correctly to calculate ΔG; however, equation 3.5 is more intuitive, since it emphasizes the fact that ΔG is a function of the extent to which a reaction is removed from equilibrium. Additionally, it is not immediately evident from equation 3.9 that ΔG° and $2.3RT\log_{10}\Gamma$ are dimensionally homogeneous terms or why apparent equilibrium constants and apparent mass action ratios (see below) can be used that make simplifying assumptions about the states of ionization of reactants and products, the pH, etc.

3.2.3 Absolute and apparent equilibrium constants and mass action ratios

To avoid confusion or ambiguity in the derivation of equilibrium constants, and hence Gibbs energy changes, a number of conventions have been adopted. Those most relevant to bioenergetics are the following:

(a) True thermodynamic equilibrium constants (K) are defined in terms of the chemical activities rather than the concentrations of the reactants and products. Generally in biochemical systems it is not possible to determine the activities of all the components, and so equilibrium constants are calculated from concentrations. This introduces no error as long as the observed mass action ratio and the equilibrium constants are calculated under comparable conditions (remember ΔG is calculated from the ratio of Γ and K).

(b) When water appears as either a reactant or product in dilute solutions, it is considered to be in its standard state (which is the pure liquid at 1 atmosphere) under both equilibrium and observed conditions. This means that the water term can be omitted from both the equilibrium and observed mass action ratio equations (once again ΔG is calculated from the ratio of Γ and K, see equation 3.6).

(c) If one or more of the reactants or products are ionizable, or can chelate a cation, there is an ambiguity as to whether the equilibrium constant should be calculated from the total sum of the concentrations of the different forms of a compound, or just from the concentration of that form, which is believed to participate in the reaction. The hydrolysis of ATP to ADP and P_i is a particularly complicated case: not only are all the reactants and products partially ionized at physiological pH, but also Mg^{2+}, if present, chelates ATP and ADP with different affinities. Thus, ATP can exist at pH 7 in the following forms:

$$[\Sigma ATP] = [ATP^{4-}] + [ATP^{3-}] + [ATP.Mg^{2-}] + [ATP.Mg^{-}] \qquad [3.10]$$

If it were known that the true reaction was:

$$ATP.Mg^{2-} + H_2O \rightleftharpoons ADP.Mg^{-} + HPO_4^{2-} + H^{+} \qquad [3.11]$$

then the true equilibrium constant would be:

$$K = \frac{[Mg.ADP^-][HPO_4^{2-}][H^+]}{[Mg.ATP^{2-}]} \; M^2 \qquad [3.12]$$

This equilibrium constant would be independent of pH or Mg^{2+}, as changes in these factors are allowed for in the equation. However, the reacting species are *not* known unambiguously and, even if they were, their concentrations would be difficult to assay, as enzymatic or chemical assay determines the total concentration of each compound (e.g. ΣATP).

In practice, therefore, an apparent equilibrium constant, K' is employed, calculated from the total concentrations of each reactant and product, ignoring any effects of ionization or chelation and omitting any protons that are involved (see equation 3.6).

The most important limitation of the apparent equilibrium constant is that K' is not a universal constant, but depends on all those factors that are omitted from the equation, such as pH and cation concentration. K' *is thus only valid for a given pH and cation concentration, and must be qualified by information about these conditions.* As the standard Gibbs energy change is derived directly from the apparent equilibrium constant, this parameter must be similarly qualified. Finally and most importantly, the apparent mass action ratio, Γ', must be calculated under exactly the same set of assumptions; if this is done, when the ratio K'/Γ' is calculated for equation 3.5, all the assumptions cancel out and a true and meaningful ΔG is obtained. In biochemistry the terms $\Delta G^{o'}$ and K' are frequently used to specify that a $[H^+]$ of 10^{-7}M is being considered, but in principle these parameters can be specified for any condition of pH, ionic strength, temperature, $[Mg^{2+}]$, etc., that is convenient – as long as Γ' is *always* calculated under exactly the same set of conditions.

3.2.4 The myth of the 'high-energy phosphate bond'

It is still possible to come across statements to the effect that the phosphate anhydride bonds of ATP are 'high-energy' bonds capable of storing energy and driving reactions in otherwise unfavourable directions. However, it should be clear from Table 3.1 that it is the extent to which the observed mass action ratio is displaced from equilibrium that defines the capacity of the reactants to do work, rather than any attribute of a single component. A hypothetical cell could utilize any reaction to transduce energy from the mitochondrion. For example, if the glucose 6-phosphatase reaction were maintained ten orders of magnitude away from equilibrium, then glucose 6-phosphate hydrolysis would be just as capable of doing work in the cell as is ATP. Conversely, the Pacific Ocean could be filled with an equilibrium mixture of ATP, ADP and P$_i$, but the ATP would have no capacity to do work. It is important that the equilibrium constant for the ATP hydrolysis reaction has about the value it does, since it means that, even in the cytoplasm under conditions where ΔG for ATP synthesis is almost $60\,kJ\,mol^{-1}$, there is still a sufficient concentration of ADP to bind to the adenine nucleotide translocator responsible for translocating the nucleotide into the matrix (Chapter 8).

3.3 OXIDATION–REDUCTION (REDOX) POTENTIALS

Review   Schafer and Buettner 2001

3.3.1 Redox couples

Both the mitochondrial and photosynthetic electron transfer chains operate as a sequence of reactions in which electrons are transferred from one component to another. While many of these components simply gain one or more electrons in going from the oxidized to the reduced form, in others the gain of electrons induces an increase in the pK of one or more ionizable groups on the molecule, with the result that reduction is accompanied by the gain of one or more protons. Cytochrome c undergoes a 1e$^-$ reduction:

$$Fe^{3+}.cyt\ c + 1e^- \rightleftharpoons Fe^{2+}.cyt\ c \qquad [3.13]$$

NAD$^+$ undergoes a 2e$^-$ reduction and gains one H$^+$:

$$NAD^+ + 2e^- + H^+ \rightleftharpoons NADH \qquad [3.14]$$

while ubiquinone undergoes a 2e$^-$ reduction followed by the addition of 2H$^+$:

$$UQ + 2e^- + 2H^+ \rightleftharpoons UQH_2 \qquad [3.15]$$

This last is sometimes incorrectly referred to as a hydrogen transfer.

Redox reactions are not restricted to the electron transport chain. For example, lactate dehydrogenase also catalyses a redox reaction:

$$Pyruvate + NADH + H^+ \rightleftharpoons Lactate + NAD^+ \qquad [3.16]$$

While all oxidation–reduction reactions can quite properly be described in thermodynamic terms by their Gibbs energy changes, electrochemical parameters can be employed, since the reactions involve the transfer of electrons. Although the thermodynamic principles are the same as for the Gibbs energy change, the origins of oxido-reduction potentials in electrochemistry sometimes obscures this relationship.

The additional facility afforded by an electrochemical treatment of a redox reaction is the ability to dissect the overall electron transfer into two half-reactions, involving respectively the donation and acceptance of electrons. Thus equation 3.16 can be considered as the sum of two half-reactions:

$$NADH \rightleftharpoons NAD^+ + H^+ + 2e^- \qquad [3.17]$$

and

$$Pyruvate + 2H^+ + 2e^- \rightleftharpoons Lactate \qquad [3.18]$$

Note that these two equations can be added together to regenerate equation 3.16. A reduced–oxidized pair such as NADH/NAD$^+$ is termed a redox couple.

3.3.2 **Determination of redox potentials**

Each of the half-reactions described above (equations 3.17 and 3.18) is reversible, and so can in theory be described by an equilibrium constant. However, it is not immediately apparent how to treat the electrons, which have no independent existence in solution. A similar problem is encountered in electrochemistry when investigating the equilibrium between a metal (i.e. the reduced form) and a solution of its salt (i.e. the oxidized form). In this case the tendency of the couple to donate electrons is quantified by forming an electrical cell from two half-cells, each consisting of a metal electrode in equilibrium with a 1 M solution of its salt. An electrical circuit is completed by a bridge which links the solutions without allowing them to mix. The electrical potential difference between the electrodes may then be determined experimentally.

To facilitate comparison, electrode potentials are expressed in relation to the standard hydrogen electrode:

$$2H^+ + 2e^- \rightleftharpoons H_2 \qquad [3.19]$$

Hydrogen gas at 1 atmosphere is bubbled over the surface of a platinum electrode, which has been coated with the finely divided metal to increase the surface area. When this electrode is immersed in 1 M H^+, the absolute potential of the electrode is defined as zero (at 25°C). The standard electrode potential of any metal/salt couple may now be determined by forming a cell comprising the unknown couple together with the standard hydrogen electrode, or more conveniently with secondary standard electrodes whose electrode potentials are known.

A similar approach has been adopted for biochemical redox couples. As with the hydrogen electrode, it is not feasible to construct an electrode out of the reduced component of the couple, so a Pt electrode is employed. Since the oxidized and reduced components of the couple (e.g. the oxidized and reduced states of a cytochrome) rarely equilibrate with the Pt electrode sufficiently rapidly for a stable potential to be registered, a low concentration of a second redox couple, capable of reacting with both the primary redox couple (e.g. the cytochrome) and the Pt electrode, is added to act as a redox mediator. As will be shown below, the primary redox couple and the redox mediator achieve equilibrium when they exhibit the same redox potential. As long as the concentration relationships of the primary couple are not disturbed, the electrode can therefore register the potential of the primary couple. The use of one or more mediating couples is of particular importance when the redox potentials of membrane-bound components are being investigated (Chapter 5).

Unlike the metal/salt and H_2/H^+ couples, both components of a biochemical couple can generally exist in aqueous solution, and standard conditions are defined in which both the oxidized and reduced components are present at unit activity, or 1 M in concentration terms, and pH = 0. Note the parallel with the conditions for ΔG^o. The experimentally observed potential relative to the hydrogen electrode is termed the standard redox potential, E^o.

3.3.3 **Redox potential and [oxidized]/[reduced] ratio**

Just as the standard Gibbs energy change ΔG^o does not reflect the actual conditions existing in the cell, the standard redox potential E^o must be qualified to take account of the relative concentrations of the oxidized and reduced species.

The actual redox potential E at pH $= 0$ for the redox couple:

$$\text{Oxidized} + ne^- \rightleftharpoons \text{Reduced}$$

is given by the relationship:

$$E = E^\circ + 2.3 \frac{RT}{nF} \log_{10}\left(\frac{[\text{oxidized}]}{[\text{reduced}]}\right) \tag{3.20}$$

where R is the gas constant and F the Faraday constant. Note that this equation is closely analogous to the 'conventional' equation involving standard Gibbs energy changes (equation 3.9).

3.3.4 Redox potential and pH

In many cases (e.g. equations 3.14 and 3.15) protons are involved in the redox reaction, in which case the generalized half-reaction becomes

$$\text{Oxidized} + ne^- + mH^+ \rightleftharpoons \text{Reduced}$$

The standard redox potential at a pH other than zero is more negative than E°, the difference being $2.3RT/F(m/n)$ mV per pH unit. This corresponds to -60 mV/pH when $m = n$ and -30 mV/pH when $m = 1$ and $n = 2$ (Table 3.2). Note that the potentials are still calculated relative to the standard hydrogen electrode at pH $= 0$.

The usual biochemical convention is to define redox potentials for pH 7. The standard redox potential under these conditions is given the symbol $E^{\circ\prime}$ and is also referred to as the mid-point potential $E_{m,7}$ because it is the potential where the concentrations of the oxidized and reduced forms are equal (Fig. 3.3).

Note that, although the potential of the standard hydrogen electrode always remains zero, $E_{m,7}$ for the $H^+/\frac{1}{2}H_2$ couple is $7 \times (-60) = -420$ mV, i.e. much more reducing, since the concentration of the oxidized component of the couple, H^+ is present at such low concentration.

So far we have concentrated on standard or mid-point potentials. However, as for the case of Gibbs energy changes, it is the actual concentrations in the cell which define the redox potential. The actual redox potential at a pH of x ($E_{h,pH = x}$) is related to the mid-point potential at that pH by the relationship:

$$E_{h,pH = x} = E_{m,pH = x} + \frac{2.3\,RT}{nF} \log_{10}\left(\frac{[\text{ox}]}{[\text{red}]}\right) \tag{3.21}$$

3.3.5 The special case of glutathione

Review   Schafer and Buettner 2001

The glutathione couple (where GSSG and GSH represent the oxidized and reduced forms):

$$\text{GSSG} + 2H^+ + 2e^- \rightleftharpoons 2\text{GSH} \tag{3.22}$$

Table 3.2   Some mid-point potentials and examples of actual redox potentials

$$\text{Oxidized} + ne^- + mH^+ = \text{reduced}$$

	n	m	$E_{m,7}$ (mV)	ΔE_m per pH	Typical ox/red ratio	$E_{h,7}$ (mV)[*]
Ferredoxin oxidized/reduced	1	0	−430	0		
$H^+/\frac{1}{2}H_2$ (at 1 atm)	1	1	−420	−60		
O_2 (1 atm[†])/(superoxide)	1	0	−330	0	10^{-5}	−30
NAD^+/NADH	2	1	−320	−30	10	−290
$NADP^+$/NADPH	2	1	−320	−30	0.01	−380
Menaquinone/menaquinol	2	2	−74	−60		
Glutathione oxidized/reduced	2	2	−172	−60	0.01	−240[‡]
(when GSH + GSSG = 10 mM)						
Fumarate/succinate	2	2	+30	−60		
Ubiquinone/ubiquinol	2	2	+60	−60		
Ascorbate oxidized/reduced	2	1	+60	−30		
Cyt c oxidized/reduced	1	0	+220	0		
Ferricyanide oxidized/reduced	1	0	+420	0		
O_2 (1 atm[†])/2H$_2$O (55 M)	4	4	+820	−60		

[*] Approximate values for mitochondrial matrix under typical conditions.
[†] 1 atm oxygen = 1.25 mM.
[‡] See equation 3.22.

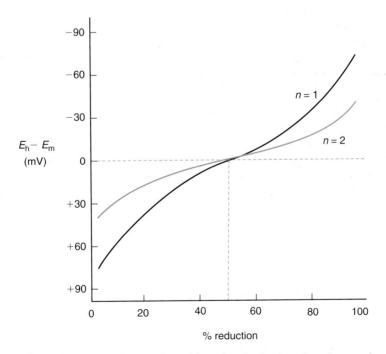

Figure 3.3   The variation of E_h with the extent of reduction of a redox couple.
$n = 1, n = 2$ refer to one- and two-electron oxido-reductions, respectively.

is one of the most important reactions defining the redox status of the cytoplasm and mitochondrial matrix. Because there are unequal numbers of substrate and product molecules, the absolute concentrations of these species, rather than merely their ratios, affect the redox potential.

In the general case, because $n = 2$:

$$E_{h,pH\,7} = E^{o\prime} + 30\log_{10}([GSSG]/[GSH]^2)\,mV \qquad [3.23]$$

Under standard conditions (1 M GSSG and 1 M GSH) $E^{o\prime} = -240\,mV$ at 25°C and pH 7.0. In the example shown in Fig. 3.4, the redox conditions required in order to maintain a redox potential of $-200\,mV$ are shown for a total glutathione pool concentration ([GSH] + 2[GSSG]) of 10 mM or 1 mM. In the former case, [GSSG] will be 1.2 mM, but this concentration must be reduced to only 20 μM if total glutathione is 1mM. This could impose kinetic problems for glutathione reductase, the enzyme reducing GSSG to GSH, and helps to explain why the pool size of glutathione is critical for preventing oxidative stress in the mitochondria.

3.3.6 Redox potential difference and the relation to ΔG

The Gibbs energy that is available from a redox reaction is a function of the difference in the actual redox potentials ΔE_h between the donor and acceptor redox couples. (Note that the difference in redox potential between two couples [redox potentials $E_{(A)}$ and $E_{(B)}$] is written in most books, as E, but we believe that use of ΔE_h clarifies that a *difference* between two couples, or a redox span in an electron transport system, is being considered). In general terms for the redox couples A and B:

$$\Delta E_h = E_{h(A)} - E_{h(B)} \qquad [3.24]$$

There is a simple and direct relationship between the redox potential difference of two couples, ΔE_h, and the Gibbs energy change ΔG accompanying the transfer of electrons between the couples:

$$\Delta G = -nF\Delta E_h \qquad [3.25]$$

where n is the number of electrons transferred, and F is the Faraday constant. From this it is apparent that an oxido-reduction reaction is at equilibrium when $\Delta E_h = 0$. Table 3.3 relates redox potential differences and Gibbs energy changes.

In the case of the mitochondrion, $E_{h,7}$ for the NADH/NAD$^+$ couple is about $-280\,mV$ and $E_{h,7}$ for the O_2/H_2O couple is $+780\,mV$ (note these values differ slightly from the mid-point potentials shown in Table 3.2 because the NADH/NAD$^+$ ratio in the matrix is only about 0.1 and because oxygen comprises only 20% of air). The redox potential difference $\Delta E_{h,7}$ of 1.16 V is the measure of the thermodynamic disequilibrium between the couples. Applying equation 3.25, the transfer of 2 mol of electrons from NADH to O_2:

$$\Delta G = -2 \times 96.5 \times 1.16 = -224\,kJ\,mol^{-1} \qquad [3.26]$$

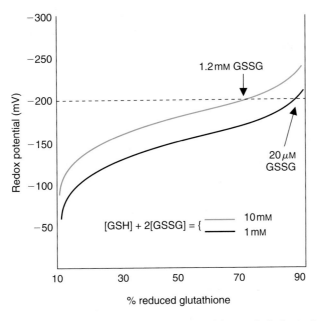

Figure 3.4   The redox potential of the GSSG/GSH couple is dependent on both the ratio of [GSH]/[GSSG] and the total concentration of glutathione. The redox potential of the glutathione couple is plotted as a function of the percentage reduction to GSH ($100 \times$ [GSH]/([GSH] + 2[GSSG]). In order to maintain a redox potential of -200 mV, [GSSG] must be maintained at 1.2 mM if total glutathione is 10 mM, but at no more than 20 μM if total glutathione is 1 mM. This could impose kinetic problems for glutathione reductase, the enzyme reducing GSSG to GSH.

Table 3.3   Interconversion between redox potential differences and Gibbs energy change for one-electron and two-electron transfers

	ΔG (kJ mol^{-1})	
ΔE_h (mV)	$n = 1$	$n = 2$
0	0	0
+100	-9.6	-19.3
+200	-19.3	-38.6
+500	-48.2	-96.5
+1000	-96.5	-193
+1200	-116	-231

One complication arises where the donor and acceptor redox couples are on the opposite sides of a membrane across which an electrical potential exists (Fig. 3.5). Examples of this occur in the respiratory chain (Chapter 5). If the electron enters via a centre at the N-side of the membrane and is transferred to a centre at the P-side, then it is intuitive that the process is energetically more downhill than if there was no $\Delta\psi$ to provide an extra 'push'

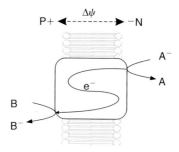

Figure 3.5 ΔE_h and ΔG for an electron transfer between redox couples located on opposite sides of a membrane sustaining a membrane potential.

$$\Delta E_h = (E_A - E_B)$$
$$\Delta G = -nF(\Delta E_h + \Delta\psi)$$

for the electron. The value of the membrane potential must be added to the redox potential difference to calculate the effective Gibbs energy change.

$$\Delta G = -nF(\Delta E_h + \Delta\psi) \tag{3.27}$$

A final note about nomenclature: a powerful *oxidizing* agent is an oxidized component of a redox couple with a relatively positive E_m (e.g. O_2), whereas a powerful *reducing* agent is the reduced component of a redox couple with a relatively negative E_m (e.g. NADH).

3.4 ION ELECTROCHEMICAL POTENTIAL DIFFERENCES

We have tried to emphasize in this chapter that the Gibbs energy change is a function of displacement from equilibrium. The disequilibrium of an ion or metabolite across a membrane can be subjected to the same quantitative treatment. As before, the derivation is not only valid for energy-transducing membranes, but has equal applicability to all membrane transport processes.

There are two forces acting on an ion gradient across a membrane, one due to the concentration gradient of the ion and one due to the electrical potential difference between the aqueous phases separated by the membrane (the 'membrane potential' $\Delta\psi$). These can initially be considered separately. It is important to remember that the term 'membrane potential' is shorthand for 'the difference in electrical potential between two aqueous compartments separated by a membrane' and says nothing about the nature of the membrane itself, or any charge on its surface.

Consider the Gibbs energy change for the transfer of 1 mol of solute across a membrane from a concentration $[X]_A$ to a concentration $[X]_B$, where the volumes of the two compartments are sufficiently large that the concentrations do not change significantly.

In the absence of a membrane potential, ΔG is given by:

$$\Delta G \ (\text{kJ mol}^{-1}) = 2.3RT \log_{10}\left(\frac{[X]_B}{[X]_A}\right) \tag{3.28}$$

Note that this equation is closely analogous to that for scalar reactions (equation 3.5). In particular, ΔG in both cases is $5.7\,\text{kJ mol}^{-1}$ for each tenfold displacement from equilibrium.

The second special case is for the transfer of an ion driven by a membrane potential in the absence of a concentration gradient. In this case, the Gibbs energy change when 1 mol of cation X^{m+} is transported down an electrical potential of $\Delta\psi$ mV is given by:

$$\Delta G \, (\text{kJ mol}^{-1}) = -mF \, \Delta\psi \qquad [3.29]$$

In the general case, the ion will be affected by *both* concentrative *and* electrical gradients, and the net ΔG when one mol of X^{m+} is transported down an electrical potential of $\Delta\psi$ mV from a concentration of $[X^{m+}]_A$ to $[X^{m+}]_B$ is given by the general electrochemical equation:

$$\Delta G \, (\text{kJ mol}^{-1}) = -mF \, \Delta\psi + 2.3RT \log_{10}\left(\frac{[X^{m+}]_B}{[X^{m+}]_A}\right) \qquad [3.30]$$

ΔG in this equation is often expressed as the ion electrochemical gradient $\Delta\tilde{\mu}_{X^{m+}}$ (kJ mol^{-1}).

In the specific case of the proton electrochemical gradient, $\Delta\tilde{\mu}_{H^+}$, equation 3.30 can be considerably simplified, since pH is a logarithmic function of $[H^+]$.

$$\Delta\tilde{\mu}_{H^+} = -F \, \Delta\psi + 2.3RT \, \Delta\text{pH} \qquad [3.31]$$

where ΔpH is defined as the pH in the P-phase (e.g. cytoplasmic) minus the pH in the N-phase (e.g. matrix). Note that this means that in a respiring mitochondrion ΔpH is usually negative. $\Delta\psi$ is defined as P-phase minus N-phase and is usually positive.

Mitchell defined the term protonmotive force (pmf or Δp), in units of voltage where:

$$\Delta p \, (\text{mV}) = -(\Delta\tilde{\mu}_{H^+})/F \qquad [3.32]$$

This facilitates comparison with redox potential differences in the electron transfer chain complexes, which generate the proton gradient, as well as emphasizing that we are dealing with a *potential* driving a proton circuit.

A $\Delta\tilde{\mu}_{H^+}$ of 1 kJ mol^{-1} corresponds to a Δp of 10.4 mV. Conversely a Δp of 200 mV corresponds to a $\Delta\tilde{\mu}_{H^+}$ of 19 kJ mol^{-1}.

Using Δp and substituting values for R and T at 25°C, the final equation is:

$$\Delta p \, (\text{mV}) = \Delta\psi - 59 \, \Delta\text{pH} \qquad [3.33]$$

3.5 PHOTONS

In photosynthetic systems, the primary source of Gibbs energy is the quantum of electromagnetic energy, or photon, which is absorbed by the photosynthetic pigments. The energy in a single photon is given by $h\nu$, where h is Planck's constant (6.62×10^{-34} J s) and ν is the frequency of the radiation (s^{-1}). One photon interacts with one molecule, and therefore N photons, where N is Avogadro's constant, will interact with 1 mol.

The energy in 1 mol (or einstein) of photons is therefore:

$$\Delta G = Nhv = Nhc/\lambda = 120\,000/\lambda \text{ kJ einstein}^{-1} \qquad [3.34]$$

where c is the velocity of light, and λ is the wavelength in nm. Thus even the absorption of an einstein of red light (600 nm) makes available 200 kJ mol^{-1}, which compares favourably with the Gibbs energy changes encountered in bioenergetics.

3.6 BIOENERGETIC INTERCONVERSIONS AND THERMODYNAMIC CONSTRAINTS ON THEIR STOICHIOMETRIES

The critical stages of chemiosmotic energy transduction involve the interconversions of ΔG between the different forms discussed in the previous sections. In the case of the mitochondrion these are: redox potential difference (ΔE_h) to protonmotive force (Δp) to ΔG for ATP synthesis. While bioenergetic systems operate under non-equilibrium conditions *in vivo*, i.e. they are 'open', with isolated preparations it is frequently possible to allow a given interconversion to achieve a true equilibrium by the simple expedient of inhibiting subsequent steps. For example, isolated mitochondria can achieve an equilibrium between the protonmotive force and ATP synthesis if reactions which hydrolyse ATP are absent.

In some cases a true equilibrium is not achievable in practice. For example, because of the inherent proton permeability of the mitochondrial membrane (Section 4.6), there is always some net leakage of protons across the membrane that results in the steady-state value of Δp lying below its equilibrium value with ΔE_h. Consequently there is always some flux of electrons from NADH to oxygen. Under these conditions it is valid to obtain a number of values for different flux rates and to extrapolate back to the static head condition of zero flux.

A test of whether an interconversion is at equilibrium is to establish whether a slight displacement in conditions will cause the reaction to run in reverse. In respiring mitochondria this test can be fulfilled by the ATP synthase and by two of the three respiratory chain proton pumps (complexes I and III, see Chapter 5). A process is, of course, at equilibrium when the overall ΔG is zero.

3.6.1 Proton pumping by respiratory chain complexes

If two electrons falling through a redox span of ΔE_h mV within the respiratory chain pump n protons across the membrane against a protonmotive force of Δp, then equilibrium would be attained when:

$$n\Delta p = 2\Delta E_h \qquad [3.35]$$

Thus the higher the H$^+$/O stoichiometry (n) of a respiratory chain complex with a particular value of ΔE_h, the *lower* the equilibrium Δp that can be attained, just as a bicycle has less ability to climb a hill in high rather than low gear.

Note that equation 3.35 will only hold if the electrons enter and leave the redox span on the same side of the membrane. If, as in the case of electron transfer from succinate dehydrogenase (on the matrix face) to cytochrome c (on the cytoplasmic face), the electrons effectively cross the membrane, they will be aided by the membrane potential (Fig. 3.4) and the relationship becomes:

$$n\Delta p = 2(\Delta E_h + \Delta\psi) \qquad [3.36]$$

3.6.2 Proton pumping by the ATP synthase

The equilibrium relationship between the protonmotive force and the Gibbs energy change ($\Delta G_{p,\,matrix}$) for the ATP synthase reaction in the mitochondrial matrix, is given by:

$$\Delta G_{p,\,matrix} = n'F\Delta p \qquad [3.37]$$

where n' is the H^+/ATP stoichiometry. Note that the higher the H^+/ATP stoichiometry (n') the *higher* the $\Delta G_{p,\,matrix}$ that can be attained at equilibrium with a given Δp. The same equation applies to the bacterial cytoplasm and the chloroplast stroma.

Since one additional proton is expended in the overall transport of P_i and ADP into, and of ATP out of, the mitochondrial matrix (Section 8.5), the relationship for the eukaryotic cytoplasmic ATP/ADP + P_i pool becomes:

$$\Delta G_{p,\,cytoplasm} = (n' + 1)\Delta p \qquad [3.38]$$

As n' lies between 3 and 4 (Chapter 7), this means that a substantial proportion of the Gibbs energy for the cytoplasmic ATP/ADP + P_i pool comes from the transport step rather than the ATP synthase itself. Naturally this occurs at a cost: the overall H^+/ATP stoichiometry is increased to supply the additional proton.

3.6.3 Thermodynamic constraints on stoichiometries

Since the above equations contain a term for the stoichiometry, it is possible to determine the thermodynamic parameters at equilibrium, substitute these values into the equations and hence calculate the stoichiometry term, without actually measuring the movement of protons across the membrane. This is known as the thermodynamic stoichiometry. Naturally such calculations are only as accurate as the determination of the thermodynamic parameters, but it does offer an alternative approach to the non-steady-state technique, which will be discussed in Chapter 4.

3.6.4 The 'efficiency' of oxidative phosphorylation

A statement of the type 'Oxidation of NADH by O_2 has a ΔG^o of $-220\,kJ\,mol^{-1}$ whilst ATP synthesis has ΔG^o of $+31\,kJ\,mol^{-1}$. Thus if three ATP molecules are synthesized for each NADH oxidized, mitochondrial oxidative phosphorylation traps approximately $93\,kJ\,mol^{-1}$ of the energy available from NADH oxidation, an efficiency of 42%' used to appear in many textbooks of biochemistry but is now mercifully rare. This analysis has at least two shortcomings. First, it refers to standard conditions that are not found in cells. Second, there is no basis in physical chemistry for dividing an output ΔG (93 kJ mol^{-1} in this case) by the input ΔG (220 kJ mol^{-1}) to calculate an efficiency. This will now be explained.

Under cellular conditions 2 mol of electrons flowing from the NADH/NAD$^+$ couple to oxygen liberate about 220 kJ. In the ideal case when there is no proton leak across the membrane, this would be conserved in the generation of a Δp of some 200 mV, while about 10 mol of H$^+$ are pumped across the membrane. The energy initially conserved in the proton gradient is thus about $10 \times 200 \times F = 200$ kJ. If 3 mol of ATP were synthesized per pair of electrons passing down the respiratory chain and the ATP is subsequently exported to the cytoplasm at a ΔG of about 60 kJ mol^{-1}, then 180 kJ would be conserved, showing that the oxidative phosphorylation machinery can closely approach equilibrium, and that there are no large energy losses between electron transport and ATP synthesis. In this sense the machine can be regarded as highly efficient. However, it is important to realize that, as ATP turnover increases, e.g. in an exercising muscle, the ΔG of the cytoplasmic ATP/ADP + P$_i$ pool will be significantly lower than 60 kJ mol^{-1}. Under these conditions the overall 'efficiency' falls as the inevitable price of running a reaction away from close-to-equilibrium conditions.

Comparison of oxidative phosphorylation in mitochondria with that in *E. coli* (Chapter 5) shows that, in the latter, NADH oxidation is coupled to the synthesis of fewer ATP molecules than in mitochondria. The reason for this appears to be that fewer protons are pumped for each pair of electrons flowing from NADH to oxygen. As all other energetic parameters (ΔG and Δp) are similar, it could be said that oxidative phosphorylation is less 'efficient' in the bacterium, owing to failure to conserve fully the energy from respiration in the form of the Δp.

In practice, all energy-transducing membranes have a significant proton leak and thus the actual output is reduced so that equilibrium between ATP synthesis and respiration is not reached. Irreversible thermodynamics, which is beyond the scope of this book, is able to calculate that the true efficiency (i.e. power output divided by power input) is optimal when the mitochondria are synthesizing ATP rapidly, since the proton leak is greatly decreased (Section 4.7) and most proton flux is directed through the ATP synthase.

3.7 THE EQUILIBRIUM DISTRIBUTIONS OF IONS, WEAK ACIDS AND WEAK BASES

The membrane potential and pH gradients across the inner mitochondrial membrane affect the equilibrium distribution of permeant ions and species with dissociable protons. These driving forces are of importance in controlling transport between the mitochondrion and its cytoplasmic environment. In addition, the equilibrium distribution of synthetic cations and dissociable species provides the basis for the experimental determination of $\Delta\psi$ and ΔpH across the inner membrane of both isolated and *in situ* mitochondria.

3.7.1 Charged species and $\Delta\psi$

As with all Gibbs energy changes, an ion distribution is at equilibrium across a membrane when ΔG, and hence $\Delta\tilde{\mu}$, for the ion transport process is zero.

At equilibrium the ion electrochemical equation (equation 3.30) becomes:

$$\Delta G \,(\text{kJ mol}^{-1}) = 0 = -mF\Delta\psi + 2.3RT\log_{10}\left(\frac{[\text{X}^{m+}]_B}{[\text{X}^{m+}]_A}\right) \qquad [3.39]$$

This rearranges to give the equilibrium Nernst equation, relating the equilibrium distribution of an ion to the membrane potential:

$$\Delta\psi = 2.3\frac{RT}{mF}\log_{10}\left(\frac{[\text{X}^{m+}]_B}{[\text{X}^{m+}]_A}\right) \qquad [3.40]$$

An ion can thus come to electrochemical equilibrium when its concentration is unequal on the two sides of the membrane. The Nernst potential is the value of $\Delta\psi$ at which an ion gradient is at equilibrium as calculated from equation 3.40. This is the diffusion potential condition (see Section 3.8).

A membrane potential is a delocalized parameter for any given membrane and acts on all ions distributed across on a membrane. It therefore follows that a membrane potential generated by the translocation of one ion will affect the electrochemical equilibrium of all ions distributed across the membrane. The membrane potential generated, for example, by proton translocation will therefore affect the distribution of a second ion. If the second ion only permeates by a simple electrical uniport, it will redistribute until electrochemical equilibrium is regained, and the resulting ion distribution will enable the membrane potential to be estimated from equation 3.40. The membrane potential will not be appreciably perturbed by the distribution of the second ion provided the latter is present at low concentration. This is because there is steady-state proton translocation and any transient drop in membrane potential following redistribution of the second ion is compensated by the proton pumping.

This is the principle for most determinations of $\Delta\psi$ across energy-transducing membranes (see Section 4.2.1). The equilibrium ion distribution varies with $\Delta\psi$ as shown in Table 3.4. Note that:

- Anions are excluded from a negative compartment (e.g. the mitochondrial matrix).
- Cation accumulation is an exponential function of $\Delta\psi$.
- Divalent cations are accumulated to much higher extents than monovalent cations.

3.7.2 Weak acids, weak bases and ΔpH

An electroneutrally permeant species will be unaffected by $\Delta\psi$ and will come to equilibrium when its concentration gradient is unity (Table 3.4). Weak acids and bases (i.e. those with a pK between 3 and 11) can often permeate in the uncharged form across bilayer regions of the membrane (Chapter 2), while the ionized form remains impermeant, even though it may be present in great excess over the neutral species. As a result the *neutral* species (protonated acid or deprotonated base) equilibrates without regard to $\Delta\psi$. However, if there is a ΔpH, the Henderson–Hasselbalch equation requires that the concentration of the *ionized* species must differ (Fig. 3.6). Weak acids accumulate in alkaline compartments

Table 3.4   The equilibrium distribution of ions permeable by passive uniport across a membrane

	$[X]_{in}/[X]_{out}$			
			Charge on ion	
$\Delta\psi$ (mV)	-1 (e.g. SCN$^-$)	0	1 (e.g. K$^+$/valinomycin)	2 (e.g. Ca^{2+})
30	0.3	1	3	10
60	0.1	1	10	100
90	0.03	1	30	1 000
120	0.01	1	100	10 000
150	0.003	1	300	100 000
180	0.001	1	1 000	1 000 000

(such as the mitochondrial matrix), while weak bases accumulate in acidic compartments. If the equilibrium gradient can be measured, for example, radioisotopically, then ΔpH can be estimated. This principle is widely used to determine ΔpH across energy-transducing membranes (Section 4.2.4).

3.8   MEMBRANE POTENTIALS, DIFFUSION POTENTIALS, DONNAN POTENTIALS AND SURFACE POTENTIALS

There are two ways in which a true, bulk-phase membrane potential (i.e. transmembrane electrical potential difference) may be generated. The first is by the operation of an electrogenic ion pump such as operates in energy-transducing membranes. The second is by the addition to one side of a membrane of a salt, the cation and anion of which have unequal permeabilities. The more permeant species will tend to diffuse through the membrane ahead of the counter-ion and thus create a *diffusion potential*. Diffusion potentials may be created across energy-transducing membranes, for example, by the addition of external KCl in the presence of valinomycin, which provides permeability for K$^+$, thus generating a $\Delta\psi$, positive inside. Under ideal conditions the magnitude of the diffusion potential can be calculated from the Nernst equation (equation 3.40).

3.8.1   Eukaryotic plasma membrane potentials

In energy-transducing organelles, diffusion potentials tend to be transient, owing to the rapid movement of counter-ions and are not in general physiologically significant. This is in contrast to eukaryotic plasma membranes where the generally slow transport processes enable potentials sometimes to be sustained for several hours. In this case the diffusion potentials owing to the maintained concentration gradients across the plasma membrane play the dominant role in determining the membrane potential. In the case where K$^+$, Na$^+$ and Cl$^-$ gradients exist across the membrane, the membrane potential is a function of the ion gradients weighted by their permeabilities, and is given by the Goldman equation:

$$\Delta\psi = 2.3\frac{RT}{F}\log_{10}\left(\frac{P_{Na}[Na^+]_{out} + P_K[K^+]_{out} + P_{Cl}[Cl^-]_{in}}{P_{Na}[Na^+]_{in} + P_K[K^+]_{in} + P_{Cl}[Cl^-]_{out}}\right) \qquad [3.41]$$

(a)   Weak acids

	Out	Membrane	In
	$H^+_{out} + A^-_{out} \rightleftharpoons HA_{out}$	$\rightleftharpoons$	$HA_{in} \rightleftharpoons H^+_{in} + A^-_{in}$

If pK of acid is the same in both compartments:

$$\frac{[H^+]_{out}[A^-]_{out}}{[HA]_{out}} = K = \frac{[H^+]_{in}[A^-]_{in}}{[HA]_{in}}$$

At equilibrium $[HA]_{out} = [HA]_{in}$

$$\therefore \frac{[H^+]_{out}}{[H^+]_{in}} = \frac{[A^-]_{in}}{[A^-]_{out}}$$

$$\Delta pH = \log_{10}\frac{[A^-]_{out}}{[A^-]_{in}}$$

(b)   Weak bases

	Out	Membrane	In
	$BH^+_{out} \rightleftharpoons A^+_{out} + B_{out}$	$\rightleftharpoons$	$B_{in} + H^+_{in} \rightleftharpoons BH^+_{in}$

$$\frac{[H^+]_{out}[B]_{out}}{[BH^+]_{out}} = K = \frac{[H^+]_{in}[B]_{in}}{[BH^+]_{in}}$$

At equilibrium $[B]_{out} = [B]_{in}$

$$\therefore \frac{[H^+]_{out}}{[H^+]_{in}} = \frac{[BH^+]_{out}}{[BH^+]_{in}}$$

$$\Delta pH = \log_{10}\frac{[BH^+]_{in}}{[BH^+]_{out}}$$

Figure 3.6   The equilibrium distribution of electroneutrally permeant weak acids and bases as a function of ΔpH.
(a) Weak acid distribution: the concentration of the protonated HA is the same on both sides of the membrane, while the anion is concentrated in the alkaline compartment. (b) Weak base distribution: the concentration of the deprotonated B is the same on both sides of the membrane, while the protonated cation is concentrated in the acidic compartment.

Note that, if only a single ion is permeant, this equation reduces to the Nernst equation (equation 3.40).

3.8.2 Donnan potentials

The limiting case of a diffusion potential occurs when the counter-ion is completely impermeant. This condition pertains in mitochondria owing to the 'fixed' negative charges of the internal proteins and phospholipids. As a result, when the organelles are suspended in a medium of low ionic strength, such as sucrose, and an ionophore such as valinomycin is added, the more mobile cations attempt to leave the organelle until the induced potential balances the cation concentration gradient. This is a stable Donnan potential.

3.8.3 Surface potentials

Surface potentials are quite distinct from the above. Owing to the presence of fixed negative charges on the surfaces of energy-transducing membranes, the proton concentration in the immediate vicinity of the membrane is higher than in the bulk phase. However, Δp is not affected, since the increased proton concentration is balanced by a decreased electrical potential. The proton electrochemical potential difference across the membrane, Δp, is thus unaffected by the presence of surface potentials, although membrane-bound indicators of $\Delta \psi$, such as the carotenoids of photosynthetic membranes (Chapter 6) might be influenced.

The proton circuit of brown adipose tissue (BAT) mitochondria has a leak which can be plugged by nucleotides such as GDP. This is part of the thermogenic mechanism enabling fatty acids to be oxidized to acetate for heat. The Cl^- permeability is puzzling

4

THE CHEMIOSMOTIC PROTON CIRCUIT

4.1 INTRODUCTION

The central concept of bioenergetics, the circuit of protons linking the primary generators of protonmotive force with the ATP synthase, was introduced in Chapter 1. The purpose of the present chapter is to discuss experimental approaches to monitoring the functioning of the proton circuit under a wide range of conditions. The close analogy between the proton circuit and the equivalent electrical circuit (Fig. 1.3) will be emphasized, not only as a simple model but also because the same laws govern the flow of energy around both circuits.

In an electrical circuit the two fundamental parameters are potential difference (in volts) and current (in amps). From measurements of these functions other factors may be derived, such as the rate of energy transmission (in watts) or the resistance of components in the circuit (in ohms). In Fig. 4.1 a simple electrical circuit is shown, together with the analogous proton circuit across the mitochondrial inner membrane (the circuit operating across a photosynthetic or bacterial membrane would be closely similar). In practice the proton circuits are a little more complicated, since multiple proton pumps feed into the proton circuit. This is equivalent to having multiple batteries connected in parallel with respect to the proton circuit and in series with respect to the pathways of electron transfer (see Fig. 4.2). These pumps, which as will be described in more detail in Chapter 5, are known as complexes I, III and IV. It should be noted that it is possible to introduce or remove electrons at the interfaces of an individual proton pump, e.g. with the redox dye TMPD, allowing a complex to be studied in isolation in the intact mitochondrion.

In an electrical open circuit (Fig. 4.1a), electrical potential (voltage) is maximal, but no current flows as the redox potential difference generated by the chemical reduction–oxidation (redox) reactions within the battery is precisely balanced by the back pressure of the electrical potential. The tight coupling of the reactions within the battery to electron flow prevents any net chemical reaction. In the case of an ideal mitochondrion with no proton leak across the inner membrane, the proton circuit is open-circuited when there is no pathway for the protons extruded by the respiratory chain to re-enter the matrix (e.g. when the ATP synthase is inhibited or when there is no turnover of ATP). As with the electrical circuit, the proton electrochemical potential, Δp, across the membrane is maximal under these conditions. As the redox reactions are tightly coupled to proton extrusion there would be no

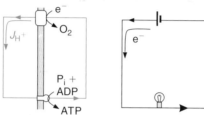

(a) Open circuit. Zero current (no respiration), potential (Δp) maximal.

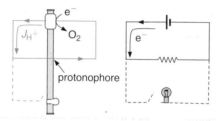

(b) Circuits completed, current flows (respiration). Useful work done (ATP synthesized). Potential (Δp) less than maximal.

(c) Short-circuit introduced. Energy dissipated, potential low, respiration maximal.

Figure 4.1   The regulation of the mitochondrial proton circuit by analogy to an electrical circuit.
(a) Open circuit, zero current (no respiration), potential (Δp) maximal. (b) Circuits completed, current flows (respiration occurs), useful work is done (ATP is synthesized). Potential (Δp) decreases slightly. (c) Short-circuit introduced, energy dissipated, potentials are low, current (respiration) is high.

respiration in this condition, and a thermodynamic equilibrium would exist between Δp and at least one of the proton-translocating regions of the respiratory chain (Section 3.6.1).

In Fig. 4.1b the electrical and proton circuits are shown operating normally and performing useful work. Both potentials are slightly less than under open-circuit conditions, as it is the slight disequilibrium between the redox potential difference available and the back-potential in the circuit which provides the net driving force enabling the battery or respiratory chain to operate. The 'internal resistance' of the battery may be calculated from the drop in potential required to sustain a given current. Analogously, the 'internal resistance' of the respiratory chain may be estimated, and is found to be very low (see Fig. 4.10).

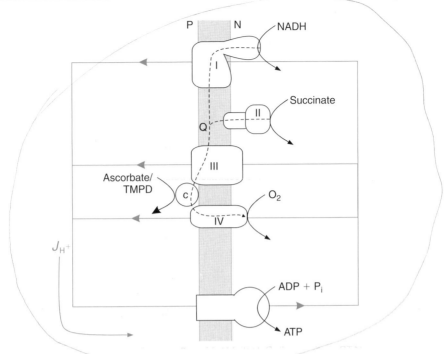

Figure 4.2 The mitochondrial respiratory chain consists of three proton pumps (complexes I, III and IV), which act in parallel with respect to the proton circuit and in series with respect to the electron flow.
Solid lines: pathway of proton flux; dotted line: pathway of electron transfer. For redox couples, see Table 3.2.

An electrical circuit may be shorted by introducing a low resistance pathway in parallel with the existing circuit, e.g. putting a copper wire across the battery terminals (Fig. 4.1c). Current can now flow from the battery without having to do useful work. Current flow is maximal, the voltage (Δp) is low and much heat can be evolved. This 'uncoupling' can be accomplished in the proton circuit by the addition of protonophores (Section 2.3.5), enabling respiration to occur without stoichiometric ATP synthesis, while a specialized class of mitochondria, in brown adipose tissue, possess a unique proton conductance pathway performing an analogous function (Sections 4.6 and 8.6).

4.2 THE MEASUREMENT OF PROTONMOTIVE FORCE

Techniques for the determination of Δp invariably involve the separate estimation of $\Delta\psi$ and ΔpH, based, for mitochondria, on the equilibrium distributions of permeant cations and weak acids, respectively (Section 3.7). Monovalent cations are accumulated by tenfold for every $60\,mV$ of $\Delta\psi$ (negative inside, e.g. in the case of mitochondria or bacterial cells), while monovalent anions are accumulated to the same extent into positive compartments (submitochondrial particles or chromatophores). Weak acids are accumulated by tenfold per pH unit into alkaline compartments (e.g. the mitochondrial matrix) and weak bases by tenfold per pH unit into acidic compartments, Fig. 3.6. To avoid disturbing the gradients, the concentration of all extrinsic probes must be kept as low as possible. Often the pH component of mitochondrial Δp is minimized by the inclusion of high P_i concentrations or addition of nigericin; the approximation is then made that $\Delta\psi$ accounts for the entire Δp or at least changes in parallel with Δp.

Early methods involved the use of radioisotopes that required the organelles to be separated from the incubation media. Currently the most common techniques either monitor the residual probe concentration in the medium at equilibrium, using ion-selective macroelectrodes or alternatively rely on the altered spectral properties of dyes when accumulated to high concentration within the organelles. In addition chloroplast and photosynthetic bacterial membranes contain carotenoid pigments, which act as intrinsic optical indicators of $\Delta\psi$. These techniques will now be described in more detail.

4.2.1 Ion-specific electrodes

Since bioenergetic organelles and bacterial cells are too small to be impaled by microelectrodes, the internal accumulation of probe cannot be determined directly. Instead ion gradients are calculated from the fall in external concentration of an indicator ion as it is accumulated. This means that both the internal volume and the activity coefficient of the ion within the organelle must be known. For mitochondria, the ions most commonly used are triphenylmethyl phosphonium ($TPMP^+$) and tetraphenyl phosphonium (TPP^+), which permeate bilayers for reasons given in Section 2.3.7.

Considerable care must be taken in the selection of appropriate indicators. To monitor $\Delta\psi$ the ion should be of the correct charge to be accumulated (a cation if the interior is negative (mitochondria or bacterial cells) or an anion if the interior is positive with respect to the medium (e.g. submitochondrial particles or chromatophores)). Secondly, the indicator must readily achieve electrochemical equilibrium and not be capable of being transported by more than one mechanism. Thirdly, the indicator should disturb the gradients as little as possible and, fourthly, the indicator should neither be metabolized nor affect enzyme activities. For the measurement of ΔpH, the above conditions should also be satisfied, with the exception that the indicator should be a weak acid to be accumulated within organelles with an alkaline interior, and a weak base to be accumulated in an acidic compartment (Fig. 3.5).

To obtain a meaningful value for $\Delta\psi$ the actual gradient of free probe across the inner membrane must be determined. It is rarely possible to do this with great precision owing to uncertainties in matrix volume and the activity coefficient of the probe in the concentrated environment of the matrix. However, precision is helped by the logarithmic nature of the Nernst equation, such that a 10% error in calculating the gradient would only affect $\Delta\psi$ by 2.5 mV, or less than 2% (equation 3.40). Lipophilic cations such as TPP^+, discussed in the next section are most accurate, as their accumulation in the matrix can be determined either isotopically or by the fall in the external concentration of the probe with a selective electrode, after allowing for binding or non-ideal behaviour in the matrix (Section 4.2.2). On the other hand, the use of fluorescent probes, either with isolated organelles or with intact cells is more empirical, since most techniques rely upon changed fluorescent properties of the probes when concentrated within the matrix. These changes are highly dependent upon the amount of probe added; an approximate calibration can be performed by varying the external K^+ concentration in the presence of valinomycin to clamp $\Delta\psi$ at values close to the corresponding K^+ Nernst potentials (equation 3.40).

Much more complexity is introduced when attempts are made to monitor $\Delta\psi$ for mitochondria *in situ* within a functioning cell. This will be discussed in Chapter 9.

4.2.2 The determination of Δp by ion-specific electrodes and radioisotopes

Further reading   Mitchell and Moyle 1969a, Nicholls 1974, Kamo *et al.* 1979, Karadjov *et al.* 1986, Dedukhova *et al.* 1993, Mootha *et al.* 1996

The first determination of Δp in mitochondria dates back to 1969, when Mitchell and Moyle employed pH- and K^+-selective electrodes in an initially anaerobic, low K^+-incubation. Valinomycin was present to allow K^+ to equilibrate, and $\Delta\psi$ was calculated from the K^+ uptake. A value of 228 mV was obtained for Δp for mitochondria respiring under 'open-circuit' conditions in the absence of ATP synthesis (state 4, see Section 4.5), and values in this range, or slightly lower, are currently accepted. The technique was modified for radioactive assay by substituting the β-emitter ^{86}Rb for K^+ and by using ^{3}H-labelled weak acids (such as acetate) and bases (e.g. methylamine) to determine ΔpH. After silicone oil centrifugation to separate the mitochondria from the medium the isotope accumulated within the matrix was calculated. In these experiments an additional indicator, such as $[^{14}C]$sucrose, is also present. Sucrose cannot permeate the inner membrane (hence the ability of sucrose to act as osmotic support during mitochondrial preparation) and the sucrose count found in the mitochondrial pellet indicates the volume of contaminating medium, which is pelleted together with the mitochondria and allows the ^{86}Rb counts to be corrected. It is also necessary to estimate the matrix volume. This is difficult to do with precision, although the difference between the sucrose-permeable and ^{3}H$_2$O-permeable spaces gives a reasonably accurate measure. A major error, however, lies in estimating the activity of the cations within the mitochondrial matrix. The matrix is enormously concentrated: about 0.5 mg of protein is concentrated into 1 μl, i.e. 500 g l^{-1}! In such a concentrated gel it is inevitable that the ions behave non-ideally.

The use of valinomycin in the above experiments has a major disadvantage in that $\Delta\psi$ is artifactually clamped at a value corresponding to the Nernst equilibrium for the pre-existing K^+-gradient across the membrane. This can be avoided by the use of phosphonium cations, such as TPP$^+$ in place of ^{86}Rb and valinomycin (Fig. 4.3). The positive charge of TPP$^+$ or the closely related TPMP$^+$ is sufficiently shielded by the hydrophobic groups to enable the cation to permeate bilayer regions (Section 2.3.7). The activity coefficients of these indicators deviate widely from unity when accumulated in mitochondria or bacteria, and accuracy can be improved by calibrating the response, e.g. by quantifying uptake in the presence of valinomycin and known K^+-gradients.

TPP$^+$ accumulation can be measured isotopically after separating mitochondria from medium as above. Alternatively a TPP$^+$-selective electrode can be constructed, allowing a continuous monitoring of the external concentration. The decrease as the mitochondria accumulate the cation allows the internal concentration and hence $\Delta\psi$ to be calculated via the Nernst equilibrium equation (equation 3.40). The combination of a TPP$^+$ electrode to

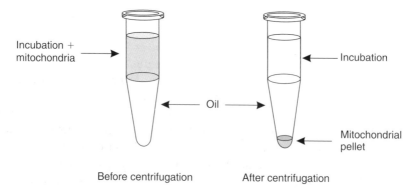

Before centrifugation   After centrifugation

Figure 4.3   Silicone-oil centrifugation of mitochondria for the determination of $\Delta\psi$.
Mitochondria are incubated under the desired conditions in media containing [^{3}H]TPP$^+$and [^{14}C]sucrose, or [^{14}C]sucrose and ^{3}H$_2$O. An aliquot of either incubation is added to an Eppendorf tube containing silicone oil and centrifuged for about 1 min at 10 000g. The mitochondria form a pellet under the oil and can be solubilized for liquid scintillation counting. An aliquot of the supernatant is also counted. The [^{14}C]sucrose in the incubation allows the sucrose-permable spaces, V_s, in the pellets to be calculated. This gives the extramatrix contamination of the pellet with incubation medium (i.e. extramitochondrial space plus intermembrane space). The difference between the ^{3}H$_2$O-permeable space in the pellet (V_h) and V_s gives the sucrose-*im*permeable space, which is taken to represent the matrix volume (as water but not sucrose can permeate the inner membrane). If the apparent [^{3}H]TPP$^+$space in the pellet (V_{TPP}) is calculated similarly, then $\Delta\psi$ can be calculated from the Nernst equation (3.40):

$$\Delta\psi = 2.3\frac{RT}{F}\log_{10}\frac{[V_{TPP} - V_s]}{[V_h - V_s]}$$

monitor $\Delta\psi$ and an oxygen electrode (see Fig. 4.7) to monitor proton current, allowing the calculation of proton conductance, will be discussed in Section 4.6. It should be borne in mind, however, in experiments involving Ca^{2+} that TPP$^+$ may inhibit the mitochondrial Ca^{2+} efflux pathway (Section 8.2.3).

Phosphonium cations can also be used to monitor changes in $\Delta\psi$ of mitochondria *in situ* within functioning cells. These techniques will be discussed in Chapter 9.

4.2.3 Intrinsic optical indicators of $\Delta\psi$

A $\Delta\psi$ of some 200 mV corresponds to an electrical field across the energy-transducing membrane in excess of 300 000 V cm^{-1}. It is not surprising, therefore, that certain integral membrane constituents respond to the electrical field by altering their spectral properties. Such *electrochromism* is due to the effect of the imposed field on the energy levels of the electrons in a molecule. The most widely studied of these intrinsic probes of $\Delta\psi$ are the carotenoids of photosynthetic energy-transducing membranes. Carotenoids are a heterogeneous class of long-chain, predominantly aliphatic pigments that are found in both chloroplasts and photosynthetic bacteria. Their roles include light-harvesting (Chapter 6) and protection against oxidative damage of the photosynthetic apparatus, since they can trap reactive excited states of oxygen molecules. A common feature of carotenoids is a central hydrophobic region with conjugated double bonds allowing delocalization of electrons and giving carotenoids their characteristic visible spectrum. The shifts in their absorption spectra in response to the membrane potentials experienced by energy-transducing membranes are only a few nanometers and so the signal is usually detected by dual-wavelength spectroscopy (Chapter 5). The carotenoids respond with extreme rapidity (nanoseconds or less) allowing primary electrogenic events in the photosynthetic apparatus to be followed (Chapter 6). The carotenoid band shift can be calibrated, especially in the case of chromatophores (Chapter 1), by monitoring the band shift in the dark with valinomycin and varying external KCl to generate defined potassium diffusion potentials (equation 3.40). One limitation of the technique, however, is that, since carotenoids are integral membrane components, they only detect the field in their immediate environment, which need not correspond to the bulk-phase membrane potential difference measured by distribution techniques, particularly as surface potential effects could be significant. In this context it is generally found that, in chromatophore membranes, the carotenoid shift gives much larger values of $\Delta\psi$ than ion distribution methods.

4.2.4 Extrinsic optical indicators of $\Delta\psi$ and ΔpH

Review and further reading   Rottenberg and Wu 1998, Nicholls and Ward 2000

Discussion here will be limited to investigation of isolated organelles, the complexity introduced by monitoring organelle function in intact cells will be reserved until Chapter 9. Lipophilic cations and anions with extensive π-orbital systems, which allow charge to be delocalized throughout the structure, and hence are membrane permeant, can achieve a Nernst equilibrium across the inner membrane and can thus be used to monitor $\Delta\psi$. These compounds frequently have characteristic absorption spectra in the visible region, and their planar structure allows them to form stacks when at high concentrations, thus reducing their ability to absorb light. The total absorbance of the mitochondrial incubation therefore decreases when the probe is accumulated into the matrix. Most of the probes currently in use are fluorescent, and under appropriate conditions a quenching of the fluorescence can be seen as the probe is accumulated into the matrix (Fig. 4.4). These techniques thus allow $\Delta\psi$ to be monitored qualitatively without separating mitochondria from the incubation. Great care must be taken both in the selection of the probe and the incubation conditions in order to obtain a reproducible signal, since many factors other than $\Delta\psi$ can interfere with the signal. Calibration is frequently performed with reference to diffusion potentials obtained with valinomycin in the presence of varying external KCl concentrations. A further problem is that many of the probes are mitochondrial inhibitors; thus while the cationic carbocyanines respond very rapidly to changes in membrane potential, at an external concentration of 40 nM they can inhibit complex I by 90%. Control experiments with an oxygen electrode are therefore important when setting up conditions.

Membrane permeant anions are accumulated by vesicles such as chromatophores and sonicated submitochondrial particles with positive-inside membrane potentials. The anionic bisoxonols are very sensitive, with a large fluorescent yield, but respond only slowly to changes in $\Delta\psi$, requiring up to 1 min to re-equilibrate after a transient. No extrinsic probe

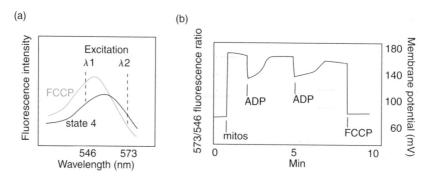

Figure 4.4   Monitoring $\Delta\psi_m$ of isolated mitochondria by dual-wavelength TMRM$^+$ fluorescence.
(a) Mitochondria were incubated in the presence of substrate plus TMRM$^+$ and excitation spectra (λ = 590 nm emission) obtained before (black trace) and after adding FCCP (blue trace) to depolarize the mitochondria. Note that the excitation peak shifts. This can be exploited to calibrate the ratio of emitted light when excited alternately at 546 and 573 nm as a function of $\Delta\psi_m$. (b) This ratiometric technique can be exploited to monitor $\Delta\psi_m$ in the fluorimeter during state 3, state 4 transitions (Section 4.5) induced by the addition of small amounts of ADP. When ADP is phosphorylated to ATP, $\Delta\psi_m$ returns to its maximal value. Data adapted from Scaduto and Grotyohann (1999).

has yet been described that acts like the intrinsic carotenoids, i.e. remaining fixed in the membrane and altering its spectrum in response to the applied field.

As discussed in Chapter 3, weak acids are accumulated by tenfold per pH unit into acidic compartments and weak bases by tenfold per pH unit into acidic compartments. ΔpH can be estimated from the fluorescence quenching of certain acridine dyes, which are weak bases and so will tend to accumulate on the acidic side of a membrane where their fluorescence may be quenched. There are often problems of quantifying such quenching in terms of a pH gradient but they can be useful qualitative probes because small amounts of membranes may be assayed without any requirement to separate the membranes from the suspending medium.

4.2.5 ΔpH determination by ^{31}P-NMR

Nuclear magnetic resonance (NMR) can be used to obtain a direct measurement of the pH inside and outside a cell or organelle, and is free of some of the drawbacks inherent in the more invasive use of weak acids or bases. The basis of the technique is that the resonance energy of the phosphorus nucleus in P$_i$ varies according to the protonation state of the latter. As the pK_a for H$_2$PO$_4^-$/HPO$_4^{2-}$ is 6.8, the technique can report pH values in the range of 6–7.6. The NMR signal is the average for the two ionization states, since proton exchange is fast on the NMR timescale. If there is phosphate in both the external and internal phases, a ΔpH can be calculated. The drawback is that NMR is an insensitive method and millimolar concentrations of P$_i$ and relatively thick cell suspensions are required, with attendant problems of supplying oxygen and substrates.

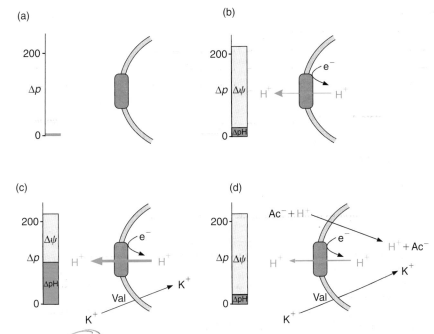

Figure 4.5 **Factors controlling the partition of Δp between Δψ and ΔpH.** Starting from a de-energized mitochondrion (a), the initiation of respiration (b) leads to a high Δψ and low ΔpH, since the electrical capacitance of the membrane is very low. In a high [K$^+$] medium, valinomycin collapses Δψ (c) and allows a high ΔpH to build up. If a permeant weak acid is additionally present (d), ΔpH will collapse and extensive swelling may occur.

Annotate diagrams from text

4.2.6 Factors controlling the contribution of Δψ and ΔpH to Δp

Some of the events which regulate the partition of Δp between Δψ and ΔpH are summarized in Fig. 4.5. The mitochondrion will be used as an example, but the discussion is equally valid for other energy-transducing systems. Starting from a 'de-energized' state of zero protonmotive force, the operation of an H$^+$-pump in isolation leads to the establishment of a Δp in which the dominating component is Δψ (Fig. 4.5a). The electrical capacity of the membrane is so low that the net transfer of 1 nmol of H$^+$ per mg of protein across the membrane establishes a Δψ of about 200 mV. The pH buffering capacity of the matrix is about 20 nmol of H$^+$ per mg of protein per pH unit, and the loss of 1 nmol of H$^+$ will only increase the matrix pH by 0.05 units (i.e. equivalent to 3 mV). Δp will thus be about 99% in the form of a membrane potential (Fig. 4.5b). In the absence of a significant flow of other ions it will stay this way in the steady state.

If an electrically permeant ion such as Ca^{2+}, or K$^+$ plus valinomycin, is now added (Fig. 4.5c), its accumulation in response to the high Δψ will tend to dissipate the membrane potential and hence lower Δp. The respiratory chain responds to the lowered Δp by a further net extrusion of protons, thus restoring the protonmotive force. Because of the pH buffering capacity of the matrix, the uptake of 20 nmol of K$^+$ (or 10 nmol of Ca^{2+}), balanced by the extrusion of 20 nmol of H$^+$, will lead to the establishment of a ΔpH of about −1 unit (equivalent to 60 mV). As the respiratory chain can only achieve the same total Δp as before, this means that the final $\Delta \psi$ must be nearly 60 mV lower than before uptake of the cation. *Thus cation uptake leads to a redistribution from $\Delta \psi$ to ΔpH.* The lowered $\Delta \psi$ means that cation uptake under these conditions becomes self-limiting, as the driving force steadily decreases until equilibrium is attained. For example, the uptake of Ca^{2+} by mitochondria in exchange for extruded protons is limited to about 20 nmol per mg of protein (Section 8.2.3), by which the time $\Delta \psi$ has decreased (and $-60\Delta pH$ increased) by about 120 mV.

The third event that can influence the relative contributions of $\Delta \psi$ and ΔpH is the redistribution of electroneutrally permeant weak acids and bases (Fig. 4.5d). Uptake of a weak acid in response to the ΔpH created by prior cation accumulation dissipates the pH gradient and allows the respiratory chain to restore $\Delta \psi$. However, the net accumulation of both cation and anion can result in osmotic swelling of the mitochondrial matrix (Section 2.5). This does not occur when the ions are Ca^{2+} and P_i, as formation of an osmotically inactive calcium phosphate complex prevents an increase in internal osmotic pressure (Section 8.2.3).

It is clear from the above discussion that $\Delta \psi$ and ΔpH indicators themselves, being ions, weak acids or weak bases, can disturb the very gradients to be measured unless care is taken. This is particularly true in the presence of valinomycin, as the ionophore brings into play the high endogenous K^+ of the matrix, with the result that $\Delta \psi$ will become clamped at the value given by the initial K^+ gradient. This risk is less apparent with cations such as TPP^+, which can be employed at very low concentrations.

4.3 THE STOICHIOMETRY OF PROTON EXTRUSION BY THE RESPIRATORY CHAIN

Further reading   Mitchell and Moyle 1967, Brand *et al.* 1976, Reynafarje *et al.* 1976

The proton current generated by the respiratory chain cannot be determined directly under steady-state conditions as there is no means of detecting the flux of protons when the rate of H^+ efflux exactly balances that of re-entry. It is, however, possible to measure the initial ejection of protons that accompanies the onset of respiration before H^+ re-entry has become established. By making a small precise addition of O_2 to an anaerobic mitochondrial suspension in the presence of substrate and monitoring the extent of H^+ extrusion with a rapidly responding pH electrode, one can thus obtain a value for the H^+/O stoichiometry for the segment of the respiratory chain between the substrate and O_2.

The practical details of an experiment to determine H^+/O are shown in Fig. 4.6. A number of precautions are necessary: first, an electrical cation permeability has to exist to allow charge compensation of the proton extrusion, which would otherwise be limited by the rapid build-up of $\Delta \psi$ (Section 4.6). Second, the pulse of O_2 must be small enough to prevent ΔpH from saturating. Third, however rapid the pulse of respiration, some protons will leak back across the membrane (and thus be undetected) before the burst of respiration is completed. These protons must be allowed for. The problem is enhanced if P_i (which is nearly always present in mitochondrial preparations) is allowed to re-enter the mitochondrion during the O_2-pulse. $H^+ : P_i^-$ symport (remember this is the same as OH^-/P_i antiport) is extremely active in most mitochondria, and the ΔpH-induced P_i uptake results in

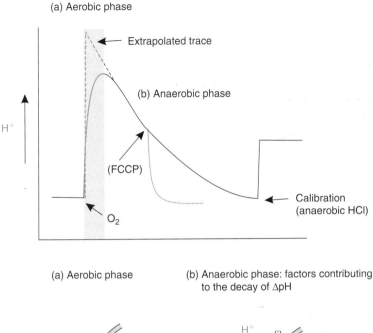

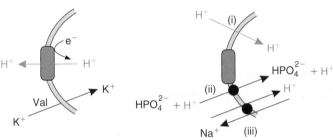

Figure 4.6 **Determination of mitochondrial H$^+$/O ratios by the 'oxygen pulse' technique.**

A concentrated mitochondrial suspension is incubated anaerobically in a lightly buffered medium containing substrate (e.g. succinate or β-hydroxybutryate from which matrix NADH can be generated), valinomycin and a high concentration of KCl. The pH of the suspension is continuously monitored with a fast-responding pH electrode. To initiate a transient burst of respiration, a small aliquot of air-saturated medium, containing about 5 nmol of O per mg protein is rapidly injected (note that some textbooks *erroneously* state that hydrogen peroxide solution was added). There is a transient acidification of the medium as the respiratory chain functions for 2–3 s while using up the added O_2. Valinomycin and K$^+$ are necessary to discharge any $\Delta\psi$, which would limit proton extrusion. When O_2 is exhausted, the pH transient decays as protons leak back into the matrix. This decay can be due to: (i) proton permeability of the membrane – note that FCCP accelerates the decay; (ii) the action of the endogenous Na$^+$/H$^+$ antiport; and (iii) electroneutral P$_i$ entry. The trace must then be corrected by extrapolation to allow for proton re-entry, which occurred before the oxygen was exhausted.

movement of protons into the matrix and hence an underestimate of H^+/O stoichiometry. Thus inhibition of the phosphate symport by N-ethylmaleimide significantly increases the observed H^+/O ratio.

Electron acceptors other than O_2 can be used to select limited regions of the mitochondrial respiratory chain, allowing a $H^+/2e^-$ ratio for the span to be obtained. Additionally, the charge stoichiometry (q^+/O or $q^+/2e^-$) may be determined instead of the proton stoichiometry by quantifying the compensatory movement of K^+ in the presence of valinomycin. Charge and proton-stoichiometry are not necessarily synonymous, since electrons can enter and leave respiratory complexes on opposite sides of the membrane (Fig. 3.5). For example, as will be described in Chapter 5 (Fig. 5.18), two electrons enter mitochondrial complex III from the matrix and are delivered to cytochrome c on the outer (P) face of the membrane; at the same time $4H^+$ are released to the P-phase; the $q^+/2e^-$ ratio is thus 2 while the $H^+/2e^-$ ratio is 4.

An alternative method for determining H^+/O ratios is based on the measurement of the initial rates of respiration and proton extrusion when substrate is added to the substrate-depleted mitochondria. This approach tends to give higher values than the oxygen pulse procedure.

Any stoichiometry determined by the above methods has to satisfy the constraints imposed by thermodynamics. In other words, the energy conserved in the proton electrochemical potential has to lie within the limits imposed by the redox span of the proton-translocating region. In addition, the proton-translocating regions of complexes I and III (which will be described in Chapter 5) are known to be in near-equilibrium as they can be readily reversed. Therefore an approximate stoichiometry for these regions can be deduced on purely thermodynamic grounds, knowing ΔE_h (Section 3.3.4) and the components of Δp.

Accurate stoichiometries have been notoriously difficult to obtain: the thermodynamic approach of equating the Gibbs energy change in reversible regions of the respiratory chain with the magnitude of the Δp under near-equilibrium conditions is beset with problems in the accurate determination of the latter parameter. Non-steady-state determinations of H^+ extrusion are also subject to controversy: it is difficult to eliminate the movement of compensatory ions masking the pH change, while some of the few protons that appear per respiratory chain complex in these experiments could conceivably originate from pK changes on the protein itself rather than as a result of transmembrane translocation. Nevertheless, most investigators agree that the NADH/O and succinate/O reactions translocate $10H^+/2e^-$ and $6H^+/2e^-$, respectively.

4.4 THE STOICHIOMETRY OF PROTON UPTAKE BY THE ATP SYNTHASE

Ultimately, the H^+/ATP ratio of the $F_1.F_0$-ATP synthase may be determined directly from the structure of the complex, which indicates that 3 mol of ATP are formed per revolution of the F_1 catalytic unit (Chapter 7). The H^+-driven rotor, which drives the rotation, is variously believed to require 10–14 protons per revolution, corresponding to between 3.3 and 4.4 protons per ATP.

The H^+/ATP ratio, Gibbs energy for ATP synthesis (ΔG_p) and Δp must be mutually consistent to satisfy thermodynamic constraints. Measured Δp values lie in the range of

170–200 mV for different systems. Comparison with Δp requires H^+/ATP values in the range 3–4 to satisfy equation 3.37 or 3.38.

A point to note is that submitochondrial particles appear to sustain a lower ΔG_p than mitochondria. This is consistent with the contribution of the adenine nucleotide translocator and phosphate transporter (Section 8.5), which utilize an additional proton for the import of ADP and P_i and export of ATP.

The stoichiometries for proton extrusion by a segment of the respiratory chain ($H^+/2e^-$) and for proton re-entry during ATP synthesis (H^+/ATP) have to be consistent with the overall observed stoichiometry of ATP synthesis related to electron flow ($ATP/2e^-$):

$$ATP/2e^- = \{H^+/2e^-\}/\{H^+/ATP\} \qquad [4.1]$$

or when O_2 is the final acceptor:

$$ATP/O = \{H^+/O\}/\{H^+/ATP\} \qquad [4.2]$$

4.5 PROTON CURRENT AND RESPIRATORY CONTROL

The previous sections have been concerned with the potential term in the proton circuit and with the gearing of the transducing complexes for the generation and utilization of this potential. This section will deal with the factors that regulate the proton current in the circuit.

The current of protons flowing around the proton circuit (J_{H^+}) may be readily calculated from the rate of respiration and the H^+/O stoichiometry:

$$J_{H^+} = \{\delta O/\delta t\}\{H^+/O\} \qquad [4.3]$$

For a given substrate, therefore, proton current and respiratory rate vary in parallel, and thus an oxygen electrode (Fig. 4.7) is an effective way of monitoring J_{H^+} as long as the H^+/O stoichiometry is known.

The oxygen electrode has for long been the most versatile tool for investigating the mitochondrial proton circuit. Although the electrode only determines directly the rate of a single reaction, the final transfer of electrons to O_2, information on many other mitochondrial processes can be obtained simply by arranging the incubation conditions so that the desired process becomes a significant step in determining the overall rate. Several such steps may be investigated (Fig. 4.8) including:

- substrate transport across the membrane
- substrate dehydrogenase activity
- respiratory chain activity
- adenine nucleotide transport across the membrane
- ATP synthase activity
- H^+-permeability of the membrane.

Three basic states of the proton circuit were shown in Fig. 4.1: open circuit, where there is no evident means of proton re-entry into the matrix; a circuit completed by proton re-entry coupled to ATP synthesis; and, thirdly, a circuit completed by a proton leak not

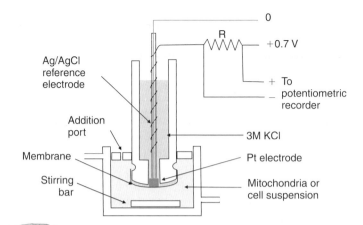

Figure 4.7 **The Clark oxygen electrode.**
O_2 is reduced to H_2O at the Pt electrode, which is maintained at 0.7 V negative with respect to the Ag/AgCl reference electrode, and a current flows which is proportional to the O_2 concentration in the medium. A thin O_2-permeable membrane prevents the incubation from making direct contact with the electrodes. Since the electrode slowly consumes O_2, the incubation must be continuously stirred to prevent a depletion layer forming at the membrane. The chamber is sealed except for a small addition port. The electrode is calibrated with both air-saturated medium and, under anoxic conditions, following dithionite addition.

coupled to ATP synthesis. These states can readily be created in the O_2-electrode chamber (Fig. 4.9).

When the oxgen electrode was first being applied to mitochondrial studies, Chance and Williams proposed a convention following the typical order of addition of agents during an experiment (Fig. 4.9):

- state 1: mitochondria alone (in the presence of P_i)
- state 2: substrate added, respiration low owing to lack of ADP
- state 3: a limited amount of ADP added, allowing rapid respiration
- state 4: all ADP converted to ATP, respiration slows
- state 5: anoxia.

Thus the addition of mitochondria to an incubation medium containing P_i (Fig. 4.9) causes little respiration as no substrate is present. Although mitochondria contain adenine nucleotides within their matrices, the amount is relatively small (about 10 nmol per mg of protein) and, when the mitochondria are introduced into the incubation, this pool will be very rapidly phosphorylated until it achieves equilibrium with Δp. That there is any respiration at all is only because the inner membrane is not completely impermeable to protons, which can therefore slowly leak back across the membrane even in the absence of net ATP synthesis. One factor that contributes to this proton leak is the slow cycling of Ca^{2+} across the membrane (Section 8.2.3), although mostly it is due to an endogenous 'non-ohmic' proton leak, which becomes most apparent at very high Δp (Section 4.6), or to the presence of a specific 'uncoupling protein' (Section 4.6.1). Only the terms state 3 and state 4 continue to be commonly used.

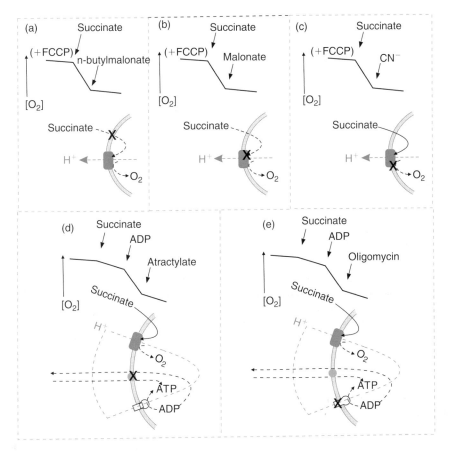

Figure 4.8   The use of the oxygen electrode to study mitochondrial energy transduction.
In the scheme, five ways of interfering (X) with energy transduction are shown. The diagrammatic oxygen electrode traces show how these perturbations might be investigated with the oxygen electrode. The incubation medium is presumed to contain osmotic support, a pH buffer and P_i. (a) Inhibition of a transport protein (in this case for succinate). (b) Inhibition of a substrate dehydrogenase (again for succinate). (c) Inhibition of a respiratory chain complex (by cyanide). (d) Inhibition of adenine nucleotide transport across the inner membrane by atractylate. (e) Inhibition of the ATP synthase by oligomycin.

4.5.1 The basis of respiratory control

How does the respiratory chain know how fast to operate? The control of respiration is a complex function shared between different bioenergetic steps and the control exerted by these steps can differ between mitochondria and under different metabolic conditions for the same mitochondria. The quantitative analysis of these fluxes is termed *metabolic control analysis* and will be derived in Section 4.7. Here we shall first give a simplified explanation based on oxygen electrode experiments of the type depicted in Fig. 4.9 to discuss respiratory control.

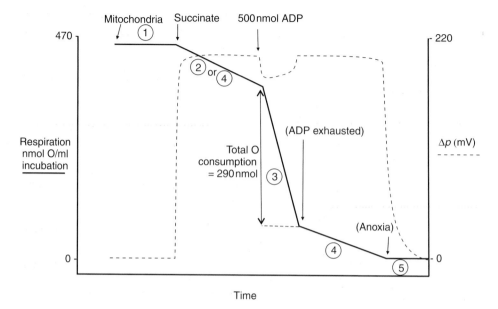

Figure 4.9 Respiratory 'states' and the determination of P/O ratios.
In this experiment mitochondria were added to an oxygen electrode chamber,
followed by succinate as substrate. Respiration is slow as the proton circuit is not
completed by H^+ re-entry through the ATP synthase. That there is any respiration
at all is because of a slow proton leak across the membrane. A limited amount of
ADP is added, allowing the ATP synthase to synthesize ATP coupled to proton
re-entry across the membrane, 'state 3' (Section 4.5). When this is exhausted,
respiration slows and finally anoxia is attained. The circled numbers refer to the
respiratory 'states'. If the amount of ADP is known, the oxygen uptake during the
accelerated 'state 3' respiration can be quantified allowing a P/O ratio to be calculated (moles ATP synthesized per mol O). Since the proton leak is almost negligible in 'state 3' (Section 4.6.2), the total oxygen uptake during 'state 3' is
effectively used for ATP synthesis. In this example, the ADP/O ratio for the substrate is found to be 500/290 = 1.72. Note the bioenergetic convention of refer-
ring to 'O', i.e. $\frac{1}{2}O_2$, which is equivalent to $2e^-$. Also, the controlled respiration
prior to addition of ADP, which is strictly termed 'state 2' is functionally the
same as state 4, and the latter term is usually used for both states. The dotted
trace reports the values of Δp during the experiment.

   A fundamental factor that controls the rate of respiration is the thermodynamic disequilibrium between the redox potential spans across the proton-translocating regions of the
respiratory chain and Δp. In the absence of ATP synthesis, respiration is automatically
regulated so that the rate of proton extrusion by the respiratory chain precisely balances the
rate of proton leak back across the membrane. If proton extrusion were momentarily to
exceed the rate of re-entry, Δp would increase, the disequilibrium between the respiratory
chain and Δp would in turn decrease and respiratory chain activity would decrease, restoring
the steady state. Once again the electrical circuit analogy is useful here.
   In the example shown in Fig. 4.9, respiration is disturbed by the addition of exogenous
ADP, mimicking an extramitochondrial hydrolysis of ATP such as would occur in an intact
cell. The added ADP exchanges with matrix ATP via the adenine nucleotide translocator,

and as a result, the ΔG_p for the ATP synthesis reaction in the matrix is lowered, disturbing the ATP synthase equilibrium. The following events then occur sequentially (but note that the gaps between them would be on the millisecond timescale):

(a) The ATP synthase operates in the direction of ATP synthesis and proton re-entry to attempt to restore ΔG_p.
(b) The proton re-entry lowers Δp.
(c) The thermodynamic disequilibrium between the respiratory chain and Δp increases.
(d) The proton current and hence respiration increases.

This accelerated 'state 3' respiration is once more self-regulating so that the rate of proton extrusion balances the (increased) rate of proton re-entry across the membrane. Net ATP synthesis, and hence state 3 respiration, may be terminated in three ways:

(a) When sufficient ADP is phosphorylated to ATP for thermodynamic equilibrium between the respiratory chain and Δp to be regained.
(b) By preventing adenine nucleotide exchange across the membrane with an inhibitor such as atractyloside (also called atractylate) (Section 8.5).
(c) By inhibiting the ATP synthase, for example, by the addition of oligomycin (Chapter 7).

Energy transduction between the respiratory chain and the protonmotive force is extremely well regulated, in that a small thermodynamic disequilibrium between the two can result in a considerable energy flux. Δp drops by less than 30% when ADP is added to induce maximal state 3 respiration. The actual disequilibrium between the respiratory chain and Δp is even less, as the ΔE values across proton translocation segments of the respiratory chain may also decrease in state 3.

Effective energy-transduction during state 3 is also apparent at the ATP synthase. A high rate of ATP synthesis can be maintained with only a slight thermodynamic disequilibrium between Δp and ΔG_p.

Proton translocators (Section 2.3.5) uncouple oxidative phosphorylation by inducing an artificial proton permeability in bilayer regions of the membrane. They may thus be used to over-ride the inhibition of proton re-entry, which results from an inhibition of net ATP synthesis. As a consequence proton translocators such as FCCP can induce rapid respiration, regardless of the presence of oligomycin or atractylate, or the absence of ADP (Fig. 4.8).

4.6 PROTON CONDUCTANCE

In an electrical circuit, the conductance of a component is calculated from the current flowing per unit potential difference. A similar calculation for the proton circuit enables the effective proton conductance of the membrane ($C_M H^+$) to be calculated

$$C_M H^+ = J_{H^+}/\Delta p \qquad [4.4]$$

This can be illustrated by a simple example. Liver mitochondria oxidizing succinate in 'state 4' might typically respire at 15 nmol of O $\text{min}^{-1}\,\text{mg}^{-1}$ and maintain a Δp of 220 mV.

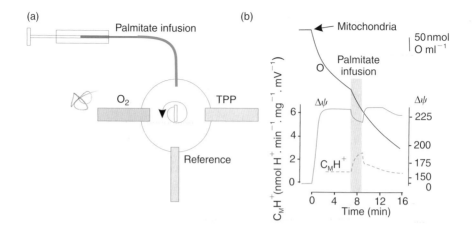

Figure 4.10 The physiological control of brown adipose tissue UCP1.
(a) Combination tetraphenylphosphonium and oxygen electrodes for the continuous monitoring of $\Delta\psi$, respiration, J_{H^+} and C_MH^+ in brown adipose tissue mitochondria during infusion of the thermogenic substrate palmitate into the stirred reaction chamber. (b) Details of an experiment in which palmitate was infused (shaded period) while respiration and membrane potential were monitored (as the mitochondria accumulate the cation in response to their membrane potential the electrode will detect the fall in external [TPP$^+$]). The incubation contained pyruvate and ATP, CoA and carnitine allowing the palmitate to be activated to palmitoyl CoA and then palmitoyl carnitine. Palmitate accumulation activates the uncoupling protein increasing C_MH^+. On the conclusion of the infusion, the mitochondria automatically recouple as the palmitate is activated and oxidized, allowing the fatty acid to leave its binding site on UCP1. *Continued* →

If the H$^+$/O ratio for the span succinate-oxygen is 6, then:

$$J_{H^+} = \{\delta O/\delta t\}\{H^+/O\} = 15 \times 6 = 90 \, \text{nmol H}^+ \text{min}^{-1}\text{mg}^{-1} \qquad [4.5]$$
$$C_MH^+ = J_{H^+}/\Delta p = 90/220 = 0.4 \, \text{nmol H}^+ \text{min}^{-1}\text{mg}^{-1}\text{mV}^{-1} \qquad [4.6]$$

If now the protonophore FCCP is added and Δp drops to 40 mV while respiration increases to 100 nmol O min^{-1}mg^{-1}, then the new values are as follows:

$$J_{H^+} = \{\delta O/\delta t\}\{H^+/O\} = 100 \times 6 = 600 \, \text{nmol H}^+ \text{min}^{-1}\text{mg}^{-1} \qquad [4.7]$$
$$C_MH^+ = J_{H^+}/\Delta p = 600/40 = 15 \, \text{nmol H}^+ \text{min}^{-1}\text{mg}^{-1}\text{mV}^{-1} \qquad [4.8]$$

The magnitude of the endogenous proton conductance of the membrane is a central parameter underlying the bioenergetic behaviour of a given preparation of mitochondria. Evidently, for an efficient transduction of energy, C_MH^+ should be as low as possible.

The respiratory chain does not distinguish between a Δp, which is lowered by the addition of a proton translocator and one which is lowered by ATP synthesis.

The combination of an oxygen electrode and a TPP$^+$ electrode (see Fig. 4.10) recording a single mitochondrial suspension is a powerful tool for continuously monitoring the proton circuit. When conditions are optimized to minimize the ΔpH component of Δp (phosphate present or nigericin addition in a KCl-based medium), $\Delta\psi$ monitored by the TPP$^+$ electrode will approximate Δp and, together with the rate of respiration, will allow J_{H^+} (equation 4.5)

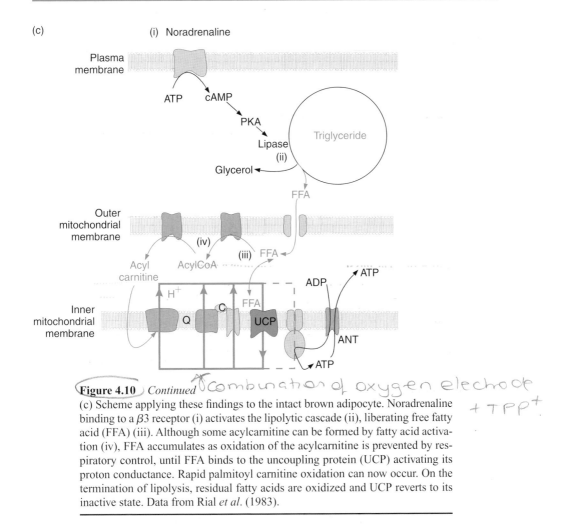

Figure 4.10 *Continued* ↑Combination of oxygen electrode + TPP+.

(c) Scheme applying these findings to the intact brown adipocyte. Noradrenaline binding to a β3 receptor (i) activates the lipolytic cascade (ii), liberating free fatty acid (FFA) (iii). Although some acylcarnitine can be formed by fatty acid activation (iv), FFA accumulates as oxidation of the acylcarnitine is prevented by respiratory control, until FFA binds to the uncoupling protein (UCP) activating its proton conductance. Rapid palmitoyl carnitine oxidation can now occur. On the termination of lipolysis, residual fatty acids are oxidized and UCP reverts to its inactive state. Data from Rial *et al.* (1983).

and $C_M H^+$ (equation 4.6) to be calculated. We shall now illustrate this approach with reference to highly specialized mitochondria in brown adipose tissue that demonstrate a regulated proton leak pathway.

4.6.1 Brown adipose tissue and the analysis of the proton circuit

Reviews Nicholls and Locke 1984, Nicholls and Rial 1999

Brown adipose tissue is the seat of non-shivering thermogenesis, the ability of hibernators, cold-adapted rodents and newborn mammals in general to increase their respiration and generate heat without the necessity of shivering. In extreme cases, whole body respiration can increase up to tenfold as a result of the enormous respiration of this tissue, which rarely accounts for more than 5% of the body weight even in a small rodent. The tissue is innervated by noradrenergic sympathetic neurons, and release the transmitter on to an unusual class of β3-receptors activates adenylyl cyclase and hence hormone-sensitive lipase. This leads to the hydrolysis of the triglyceride stores, which are present in multiple small droplets, giving the cell a 'raspberry' appearance. The brown adipocytes are packed with mitochondria, whose extensive inner membranes indicate a high capacity for respiration. However, the chemiosmotic theory now poses a problem: how can the fatty acids liberated by lipolysis be oxidized by the mitochondria when the main rate limitation in the proton circuit is the re-entry of protons into the mitochondrial matrix? The problem is compounded by the relatively low amount of ATP synthase and by the absence of any significant extra-mitochondrial ATP hydrolysis activity.

Two solutions are possible from first principles: either the brown fat mitochondrial respiratory chain is modified so that it does not expel protons, or the membrane is modified to allow re-entry of protons in the absence of ATP synthesis. The latter turns out to be the case. The mitochondrial inner membrane contains a unique 32 kDa uncoupling protein (UCP1), which binds a purine nucleotide to its cytoplasmic face and is inactive until the free fatty acid concentration in the cytoplasm starts to rise. The protein then binds a fatty acid and alters its conformation to become proton conducting. The uncoupling protein thus acts as a self-regulating endogenous uncoupling mechanism, which is automatically activated in response to lipolysis allowing uncontrolled oxidation of the fatty acids (Fig. 4.10). The low conductance state is restored when lipolysis is terminated, and the mitochondria oxidize the residual fatty acids.

The physiological regulation seen in intact brown adipocytes can be mimicked with isolated mitochondria in a combined oxygen electrode/TPP$^+$ electrode chamber by the infusion of fatty acid (mimicking lipolysis) in the presence of coenzyme A, carnitine and ATP to allow the fatty acid to be activated. The increase in $C_M H^+$ correlates with the steady-state concentration of free fatty acid during the infusion and on the termination of lipoly-sis (Fig. 4.10). The structures of UCP1 and related candidate uncoupling proteins are homologous to those of inner membrane metabolite transporters and will be discussed in Chapter 8.

Expression of the uncoupling protein occurs in response to the adaptive status of the animal: it is present in the mitochondria at high concentration at birth, but is then repressed so that the mitochondria lose the protein and the capacity for non-shivering thermogenesis. Cold-adaptation or, interestingly, overfeeding, can under certain conditions lead to re-expression of the protein.

4.6.2 The basal proton leak

A proton leak can also be found in mitochondria that do not express uncoupling proteins. This basal leak is responsible for the state 4 respiration seen in even the most carefully prepared mitochondria and accounts for a high proportion of the basal metabolic rate in tissues such as skeletal muscle. The molecular basis of the leak is still unclear, it cannot be correlated with the presence of a specific protein and does not correlate with the phospholipid composition of the bilayer. It is possible that protons leak across the membrane at the junctions between protein and lipid. Within the context of electrical analogy for the proton circuit, the basal leak displays a non-ohmic current/voltage relationship, which can be investigated by progressively restricting respiration. Proton conductance is most apparent at very high Δp and decreases more than proportionately with Δp (Fig. 4.11). This 'non-ohmic' I/V relationship suggests that the endogenous leak may have evolved to limit the maximal value of Δp attainable by mitochondria. Since the production of reactive oxygen species by

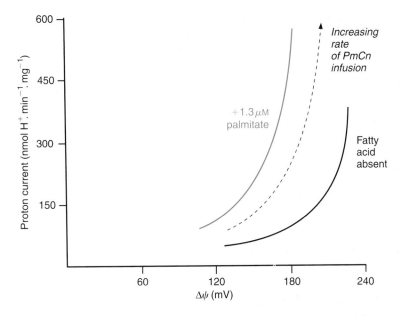

Figure 4.11   Determining the current voltage (*I/V*) relationship for the brown adipose tissue (BAT) UCP1 for BAT mitochondria in the presence or absence of fatty acids.
Brown adipose tissue mitochondria were incubated in state 4 in the presence of excess ATP to inhibit the uncoupling protein. Palmitoyl carnitine (PmCn) was infused as substrate at varying rates and respiration and $\Delta\psi$ determined as in Fig. 4.10 to generate an *I/V* curve for the inherent inner membrane proton leak in the absence of UCP1. The experiment was then repeated in the presence of 1.3 μM palmitate to activate the uncoupling protein. Note that the conductances are 'non-ohmic' and that fatty acid lowers $\Delta\psi$. Data from Rial *et al.* (1983).

mitochondria is highly dependent on Δp (Chapter 5), proton leaks, whether endogenous or uncoupler protein mediated, may serve the purpose of restricting oxidative damage.

4.7 MITOCHONDRIAL RESPIRATION RATE AND METABOLIC CONTROL ANALYSIS

Review and further reading   Davey *et al.* 1998, Murphy 2001

In the previous section we have explained a simplified model of the connections between mitochondrial respiration, Δp and $C_M H^+$. However, in practice there are many more steps interacting with the proton circuit, including the supply and transport across the inner membrane of substrate, the supply of electrons to the respiratory chain via the metabolite dehydrogenases, and at the other end of the circuit the activity of the adenine nucleotide translocator (Section 8.5) and the rate of ATP turnover. One approach to this complexity is to invoke non-equilibrium thermodynamics where fluxes are described in terms of the net thermodynamic driving forces under near-equilibrium conditions, but the most useful technique is to apply quantitative *metabolic control analysis* (MCA) to provide a simple description of how control is distributed between multiple steps.

'Control' has a precise meaning in MCA. Consider a simple metabolic pathway comprising two enzymes, E_1 and E_2, where the overall flux through the pathway in steady state is J, i.e.

$$A \xrightarrow{\ E_1\ } B \xrightarrow{\ E_2\ } C \longrightarrow J \qquad [4.9]$$

The *flux control coefficient* C relates changes in the overall flux through the pathway to changes in the activity of an enzyme or transport process. Strictly it is defined as the fractional change in flux divided by the fractional change in the amount of the enzyme as the change tends to 0, i.e. for E_1 in the above example the control coefficient $C_{E_1}^J$ equals:

$$C_{E_1}^J = \lim_{\delta E \to 0} \frac{\delta J / J}{\delta E_1 / E_1} \qquad [4.10]$$

We can illustrate this with a simple example. Consider a mitochondrion respiring in state 3. If we deliberately alter the activity of a single step in the overall sequence, for example, the adenine nucleotide translocator, by a small fraction, say 1%, what effect does this have on the overall respiration rate? Two extreme results are possible in this type of experiment. First, the change in flux through the entire pathway may be the same percentage as the change in activity of the single step, i.e. 1%. In this case, the *flux control coefficient* of the adenine nucleotide translocator would be said to be 1. The second extreme would be when a 1% change of the translocator activity had no effect on the overall flux. In this case, the step would have a flux control coefficient of 0. In practice flux control coefficients of one are rare; the idea of a single rate-determining step, to which a flux control coefficient of 1 corresponds, although often encountered in chemical reactions, rarely applies to metabolic sequences. Instead there is an interplay between many steps, each of which may have significant flux control coefficients, with values in non-branched pathways between 0 and 1. The *summation theorem* states that, in any pathway, the sum of all the individual flux control coefficients is always one, i.e. for the pathway in equation 4.9:

$$C_{E_1}^J + C_{E_2}^J = 1 \qquad [4.11]$$

The *elasticity coefficient* ϵ is the fractional change in activity of an enzyme or transport process in response to a small change in its substrates, products or other effectors. For the example in equation 4.9, consider how the activity, V_{E_2}, of enzyme E_2 responds to changes in concentration of its substrate B:

$$\varepsilon_B^{E_2} = \lim_{\delta B \to 0} \frac{\delta V_{E_2} / V_{E_2}}{\delta B / B} \qquad [4.12]$$

Finally, the *connectivity theorem* states that the products of the flux control coefficient and the elasticity to a given substrate for all enzymes connected by that substrate add up to 0.

If all this seems a little dry and theoretical, we shall now apply MCA to mitochondrial oxidative phosphorylation, which is especially suited to this type of analysis. For an isolated mitochondrion, some processes that can be analysed are summarized in Fig. 4.12. Two

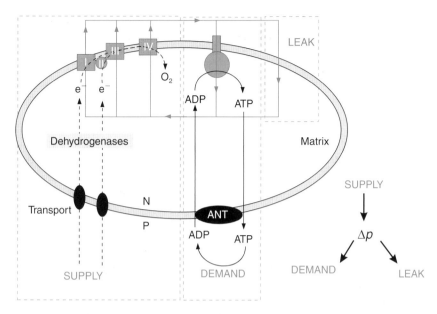

Figure 4.12   Modules for metabolic control analysis.
The complexity of the bioenergetic pathways may be lessened for metabolic control analysis by grouping processes together that are linked by a single common intermediate (here Δp). ANT is the adenine nucleotide translocator (see Chapter 8).

approaches are possible – bottom-up and top-down. *Bottom-up* analysis examines the effects of titrating specific mitochondrial enzymes and transporters with irreversible inhibitors and determining the effects on respiratory rate and ATP synthesis.

As will be discussed in Chapter 9, even modest restrictions in respiratory chain capacity *in vivo* greatly sensitize neurons to damaging stimuli. Indeed, chronic complex I restriction in animal models can simulate the neurodegenerative characteristics of Parkinson's disease, while complex II inhibition reproduces the damage to the striatum found in Huntington's disease. A bottom-up approach has been made for isolated brain mitochondria by titrating mitochondria with high-affinity inhibitors acting on individual complexes. For example, flux control coefficients of 0.29, 0.2 and 0.13 were determined in state 3 for complexes I, III and IV, respectively, of presynaptic mitochondria. The high coefficient seen at complex I is consistent with the sensitivity of the neuron to even slight inhibition of this complex (Chapter 9).

A limitation of the bottom-up technique is the requirement for irreversible inhibitors and the need to know each system component, which is difficult for more complex systems such as cells or tissues. An alternative approach is the *top-down* approach (or top-down elasticity analysis), in which processes are grouped into blocks linked to a common intermediate. The elasticities (equation 4.12) of each block to the common intermediate allows the overall flux control coefficients of each block to be determined. For the mitochondrion, $\Delta\psi$ is usually considered the common intermediate, fed by a block comprising substrate transport, metabolism and respiratory chain and drained by two blocks: the proton leak and a second comprising ATP synthesis, transport and turnover. The elasticities of each block towards $\Delta\psi$ can be determined by titrating with uncouplers or inhibitors. Finally, the connectivity

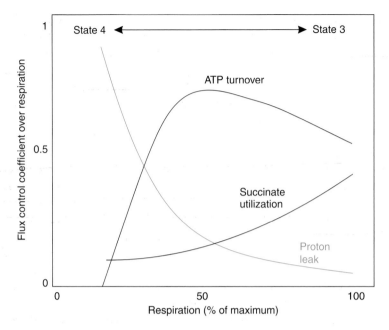

Figure 4.13 Metabolic control analysis of steps in mitochondrial oxidative phosphorylation during a progressive transition from state 4 to state 3.
As described in the text, as the respiration rate increases from state 4 to state 3, with increasing addition of hexokinase to increase the rate of the ATP turnover reactions, the flux control coefficients alter as shown (adapted from Hafner *et al.* 1990).

theorem can be used to determine the flux control coefficients of the blocks over respiration, ATP turnover and proton leak.

Figure 4.13 shows the control exerted by the three blocks in isolated mitochondria over respiration. Control is shared between multiple steps and the distribution changes with metabolic state during a transition from state 4 to state 3. The mitochondria are supplied with succinate as respiratory substrate together with ADP and P_i. The initial state 4 is attained when the net conversion of ADP and P_i to ATP ceases. As intuitively expected, the overall flux control coefficient of the set of reactions: adenine nucleotide translocation, ATP synthesis and consumption of ATP ('ATP turnover'), is 0 in state 4. In the previous sections we have made the simplification that the proton leak across the mitochondrial membrane completely controls the respiration in state 4 (i.e. has a flux control coefficient of 1). However, more careful analysis shows that, although the proton leak is indeed dominant (flux control coefficient 0.9), there is also significant control (coefficient of 0.1) in the set of reactions catalysing transport of succinate into the mitochondrion and its oxidation by the electron transport chain.

If glucose and incremental amounts of hexokinase are now added, to accelerate ATP turnover, respiration will steadily increase until the rate of ATP synthesis reaches a maximum and the mitochondria are in state 3. The first additions of hexokinase each cause a marked increase in the respiration rate and thus the flux control coefficient of the 'ATP turnover' reactions is high, corresponding to the classic respiratory control. As further hexokinase is added, other components of the 'ATP turnover' reactions, particularly the adenine nucleotide translocator, assume an increasing share of the control. Concomitantly, the control by hexokinase becomes a progressively smaller component of the control exerted by the 'ATP turnover' reactions. At the limit of state 3 respiration, further additions of hexokinase are without effect on the respiration rate and thus its flux control coefficient falls to 0, the classic 'uncontrolled respiration' (state 3).

In state 3 control is shared almost equally (Fig. 4.13) between 'ATP turnover' reactions and those of 'succinate utilization'. More detailed analysis shows that it is distributed between the adenine nucleotide translocator, the dicarboxylate translocator, the cytochrome bc_1 complex and cytochrome oxidase. As the respiration rate alters between the extremes of state 3 and state 4, the quantitative contribution of each of these components varies; for example, control due to the adenine nucleotide translocator rises to its greatest flux control strength at 75% of the maximum respiration rate. *The important outcome of this analysis is that neither in state 3 nor state 4 is a single step responsible for the control of the mitochondrial respiration rate.* Traditional attempts to correlate respiratory control with the [ATP]/[ADP][P_i] ratio or a single irreversible step in the electron transport chain (e.g. a step in the cytochrome oxidase reaction) are thus not tenable.

Top-down analysis can be used for more complex systems such as intact cells and can include cytoplasmic metabolic blocks such as glycolysis and cellular ATP turnover. In an intact cell the factors controlling mitochondrial respiration rate will be more varied and complex than those considered above. The major respiratory substrate will not be succinate but rather NADH generated in the matrix. There will also be important differences between mitochondria from different cell types. Mitochondria in a liver cell respire at a rate intermediate between state 3 and state 4. Control analysis shows that this rate is controlled by processes (such as glycolysis, fatty acid oxidation and the tricarboxylic acid cycle) that supply mitochondrial NADH (flux control coefficient 0.15 to 0.3) by the proton leak (flux control coefficient 0.2) and by the 'ATP turnover' reactions (flux control coefficient 0.5). Oxidation of NADH is less important with a flux control coefficient between 0 and 0.15. Fluctuations in rate can be caused by hormones or increases in cytoplasmic and matrix Ca^{2+} via three separate effects; alteration of either ATP turnover, NADH supply or proton leak. Each of these effects may be important. Muscle mitochondria can experience periods of resting activity when they may be close to the state 4 respiration rate but upon initiation of contraction the ATP demand and raised Ca^{2+} may be such as to cause transition to state 3. If anaerobiosis approaches, then the rate of respiration could conceivably pass transiently through a stage where cytochrome oxidase has a higher flux control coefficient, owing to restriction on the supply of oxygen.

4.8 OVERALL PARAMETERS OF ENERGY TRANSDUCTION

These are independent of the chemiosmotic theory and indeed predate it by many years, but are frequently still found in the current literature owing to the ease of their determination, for example, with an oxygen electrode.

4.8.1 Respiratory control ratio

This is an empirical parameter frequently used for assessing the 'integrity' of a mitochondria preparation. It is based on the observation that 'damaged' mitochondrial preparations tend to show an increased proton leak above and beyond that which is now known to be an inherent physiological property of the mitochondrion. It is defined as the state 3 respiratory rate attained during maximal ATP synthesis (i.e. in the presence of ADP), or in the presence of a proton translocator, divided by the rate in the absence of ATP synthesis or proton translocator. It is therefore a hybrid parameter depending on a number of primary parameters, notably the endogenous proton leak. Typical values for the ratio vary from 3 to 15 in different preparations. Note that most bacterial cells do not show significant respiratory control owing, presumably, to the continuous activity of the ATP synthase and other protonmotive force-driven reactions. Respiratory control can be observed in some inside-out vesicle preparations. The rate of electron transport in isolated thylakoids does accelerate when ATP synthesis is occurring.

4.8.2 P/O (ATP/O, ADP/O) and P/2e⁻ (ATP/2e⁻, ADP/2e⁻) ratios

While the stoichiometries of proton translocation by the respiratory chain and ATP synthase appear to be fixed, even if the actual values are still a matter for contention, the overall stoichiometry of mitochondrial ATP synthesis in relation to respiration can vary from a theoretical maximum of about 2.5 ATP per $2e^-$ passing from NADH to oxygen down to zero, depending on the activity of the parallel proton leak pathway bypassing the ATP synthase. The variety of terms in the heading are used to describe this empirical ratio; they are roughly synonymous.

The P/O ratio is the number of moles of ADP phosphorylated to ATP per $2e^-$ flowing through a defined segment of an electron transfer to oxygen. If the terminal acceptor is not oxygen then the term P/$2e^-$ ratio is used. P/O ratios can be determined from the extent of the burst of accelerated state 3 respiration obtained when a small measured aliquot of ADP is added to mitochondria respiring in state 4. Almost all the added ADP is phosphorylated to ATP, the ATP : ADP ratio being typically at least 100 : 1 when state 4 is regained, and the ratio 'moles of ADP added/moles of O consumed' can be calculated. It is the convention to assume that the proton leak ceases during state 3 respiration, which is largely true, due to its 'non-ohmic' nature. Thus the total oxygen consumed is used in the calculation (Fig. 4.9).

Values for ADP/$2e^-$ ratios are, as with all stoichiometries, a source of debate. The 'classic' value of 1 ATP per $2e^-$ per proton-translocating complex is no longer tenable. For example, complex III (UQH_2–cyt c) has a $H^+/2e^-$ ratio of 4, but because electrons enter the complex from the matrix side and leave from the cytoplasmic face, the charge/$2e^-$ ($q^+/2e^-$) ratio is only 2. Conversely, complex IV (cyt c–O_2) has a $H^+/2e^-$ ratio of 2 but a $q^+/2e^-$ ratio of 4 (Fig. 5.18). To maintain overall electroneutrality, positive charge, in the form of protons, must flow back through the ATP synthase to balance the charge displacement and to make ATP, it follows that the P/$2e^-$ ratio for complex IV should be twice that for complex III. This is more than a semantic argument for many bacteria, since electron transfer frequently terminates at the level of cyt c and so a dissection of P/$2e^-$ ratios for individual parts of an electron transport system is important.

Taking the consensus view for mitochondria that the $H^+/2e^-$ ratio is 6 for the span succinate-O_2, that the H^+/ATP ratio at the ATP synthase is about 3 (Chapter 7), and that one additional proton is consumed in the transport of P_i and translocation of adenine nucleotides, it follows that the theoretical maximum P/O ratio for succinate oxidation would be 1.5, close to what is observed, rather than the 'classic' value of 2. The $H^+/2e^-$ stoichiometry of complex I (NADH-UQ) is thought to be 4. If so then the maximum P/O ratio for the span NADH-O_2 would be 2.5. P/O values of 2.5 and 1.5 for mitochondrial oxidation of NADH and succinate (or of other substrates from which electrons enter the chain at the level of ubiquinone) mean that the standard textbook stoichiometries of ATP synthesis associated with the total oxidation of carbohydrates and fats need revising downwards.

Similar considerations apply to ATP synthesis in bacterial and thylakoid systems; the maximum $P/2e^-$ ratio will be determined by the relative values of $H^+/2e^-$ and H^+/ATP. Note that the translocation step for adenine nucleotide and phosphate is not involved in the synthesis of ATP by bacterial and thylakoid membranes, nor indeed in submitochondrial particles, and thus these systems should therefore have higher P/O ratios for a given $H^+/2e^-$ stoichiometry.

4.9 REVERSED ELECTRON TRANSFER AND THE PROTON CIRCUIT DRIVEN BY ATP HYDROLYSIS

The ATP synthase is reversible and is only constrained to run in the direction of net ATP synthesis by the continual regeneration of Δp and the use of ATP by the cell. If the respiratory chain is inhibited and ATP is supplied to the mitochondrion, or if sufficient Ca^{2+} is added to depress Δp below that for thermodynamic equilibrium with the ATP synthase reaction, the enzyme complex functions as an ATPase, generating a Δp comparable to that produced by the respiratory chain. The proton circuit generated by ATP hydrolysis must be completed by means of a proton re-entry into the matrix. Proton translocators therefore accelerate the rate of ATP hydrolysis, just as they accelerate the rate of respiration; this is the 'uncoupler-stimulated ATPase activity'. This is of particular importance in the cellular context, where mitochondrial dysfunction can cause ATP synthase reversal and drain glycolytically generated ATP (Section 9.1.1).

The classic means of discriminating whether a mitochondrial energy-dependent process is driven directly by Δp or indirectly via ATP, is to investigate the sensitivity of the process to the ATP synthase inhibitor oligomycin. A Δp-driven event would be insensitive to oligomycin when the potential was generated by respiration, but sensitive when Δp was produced by ATP hydrolysis. The converse would be true of an ATP-dependent event. If Δp or $\Delta \psi$ is being monitored, mitochondria (isolated or *in situ*), which are net generators of ATP, will hyperpolarize on addition of oligomycin, while those whose Δp is supported by ATP hydrolysis will depolarize. This 'null-point' assay is a simple way of monitoring mitochondrial function within cells.

The near-equilibrium in state 4 between Δp and the redox spans of complexes I and III suggests that conditions could be devised in which these segments of the respiratory chain could be induced to run backwards, driven by the inward flux of protons. It should be noted that this does not apply to complex IV, which is essentially irreversible. Reversed electron transfer may be induced in two ways, either through generating a Δp by ATP hydrolysis, or by using the flow of electrons from succinate or cytochrome c to O_2 to reverse electron transfer through complexes I or I and III, respectively (Fig. 4.14). Such flow of electrons, e.g. from succinate, involves the majority of the electron flux passing to O_2 and thereby generating Δp, whilst a minority is driven energetically uphill to reduce NAD^+ at the expense of Δp.

Under physiological conditions the mitochondrial ATP synthase will not normally be called upon to act as a proton-translocating ATPase, except possibly during periods of anoxia when glycolytic ATP could be utilized to maintain the mitochondrial Δp. However, some bacteria, such as *Streptococcus faecalis* when grown on glucose, lack a functional respiratory chain and rely entirely upon hydrolysis of glycolytic ATP to generate a Δp across their membrane and enable them to transport metabolites. Reversed electron transport driven by Δp generated through respiration is an essential process in some bacterial species (see Chapter 5).

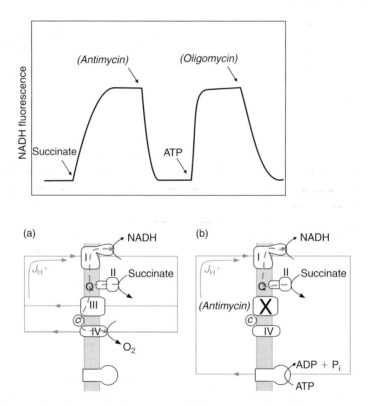

Figure 4.14   Reversed electron transfer in the mitochondrial respiratory chain.
Schematic response of SMPs incubated in the presence of NAD^+. In (a), a Δp is generated by succinate oxidation. Δp then drives complex I in reverse, causing NAD^+ reduction, i.e. succinate acts as both donor of electrons for reversed electron transfer and as substrate for complexes III and IV. In (b) complex III is inhibited by antimycin A and the Δp is generated by ATP hydrolysis. Succinate merely donates electrons for reversed electron transfer through complex I.

4.10 ATP SYNTHESIS DRIVEN BY AN ARTIFICIAL PROTONMOTIVE FORCE

An artificially generated Δp must be able to cause the net synthesis of ATP in any energy-transducing membrane with a functional ATP synthase. The first demonstration that this was so came from thylakoids after equilibration in the dark at acid pH. They could be induced to synthesize ATP when the external pH was suddenly increased from 4 to 8, creating a transitory pH gradient of 4 units across the membrane, the *acid bath experiment* (Fig. 4.15). For many years this experiment has been interpreted on the basis that thylakoids normally operate with ΔpH as the main component of Δp, owing to the ease with which Cl^- redistributes across the thylakoid membrane to collapse $\Delta\psi$ (Chapter 6).

An important corollary of this experiment is that the ATP synthase can be driven by ΔpH alone. There is no thermodynamic objection to this, but it is currently being argued (Chapter 7) that, mechanistically, a $\Delta\psi$ is required. In this context it has been recently proposed that the acid bath experiment should be re-interpreted. It is argued that the transition to higher external pH was accompanied by the efflux of the succinate monoanion (towards which the membrane is claimed to be permeable), thus generating a diffusion potential (Chapter 3), positive inside, for this ion. Thus an induced $\Delta\psi$ would be at least part of the driving force for the ATP synthesis seen in this type of experiment (Fig. 4.15). It remains to be finally decided if this re-interpretation is valid but it is important to understand that it makes no difference to the validity of the experimental approach for showing that an imposed protonmotive force, independent of electron transport reactions, can drive ATP synthesis.

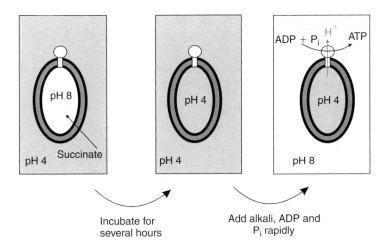

Figure 4.15   The 'acid bath' experiment: a ΔpH can generate ATP.
Thylakoid membranes were incubated in the dark at pH 4 in the presence of electron transport inhibitors in a medium containing succinate, which slowly permeated into the thylakoid space liberating protons and lowering the internal pH to about 4. The external pH was then suddenly raised to 8, creating a ΔpH of 4 units across the membrane. Traditionally H^+ efflux through the ATP synthase has been regarded as charge compensated by Cl^- efflux and/or Mg^{2+} influx. A $\Delta\psi$ may also be induced, see text. ADP and P_i were simultaneously added and proton efflux through the ATP synthase led to the synthesis of about 100 mol of ATP per mol of synthase. Protonophores such as FCCP inhibited the ATP production.

For an analogous acid bath experiment with mitochondria or bacteria, an ionophore such as valinomycin is needed to allow movement of compensating charge. Submitochondrial particles, which are inverted relative to intact mitochondria, are treated with valinomycin to render them permeable to K^+, incubated at low pH in the absence of K^+ to acidify the matrix, and then transferred to a medium of higher pH containing K^+. K^+ entry creates a diffusion potential, positive inside, and this, together with the artificial ΔpH that has just been created, generates a short-lived Δp. Protons exit through the ATP synthase, generating a small amount of ATP. K^+ enters on valinomycin to maintain charge balance. Eventually the K^+ and H^+ gradients run down to the extent that ATP synthesis ceases. An analogous approach has been used to demonstrate Δp-driven secondary active transport (Chapter 8).

4.11 KINETIC COMPETENCE OF Δp IN THE PROTON CIRCUIT

4.11.1 Proton utilization

If Δp is the intermediate between electron transport and ATP synthesis, then the sudden imposition of an artificial Δp of comparable magnitude to that normally produced by the respiratory chain should lead to ATP synthesis with minimal delay and at an initial rate comparable to that seen in the natural process. In other words, the proton circuit requires a cause and effect relationship.

Tests of kinetic competence have been made for both the thylakoid and submitochondrial particle systems as described above, except that the protonmotive force was imposed by rapid mixing. The subsequent reaction period can be altered by varying the length of tubing between the mixing and quenching points (where the reaction is terminated by concentrated acid). In this way ATP synthesis on the millisecond timescale can be followed. In both preparations ATP synthesis was initiated with no significant lag and at initial rates comparable to those seen for the normal energy transduction. Indeed, in the case of the submitochondrial particles, the onset of ATP synthesis was more rapid than following initiation of respiration.

4.11.2 Proton movements driven by electron transport

While the experiments described above are clearly consistent with the kinetic competence of Δp as the intermediate, an important complementary experiment would be to show that the generation of Δp by electron transport preceded ATP synthesis. This requires a method with a high time resolution for detection of Δp. The carotenoid band shift, an indicator of membrane potential in thylakoid membranes and bacterial chromatophores (Chapter 4) has an almost instant response to an imposed membrane potential, and responded within microseconds to the initiation of light-driven electron transport initiated by a laser flash. Furthermore, the subsequent decay of the membrane potential was accelerated by the presence of ADP and P_i. As the increased decay is due to the passage of protons through the ATP synthase to make ATP, it follows that ATP synthesis occurs after the formation of $\Delta\psi$.

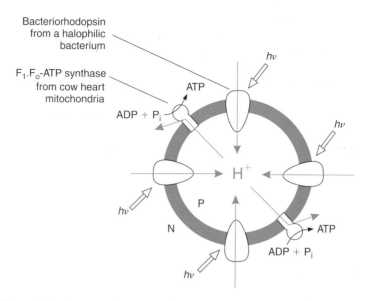

Figure 4.16   A proton circuit between a light-driven proton pump (bacterio-rhodopsin) and ATP synthase from mitochondria.
The establishment of the proton circuit depends on the majority of the bacterio-rhodopsin molecules adopting (for poorly understood reasons) the orientation in which they pump protons inward. Similarly, the ATP synthase had to incorporate predominantly with the topology shown. Opposite orientations of both bacteriorhodopsin and ATP synthase would in principle also have permitted an H^+ circuit, in the opposite direction, to be established. In practice, this would have meant that added ADP and P_i (both membrane impermeant) would not have been able to reach the active site of the ATP synthase.

4.12 LIGHT-DEPENDENT ATP SYNTHESIS BY BOVINE HEART ATP SYNTHASE

An important qualitative demonstration of the proton circuit comes from an instructive reconstitution experiment. The ATP synthase of mitochondria ought to be able to drive ATP synthesis if it is incorporated in phospholipid vesicles, with the correct relative orientation, along with another protein that generates protonmotive force of the correct polarity. A dramatic substantiation of this point was achieved when this experiment was performed using the light-driven proton pump, bacteriorhodospin (Chapter 6), from a halophilic bacterium (Fig. 4.16). An important point about this experiment is that the ATP synthase and the bacteriorhodopsin originate from such disparate sources: it is inconceivable that the coupling between them occurred through any other mechanism than the proton circuit shown in Fig. 4.16.

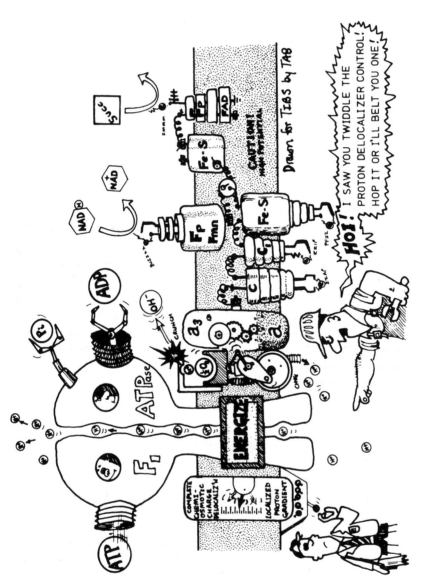

The mitochondrial respiratory chain and the ATP synthase: the 1970s and 1980s debate as to whether protons travel on a delocalized chemiosmotic circuit or via more localized pathways

5 RESPIRATORY CHAINS

5.1 INTRODUCTION

This chapter will describe our knowledge of the respiratory chains of mitochondria and selected bacteria, along with a brief outline of some of the approaches that have been taken to investigate these systems. The respiratory chain of mammalian mitochondria is an assembly of more than 20 discrete carriers of electrons that are mainly grouped into several multi-polypeptide complexes (Fig. 5.1). Three of these complexes (I, III and IV) act as oxidation–reduction-driven proton pumps. There are now detailed crystal structures for two of these complexes, and the sequences of all the constituent polypeptides are available. This information has advanced functional understanding considerably, but many aspects still remain to be understood at the molecular level. We shall illustrate methods for studying electron transport by reference to mitochondria, although comparable approaches are applied to bacteria and photosynthetic systems.

5.2 COMPONENTS OF THE MITOCHONDRIAL RESPIRATORY CHAIN

Reviews   Barker and Ferguson 1999, Saraste 1999, Schultz and Chan 2001

The respiratory chain transfers electrons through a redox potential span of 1.1 V, from the $NAD^+/NADH$ couple to the $O_2/2H_2O$ couple. Much of the respiratory chain is reversible (Section 3.6.4) and, to catalyse both the forward and reverse reactions, it is necessary for the redox components to operate under conditions where both the oxidized and reduced forms exist at appreciable concentrations. In other words, the operating redox potential of a couple, E_h (Section 3.3.3), should not be far removed from the mid-point potential of the couple, E_m. As will be shown later (Section 5.4.1), this constraint is generally obeyed, and this in turn gives some rationale to the selection of redox carriers within the respiratory chain.

The initial transfer of electrons from the soluble dehydrogenases of the citric acid cycle requires a cofactor that has mid-point potential in the region of $-300\,mV$, and is

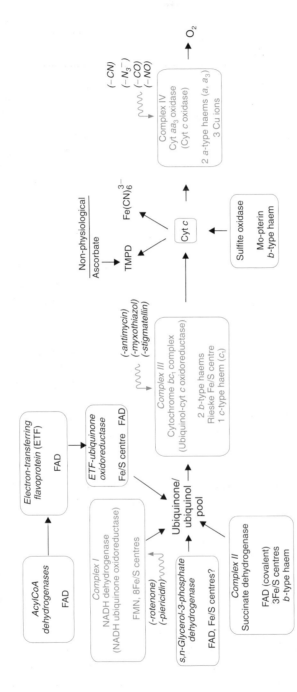

Figure 5.1 An overview of the redox carriers in the mitochondrial respiratory chain and their relation to the four respiratory chain complexes.

A wavy arrow indicates a site of action of an inhibitor. Sulfite oxidase and cyt c are in the intermembrane space which is topologically equivalent to the bacterial periplasm which is where other c-type cytochromes and molybdenum proteins related to sulfite oxidase are found (cf. Figs 5.23 and 5.25). Note that components are not all present at equal stoichiometry (see text).

sufficiently mobile to shuttle between the matrix dehydrogenases and the membrane-bound respiratory chain. This function is filled by the NAD^+/NAD couple, which has an $E_{m,7}$ of $-320\,mV$.

While the majority of electrons are transferred to the respiratory chain in this way, a group of enzymes catalyse dehydrogenations where the mid-point potential of the substrate couple is close to $0\,mV$, and are thus not thermodynamically capable of reducing NAD^+. These, succinate dehydrogenase, s,n-glycerophosphate dehydrogenase and the 'electron-transferring flavoprotein (ETF)-ubiquinone oxidoreductase' (transferring electrons via ETF from the flavin-linked oxidation step in the catabolism of fatty acids by β-oxidation), feed electrons directly into the respiratory chain at a potential close to $0\,mV$ independently of the $NAD^+/NADH$ couple (Fig. 5.1). This direct transfer requires that these enzymes be membrane bound. A third site of electron entry from sulfite oxidase (the vital final step of degradation of sulfur-containing amino acids in liver, which occurs at a very low rate compared with other inputs) is at cyt c, which is also where electrons can be donated artificially from chemicals such as tetramethyl-p-phenylene diamine (TMPD).

The redox carriers within the respiratory chain consist of: *flavoproteins*, which contain tightly bound FAD or FMN as prosthetic groups (note, unlike $NAD^+/NADH$ these flavins do not diffuse from one enzyme to another) and undergo a $(2H^+ + 2e^-)$ reduction; *cytochromes*, with porphyrin prosthetic groups undergoing a $1e^-$ reduction; *iron–sulphur (non-haem iron) proteins*, which possess prosthetic groups also reduced in a $1e^-$ step; *ubiquinone*, which is a free, lipid-soluble cofactor reduced by $(2H^+ + 2e^-)$; and, finally, *protein-bound Cu*, reducible from Cu^{2+} to Cu^+.

Cytochromes are classified according to the structure of their porphyrin prosthetic group, which in a b-type cytochrome is just the same as in many other proteins, e.g. haemoglobin. In a c-type cytochrome the haem is covalently attached to the polypeptide chain via two cysteine groups. Curiously it is not really understood what advantage the c-type cytochromes alone gain by undergoing a post-translational modification to generate the covalent attachment. The haem in an a-type cytochrome has undergone a modification relative to that in a b-type so as to have (i) a 15-carbon atom farnesyl side chain instead of a vinyl group at carbon atom 2, and (ii) replacement of the methyl group on carbon atom 8 by a formyl group. Again the rationale for these modifications is surprisingly unclear. In bacterial respiratory chains, other types of haem groups are encountered, notably those called d, d_1 and o. The former two involve quite significant modifcation to the standard haem structure, but the o-type has proved to be half way between the b- and a-types, possessing the farnesyl group but lacking the formyl group. The purposes of these other modifications are still not understood.

5.2.1 Fractionation and reconstitution of mitochondrial respiratory chain complexes

Although the mitochondrial electron transport chain contains approximately 20 discrete electron carriers, they do not all function independently in the membrane. The only mobile components are ubiquinone (also called coenzyme Q, UQ or simply Q), which is found in mammalian mitochondria as UQ_{10}, i.e with a side chain of ten 5-carbon isoprene units (see Fig. 5.6), its reduced form ubiquinol (UQH_2) and the water-soluble cytochrome c that is located on the P-side of the membrane.

Certain detergents, when employed at low concentrations, disrupt lipid–protein interactions in membranes, leaving protein–protein associations intact. Using these, the mitochondrial respiratory chain can be fractionated into four complexes, termed complex I (or NADH-UQ oxidoreductase), complex II (succinate dehydrogenase), complex III (UQH_2–cyt c oxidoreductase, or bc_1 complex) and complex IV (cytochrome c oxidase). Complex V is another name for the ATP synthase (Chapter 7). It is a source of confusion that the s,n-glycerophosphate dehydrogenase and ETF-ubiquinone oxidoreductases do not have the 'complex' nomenclature, even though they are connected to the respiratory chain in similar fashion to complexes I and II.

The electron-transfer activity of each complex is retained during this solubilization, and when complexes I, III or IV are reconstituted into artificial bilayer membranes, their ability to translocate protons is restored. Fractionation and reconstitution of the complexes has served a number of purposes:

(1) The complexity of the intact mitochondrion is reduced.
(2) It is possible to establish the minimum number of components that are required for its function.
(3) During the period in which the chemiosmotic theory was being tested, reconstitution proved to be one of the most persuasive techniques for eliminating the necessity of a direct chemical or structural link between the respiratory chain and the ATP synthase.

For example, it proved possible to 'reconstitute' ATP synthesis by combining complex IV and the bovine heart mitochondrial ATP synthase in bilayer membranes. However, the technical problems surrounding reconstitution are considerable. First, the complex must be incorporated into a bilayer in a way that retains catalytic activity. Secondly, allowance has to be made for the possibility of a random orientation of the reconstituted proton pumps that would prevent the detection of net transport. Incorporation into vesicles is normally by dissolving the proteins in a detergent together with phospholipid and then slowly removing the detergent by dialysis. Alternatively, the proteins can be sonicated together with phospholipid. An example of a reconstituted system was given in Section 4.12.

There are approximately 2 mol of complex IV per mol of complex III, while complex I and complex II are present at a substantially lower stoichiometry. However, there is a considerable molar excess of ubiquinone (Fig. 5.1), consistent with its role as a diffusing connector in the respiratory chain. The isolated complexes readily reassemble, for example, complex I and complex III reassemble spontaneously in the presence of phospholipid and UQ_{10} to reconstitute NADH–cyt c oxidoreductase activity. Such reconstitution experiments provided an important approach to establishing the order of the components in the chain.

5.2.2 Methods of detection of redox centres

(a) Cytochromes

The cytochromes were the first components to be detected, thanks to their distinctive, redox-sensitive, visible spectra. An individual cytochrome exhibits one major absorption band in its oxidized form, while most cytochromes show three absorption bands when reduced. Absolute spectra, however, are of limited use when studying cytochromes in intact mitochondria or bacteria, owing to the high non-specific absorption and light scattering of the organelles. For this reason, cytochrome spectra are studied using a sensitive differential, or split-beam, spectroscopy in which light from a wavelength scan is divided between two cuvettes containing incubations of mitochondria identical in all respects except that an addition is made to one cuvette to create a differential reduction of the cytochromes (Fig. 5.2). The output from the reference cuvette is then automatically subtracted from that of the sample cuvette, to eliminate non-specific absorption. Figure 5.3 shows the reduced, oxidized and reduced-minus-oxidized spectra for isolated cyt c, together with the complex reduced-minus-oxidized difference spectra obtained with submitochondrial particles in which the peaks of all the cytochromes are superimposed.

The individual cytochromes may most readily be resolved on the basis of their α-absorption bands in the 550–610 nm region. The sharpness of the spectral bands can be enhanced by running spectra at liquid N_2 temperatures (77°K), owing to a decrease in line broadening, resulting from molecular motion, and to an increased effective light path through the sample, resulting from multiple internal reflections from the ice crystals (Fig. 5.3).

Room-temperature difference spectroscopy can only clearly distinguish single a-, b- and c-type cytochromes. However, each is now known to comprise two spectrally distinct components. The a-type cytochromes can be resolved into a and a_3 in the presence of CO, which combines specifically with a_3. a and a_3 are chemically identical but are in different environments. The b-cytochromes consist of two components with different E_m values (high $- b_H$ and low $- b_L$). These respond differently when a Δp is established across the membrane (Section 3.6.2). It is now clear (Section 5.8) that the two components reflect the presence on one polypeptide chain of two b-type haems; the different local environments provided by the polypeptide chain account for the differences in spectral and redox properties. The two c-type cytochromes, cyt c and cyt c_1, can be resolved spectrally at low temperatures. Cyt c_1 is an integral protein within complex III (Section 5.8), while cyt c is a peripheral protein on the P-face of the membrane and links complex III with complex IV (cytochrome c oxidase).

(b) Fe/S centres

While their distinctive visible spectra aided the early identification and investigation of the cytochromes, the other major class of electron carriers, the iron–sulfur (Fe/S) proteins (Fig. 5.4) have ill-defined visible spectra but characteristic electron spin resonance spectra (ESR or EPR) (see Fig. 5.5). The unpaired electron, which may be present in either the oxidized or reduced form of different Fe/S proteins, produces the ESR signal. Each Fe/S group that can be detected by ESR is termed a centre or cluster. A single polypeptide may contain more than one centre. Currently, complexes I, II and III between them are thought to have 12 such centres (see below).

Fe/S proteins contain Fe atoms covalently bound to the apoprotein via cysteine sulfurs and bound to other Fe atoms via acid labile sulfur bridges (Fig. 5.4). Fe/S centres may contain two or four Fe atoms, even though each centre only acts as a $1e^-$ carrier. Fe/S proteins are widely distributed among energy-transducing electron-transfer chains and can have widely different $E_{m,7}$ values from as low as -530 mV for chloroplast ferredoxin (Section 6.4) to $+360$ mV for a bacterial HiPIP ('high-potential iron–sulfur protein'). This emphasizes the general point that the redox potential of a particular type of centre can be considerably 'tuned' by the environment provided by the protein.

(a) Split-beam spectrophotometer

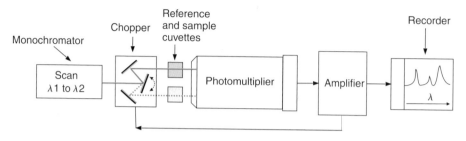

(b) Dual-wavelength spectrophotometer

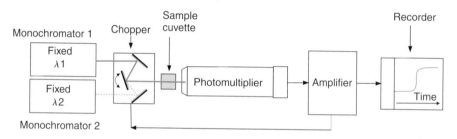

(c) Rapid kinetics: stopped flow in combination with dual wavelength

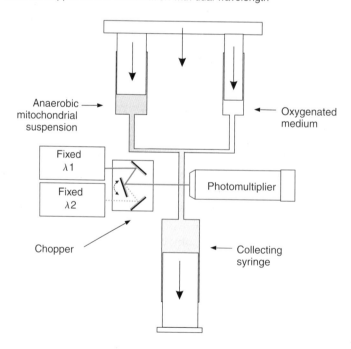

Figure 5.2   Spectroscopic techniques for the study of the respiratory chain.
(a) The split-beam spectrophotometer uses a single monochromator, the output from
which is directed alternately (by means of a chopper oscillating at about 300 Hz)
into reference and sample cuvettes. A single large photomultiplier is used and the
alternating signal is amplified and decoded so that the output from the amplifier
is proportional to the difference in absorption between the two cuvettes. If the
monochromator wavelength is scanned, a difference spectrum is obtained. The
split beam is therefore used to plot difference spectra that do not change with time.
(b) The dual-wavelength spectrophotometer uses two monochromators, one of
which is set at a wavelength optimal for the change in absorbance of the species
under study and one set for a nearby isosbestic wavelength at which no change is
expected. Light from the two wavelengths is sent alternately through a single cuvette.
The output plots the difference in absorbance at the two wavelengths as a function
of time, and is therefore used to follow the kinetics or steady-state changes in the
absorbance of a given spectral component, particularly with turbid suspensions.
(c) To improve the time resolution of the dual-wavelength spectrophotometer,
a rapid-mixing device can be added. The syringes are driven at a constant speed
and the 'age' of the mixture will depend on the length of tubing between the mixing chamber and the cuvette. When the flow is stopped, the transient will decay
and this can be followed.

Quinones and quinols

The 50-carbon hydrocarbon side chain of ubiquinone renders UQ_{10} highly hydrophobic
(Fig. 5.6). UQ undergoes an overall $2H^+ + 2e^-$ reduction to form UQH_2 (ubiquinol),
although in general the reaction will take place in two one-electron steps (Fig. 5.6); the partially reduced free radical form $UQ^{\bullet-}$ (ubisemiquinone) plays a defined role in both the photosynthetic reaction centre and the cytochrome bc_1 complex, where it is stabilized by binding
sites in the proteins. Ubiquinone reduction and ubiquinol oxidation will always occur at catalytic sites provided by the membrane proteins for which they are substrates; the (de)proton-
ation and oxidation–reduction steps are believed to proceed as shown in Fig. 5.6b.

The radical form can be detected by its ESR spectrum or a characteristic absorption band
in the visible region, but ubiquinone and ubiquinol are more difficult to detect because, in
common with proteins, they absorb around 280 nm, although the absorbance of the oxidized and reduced forms differ.

The simplest postulate for the role of UQ is as a mobile redox carrier linking complexes
I and II with complex III, although the 'Q-cycle' of electron-transfer in complex III involves
a more integral role (Section 5.8). While UQ_{10} is the physiological mediator, its hydrophobic
nature makes it difficult to handle, and ubiquinones with shorter side chains, and consequently greater water solubility, are usually employed *in vitro*. Some anaerobic bacterial
respiratory chains employ menaquinone in place of UQ (Sections 5.15.2 and 5.15.4), while
in the chloroplast the corresponding redox carrier (Chapter 6) is plastoquinone (Fig. 5.6).

5.3 THE SEQUENCE OF REDOX CARRIERS IN THE RESPIRATORY CHAIN

The sequence of electron carriers in the mitochondrial respiratory chain (Fig. 5.1) was
largely established by the early 1960s as a result of the application of oxygen electrode

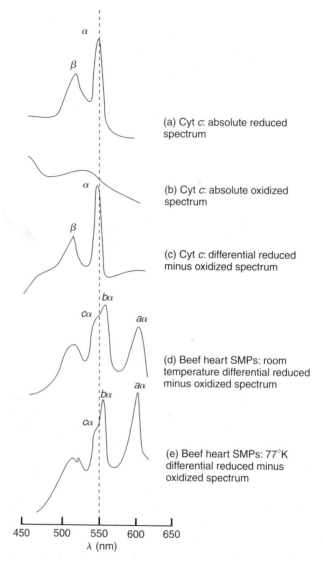

Figure 5.3   Cytochrome spectra.
The absolute oxidized (a) and reduced (b) spectra were obtained with purified cyt c in a split-beam spectrophotometer with water in the reference cuvette. The reduced minus oxidized spectrum (c) was obtained with reduced cyt c in one cuvette and oxidized cyt c in the other. (d) shows the reduced (with dithionite) minus oxidized (with ferricyanide) spectrum from beef heart SMPs. In (e) the scan was repeated at 77°K, note the greater sharpness of the α-bands.

(Fig. 4.7) and spectroscopic techniques. This work was greatly facilitated by the ability to feed in and extract electrons at a number of locations along the respiratory chain, corresponding to the junctions between the respiratory complexes. Thus NADH reduces complex I, succinate reduces complex II, and tetramethyl-p-phenylenediamine reduces cytochrome oxidase via cyt c. In this last case, ascorbate is usually added as the reductant

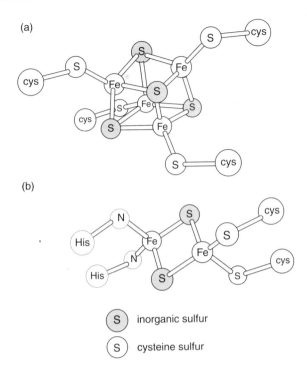

(a)

(b)

S   inorganic sulfur

S   cysteine sulfur

Figure 5.4   Iron–sulfur centres.
(a) A centre with four Fe and four acid-labile sulfurs is shown. On treatment with acid, these sulfurs (shaded) are liberated as H_2S. Although there are four Fe atoms, the entire centre undergoes only a $1e^-$ oxido-reduction. (b) The structure of the 2Fe/S Rieske centre in complex III with two histidine ligands is shown; other 2Fe/S structures will have four cysteine ligands to the Fe.

to regenerate TMPD from its oxidized form known as Wurster's blue (WB). Ferricyanide (hexacyanoferrate (III)) is a non-specific, but impermeant, electron acceptor and can be used not only to dissect out regions of the respiratory chain but also to provide information on the orientation of the components within the membrane. The reconstitution approach also showed that the complexes could not interact randomly; for example, complex I could only transfer electrons to complex III and the transfer depended on the presence of ubiquinone.

It should be emphasized that it is now clear that the electron carriers do not operate in a simple linear sequence, but that electrons may divide between carriers in parallel (as happens with complex III) and that there have to be mechanisms for permitting a switch from one to two electron steps.

The discovery of specific electron-transfer inhibitors enabled the relative positions of sites of electron entry and inhibitor action to be determined (Figs 4.8 and 5.1). Armed with this information, it was possible to proceed to a spectral analysis of the location of each redox carrier relative to these sites.

An independent approach to the ordering of the redox components came with the development of techniques for studying their kinetics of oxidation following the addition of oxygen to an anaerobic suspension (Fig. 5.2). The sequence with which the components become oxidized can reflect their proximity to the terminal oxidase and also whether they are kinetically competent to function in the main pathway of electron transfer. The rapidity

(a) Apparatus

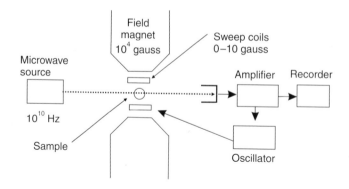

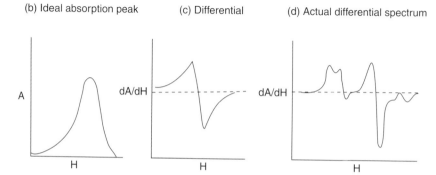

(b) Ideal absorption peak   (c) Differential   (d) Actual differential spectrum

Figure 5.5   Electron spin resonance and the detection of Fe/S centres.
(a) Apparatus: a microwave source produces monochromatic radiation at about
10^9 Hz, 30 cm. Unpaired electrons in the sample absorb the radiation when a magnetic field is applied, the precise value of the field required for absorption depending on the molecular environment of the electron, according to the formula:

$$h\nu = g\beta H$$

where h is Planck's constant, ν the frequency of the radiation, β a constant, the
Bohr magneton, H the applied magnetic field, and g the spectroscopic constant,
which is diagnostic of the species. The spectrum is obtained by keeping the microwave frequency constant and varying the magnetic field, trace (b). In practice,
a differential spectrum, trace (c), is obtained by superimposing upon the steadily
increasing field a very rapid modulation of small amplitude obtained with auxil-
liary sweep coils. The change in microwave absorption across each of these
sweeps enables the differential to be obtained. Spectra of energy-transducing
membranes are complex, trace (d); g values are obtained from either the peaks,
the troughs or the points of inflexion of the trace. Samples must be frozen and
generally present at a high protein concentration (10–50 mg protein ml^{-1}).

of the oxidations observed under these conditions requires the use of stopped-flow
techniques (Fig. 5.2).

The carriers in the respiratory chain must be ordered in such a way that their operating
redox potentials, E_h (Chapter 3) form a sequence from NADH to O_2. E_h is determined from

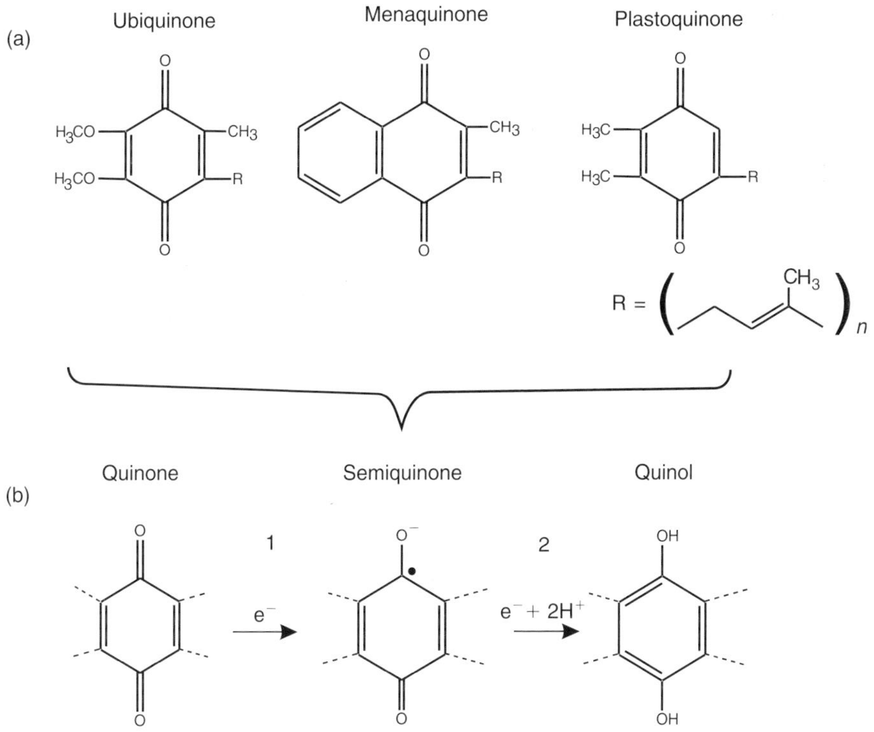

Figure 5.6 Quinones and quinols.
(a) Structures of the common quinone and quinols found in energy-transducing membranes. The length of the side chain can vary, for example, in ubiquinone $n = 10$ in mammals but $n = 6$ in yeast. (b) The two steps of quinol oxidation–quinone reduction. Note that $E^{o'}$ for the $Q^{\bullet -}/Q$ couple (step 1) is usually approximately at least 100 mV more negative than $E^{o'}$ for the $QH_2/Q^{\bullet -}$ couple (step 2), i.e. $Q^{\bullet -}$ is a stronger reductant than QH_2.

the mid-point potential, E_m, and the extent of reduction (equation 3.21). Although the extent of reduction of a component in the respiratory chain can be measured spectroscopically, indirect methods are needed to measure the mid-point potential *in situ*. It should be noted that the mid-point potential of a component in the respiratory chain can be different from that of the purified solubilized component.

5.4 THE MECHANISM OF ELECTRON TRANSFER

Review and further reading   DeVault and Chance 1966, Page *et al.* 1999

A fundamental process in bioenergetics is the transfer of electrons from one centre within a protein to another. We know from many protein structures that these centres are rarely directly adjacent to each other; frequently they are separated by protein. In this context, protein means any aspect of a polypeptide chain (e.g. peptide bonds or side chains) or water that lies between the two redox centres. In such circumstances the electron-transfer process cannot be similar to that which occurs when two inorganic ions encounter each other in aqueous solution, the so-called inner sphere mechanism. As we shall see later in this chapter and in Chapter 6, protein crystal structures have shown that electrons are transferred over distances of up to 14 Å. Simple intuition might suggest that the intervening protein structure would be organized so as to provide a pathway for the electron transfer. However, it is generally accepted that this is not the case and that the electron passes from one centre to another by a process known as *tunnelling*; the latter is a prediction of quantum mechanics. More specifically, the wavefunction for an electron held in an energy well on a donor shows that there is a finite, if very low, probability that the electron will be found at a potential acceptor. In effect, the electron can tunnel through a barrier.

A diagnosis of tunnelling is an insensitivity of electron transfer rate to temperature, even down to liquid helium temperatures. Such behaviour has been observed in proteins. Development of the theory of electron transfer within and between proteins indicates that proteins present a rather uniform barrier through which the electron tunnels, and that three factors influence the rate.

(a) The distance between electron donor centre and the acceptor (which is relatively easy to define if the transfer is between two ions, e.g. copper, but difficult to define if, say, the donor and acceptors are both haems, but generally taken to be haem edge to edge and not iron to iron). This is often the most important factor.

(b) The size of the free energy (or redox potential) difference between the donor and the acceptor.

(c) The response of the donor and acceptor, plus their environments, to the increased positive charge on the donor and the increased negative charge on the acceptor that follow the transfer of the electron. The latter term is the reorganization energy. It is important to note that this does not include what might be termed chemical events such as a concomitant transfer of a proton or dissociation of a ligand from one or both of the centres between which electrons are exchanged. Such events can limit, or 'gate', the rate of an electron transfer process.

For most individual steps of biological electron transfer, the driving force is between 0 and 100 mV, and the rearrangement energy varies between proteins by only a factor of two or three. It is in this context that distance between centres correlates closely with rate; edge to edge distances of 3, 10 and 14 Å are predicted to allow rates of approximately 10^{10}, 10^7 and $10^4 \, \text{s}^{-1}$ with a driving force of 100 mV. This last figure is significant because enzymes, to which electron transport chains act as donors or acceptors, tend to turn over on at least the millisecond timescale. Separations between redox groups of more than 14 Å are predicted to give electron transfer rates that are too slow for normal biological functions, for example, at 25 Å the rate would be approximately on the timescale of hours.

It is striking that, for all the known protein structures with more than one redox group, the edge to edge distance between centres that exchange electrons is always below 14 Å in the structural state where transfer occurs; change of distance linked to conformational change can be a means of controlling electron transfer as exemplified by the cytochrome bc_1 complex (Section 5.8.6). Thus specificity and directionality in electron transfer is achieved as a consequence of the spatial relationships of the participating groups; transfer of an electron from a donor to the 'wrong' acceptor will occur so slowly as to be insignificant.

Although distance is the usual dominant term in determining biological electron transfer rates, the free energy difference between a donor and an acceptor can be a significant factor. In particular, the theory of electron transfer predicts a counterintuitive, but subsequently experimentally verified, *decrease* in electron transfer rate as the driving force reaches very high values. This phenomenon contributes to the effectiveness of photosynthesis (Chapter 6).

A puzzling feature of many proteins involved in biological electron transfer is that they often have at least one cofactor for which the redox potential is dissimilar to the others. For example, suppose that a protein system contains five redox centres, B, C, D, E and F, with standard potentials of -50 mV, 300 mV, $+50$ mV, $+350$ mV and $+500$ mV, respectively, and accepts electrons from protein A ($+280$ mV). It would be reasonable to assume that the direction of electron flow through the protein would be from A to F, but the redox potentials of B and D suggest that they might not be involved in the overall electron transfer process and might have some other role in the protein (Fig. 5.7).

Several structures are available for proteins containing redox centres that are at first sight unsuitable, on grounds of redox potential, for a role in electron transfer. It is clear that

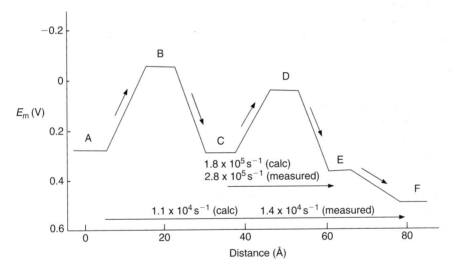

Figure 5.7   Electron transfer over large distances in proteins is catalysed by chains of redox centres with both uphill (endergonic, A to B and C to D) and downhill (exergonic) steps.
This energy diagram corresponds to part of the reaction centre of *Rhodopseudomonas viridis* (see Chapter 6). A is cytochrome c_2, B, C, D and E are, respectively, haems 4 to 1 of the cytochrome subunit, while F is the special pair of bacteriochlorophyll molecules $(Bchl)_2$. The cation form of the latter accepts electrons from haem 1 and haem 4 accepts electrons, in an intermolecular step, from the cytochrome c_2. Note the good agreement between measured rates and those calculated according to current electron transfer theory (see text). If electron transfer did not involve B and D, then calculations show that each of the two (A to F and C to E) rates shown would slow to a few per hour. (Adapted from Page *et al.* (1999), where further details of the calculation procedures can be found.)

these centres often lie spatially on the path of electron transfer. In the example considered here, B would be between A and C. If B were to be absent, the distance between A and C would be so large as to render the rate of electron transfer inadequate. Thus B is deduced to lie on the electron transfer pathway. Thermal energy is sufficient for the electron to reach B from A, while the overall movement of an electron from A to C is thermodynamically favourable and thus provides the driving force. (The thermodynamically uphill reaction from A to B means, of course, that, in the steady state, only a very small fraction of B can be in the reduced state.) Similar arguments apply to the role of D in electron transfer from C to E.

Electron transfer from C to E via D will not be as fast as it would be if the redox potential of D were to be, say, 325 mV, but such more positive redox potential is not necessary as the overall rate of electron transfer will still be fast enough to more than match the rates of chemical steps (i.e. catalysis) at the beginning or end of any chain of electron transfer reactions. Nevertheless, it is still puzzling why these redox centres with potentials very different from their neighbours should occur. There may be two reasons. First, the uphill reactions A to B and C to D could be a locus of control, although there is no documentation of such an effect. Second, many electron transfer proteins are evolutionarily related to one another and it may be that the recruitment of a particular redox centre for a new reaction has not exerted any pressure for change in its redox potential. An example is an Fe/S centre that is found in both a nitrate reductase and a formate dehydrogenase in bacteria. Its redox potential is much more negative than either the donor protein or the active site cofactor of the nitrate reductase, but it is much better tuned to its role as the electron acceptor from the active site of formate dehydrogenase.

Electron transfer occurs not only within a protein, or proteins that form a permanently associated complex, but also between proteins that interact transiently, for example, transfer from A to B in Fig. 5.7. The requirement here is that a transiently formed complex places the redox groups of the donor and acceptor proteins at an edge to edge distance no more than 14 Å. This means that the redox groups of the two proteins cannot be deeply buried. The much studied mitochondrial cytochrome c (Plate A) illustrates this point. The asymmetrically positioned haem group has one edge within 5 Å of one surface of the protein at which a group of lysines (especially residues 13, 86 and 87) are found. These have been implicated in the interaction of cytochrome c with its partners on the basis that:

- their chemical modification inhibits electron transfer, both from the donor, the cytochrome bc_1 complex, and to the acceptor, cytochrome aa_3; and
- complexation of cytochrome c with either cytochrome bc_1 or cytochrome aa_3 protects against the chemical modification of the patch of lysines.

The haem of cytochrome c_1 (Section 5.8) and the Cu_A centre of cytochrome oxidase are each close to a surface of their respective polypeptides, which both have sites with complementary charges suitable for docking with the lysine patch on cytochrome c. Involvement of the same region of the cytochrome c surface for interaction with both its electron donor and acceptor partners indicates a single route for electron transfer into and out of the haem of cytochrome c, with the rest of the surface of the 30 Å diameter cytochrome c molecule insulated from adventitious electron transfer to or from the haem. The single route involves electron tunnelling through the relatively uniform protein dielectric and does not require specifically positioned amino acid side chains; the idea that the aromatic side chain of highly conserved phenylalanine residue at position 82 plays a key role in electron transfer as an aromatic conductor was disproved by the demonstration that its replacement by alanine had no effect on the rate of biological electron transfer by cytochrome c.

Clearly, if a single patch on the surface of the cytochrome is responsible for the protein–protein interaction with the redox partner, it follows that after reduction by the bc_1 complex, the cytochrome c must dissociate before forming a productive complex in which to pass the electron to cytochrome c oxidase. This is in accord with the current view of the electron transport system in which the cytochrome bc_1 and oxidase complexes are thought to diffuse relatively slowly in the plane of the membrane whilst the peripheral protein, cytochrome c, undergoes more rapid lateral diffusion along the surface.

Finally, we note that, although the general rule is that proteins present a fairly uniform dielectric through which electrons tunnel from one site to the next, there are rare instances (Section 5.12 and Chapter 6) where the side chain of an amino acid participates as an oxidation–reduction centre. This requires the involvement of a very electropositive redox centre to be the acceptor of an electron from the amino acid because side chains are thermodynamically difficult to oxidize. But these are the exceptions, and in general the idea of a 'best pathway' for electrons through particular bonds or amino acids that happen to lie between redox centres is incorrect; even bound water molecules within proteins can support electron tunnelling.

5.4.1 Redox potentiometry

The technique of redox potentiometry (Fig. 5.8) combines dual-wavelength spectroscopy with redox potential determinations. As with redox potentiometry of most biological couples, it is necessary to add a low concentration of an intermediate redox couple in order to speed the process of equilibrium between the Pt electrode and the redox centres. As a secondary mediator will only function effectively in the region of its mid-point potential (so that there are appreciable concentrations of both its oxidized and reduced forms), a set of mediators is required to cover the whole span of the respiratory chain, with mid-point potentials spaced at intervals of about $100\,mV$. Mediators are usually employed at concentrations of 10^{-6} to $10^{-4}\,M$. Many mediators are autoxidizable, and the incubation has to be maintained anaerobic, both for this reason and to prevent a net flux through the respiratory chain from upsetting the equilibrium.

A second requirement for membrane-bound systems is that the mediators must be able to permeate the membrane in order to equilibrate with all the components. This introduces a considerable complication if the mitochondria are studied in the presence of a $\Delta\psi$, as $\Delta\psi$ (or indeed ΔpH) may affect the distribution of the oxidized and reduced forms of the mediators across the membrane, and the oxidized–reduced ratio of the mediator at the site of the component will differ from that at the platinum electrode. Furthermore, redistribution of electrons across the membrane will occur upon induction of a membrane potential; this effect can be informative but leads to complications in the analysis. The simplest redox potentiometry is therefore performed with mitochondria or submitochondrial particles at zero Δp; it is also necessary to use anaerobic conditions so as to prevent steady-state transfer of electrons from the mediators to oxygen.

(a) Apparatus

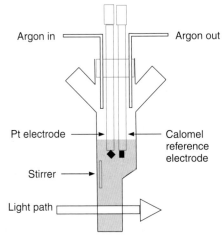

(b) Redox difference spectra succinate – cytochrome c reductase (complexes II + III)

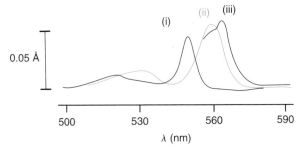

Figure 5.8   Redox potentiometry of respiratory chain components.
(a) Apparatus for the simultaneous determination of redox potential and
absorbance. (b) Difference spectra obtained with a suspension of succinate-
cytochrome c reductase (i.e. complexes II + III). The complex, held in solution
by a low concentration of detergent, was added to an anaerobic incubation containing redox mediators. The ambient redox potential was varied by the addition
of ferricyanide. (i) Reference scan (baseline) at $+280\,mV$ (all cytochromes
oxidized), second scan at $+145\,mV$ (cyt c_1 now reduced). (ii) Baseline at
$+145\,mV$ (cyt c_1 reduced), second scan at $-10\,mV$ (cyt b_L additionally
reduced). (iii) Baseline at $-10\,mV$ (c_1 and b_L reduced), second scan at $-100\,mV$
(b_H additionally reduced).

The practical determination of the E_m of a respiratory chain component (Fig. 5.8)
involves incubating mitochondria anaerobically in the presence of the secondary mediators. The state of reduction of the relevant component is monitored by dual-wavelength
spectrophotometry (Fig. 5.2), while the ambient redox potential is monitored by a Pt or Au
electrode. The electrode allows the secondary mediators and the respiratory chain components all to equilibrate to the same E_h. This potential can then be made more electronegative
(by the addition of ascorbate, NADH, or dithionite) or more electropositive by the addition
of ferricyanide. E_h and the degree of reduction of the component, by spectrophotometry,

are monitored simultaneously. In this way a redox titration for the component can be established.

Considerable information can be gathered from such a titration. Besides E_h itself, the slope of $\log_{10}[\text{ox}]/[\text{red}]$ establishes whether the component is a $1e^-$ carrier (60 mV per decade) or a $2e^-$ carrier (30 mV per decade), Table 3.2. By repeating the titration at different pH values, it can be seen whether the mid-point potential is pH dependent, implying that the component is a $(H^+ + e^-)$ carrier. Finally, the technique frequently allows the resolution of a single spectral peak into two or more components based on differences in E_m. In this case the basic Nernst plot (Fig. 3.3) is distorted, being the sum of two plots with differing E_m values, which can then be resolved. One of the most interesting findings with this technique was that cyt b in complex III can be resolved into two components (Section 5.8). Redox potentiometry can also be employed for Fe/S proteins, in which case the redox state of the components is monitored by ESR.

5.4.2 E_h values for respiratory chain components fall into isopotential groups separated by regions where redox potential is coupled to proton translocation

The mid-point potentials of some of the identifiable components of the respiratory chain are depicted in Fig. 5.9. Once the E_m values have been established for non-respiring mitochondria, an E_h for a component can be assigned to any component in respiring mitochondria simply by determining the degree of reduction. The results for mitochondria respiring in state 4 are shown. The oxido-reduction components fall into four equipotential groups, the gaps between which correspond to the regions where proton translocation occurs. The drop in E_h of the electrons across these gaps is conserved in Δp.

5.5 PROTON TRANSLOCATION BY THE RESPIRATORY CHAIN: 'LOOPS', 'CONFORMATIONAL PUMPS' OR BOTH?

The original formulation of the chemiosmotic mechanism envisaged that electron transport chains would achieve the net translocation of protons by providing a series of alternate hydrogen and electron carriers. Charge movement across the membrane would be achieved by migration of electrons rather than protons; this was called a loop mechanism. It has the implication of a one to one stoichiometry between electron movement and net proton translocation (Fig. 5.10). It turns out that electron movement across the membrane is an important feature of many proton translocation mechanisms and that a loop mechanism does operate in several bacterial electron transport reactions (Section 5.13).

In the mitochondrial electron transport system, elements of the loop mechanism are used, as we shall see later in this chapter. However, it is also now clear that *direct* proton pumping across a membrane can be driven by some electron transfer proteins. What this means is that the oxidation–reduction reaction at one or more centres in the protein is linked with conformational changes that result in net proton translocation across the membrane. This type of mechanism places no limits on the stoichiometry of proton translocation, although there is, of course, the thermodynamic constraint imposed by the relationship between the available driving force and the size of Δp (Section 3.6).

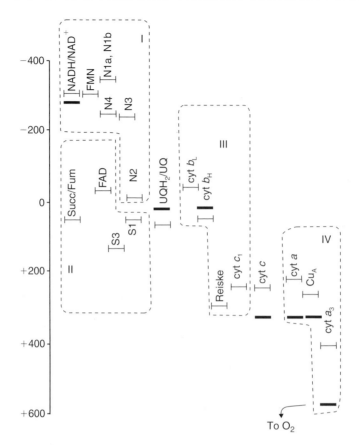

Figure 5.9 E_m **values for components of the mitochondrial respiratory chain and** E_h **values for mitochondria respiring in state 4.**
Values are consensus values for mammalian mitochondria. ($\vdash\!\!\!-\!\!\!\dashv$) $E_{m,7}$ values obtained with de-energized mitochondria; (──) $E_{h,7}$ values for mitochondria in state 4.

Understanding the mechanisms underlying proton-pumping is not trivial because one must identify the steps which ensure that a proton in transit across the membrane, through the protein, cannot retrace its steps to the side of the membrane from which it originated. This type of mechanism clearly contributes to protonmotive force generation in complex IV (Section 5.9), but the mechanistic understanding is most advanced for the protein bacterio-rhodospin (Chapter 6).

It was sometimes postulated that the 'loop' hypothesis requires that the loops should span the membrane and that the appropriate redox centres should be located at the two sides (P and N) of the membrane. However, all that is required for a functional loop is that there should be a means for taking up and releasing the protons at the two sides. Thus specific pathways through a protein from a site of oxidation or reduction to a surface can contribute to the operation of a loop mechanism; such pathways clearly occur, for example, in complex IV (Section 5.9).

In a purely *conformational pump* model a redox carrier anchored within a flexible protein is proposed to undergo redox-induced changes in pK_a, the directionality of proton

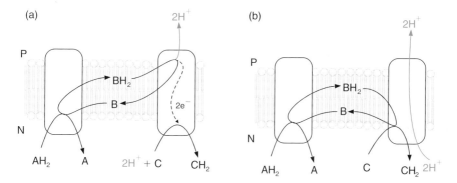

Figure 5.10   Net proton translocation by (a) a redox loop and (b) a proton pump.
In (a) most of the charge transfer across the membrane will be achieved by the inward movement of electrons to reduce D, but there will be some contribution from the outward movement of H^+, depending upon the depth in the membrane of the site of CH_2 oxidation. In (b) all the charge translocation will be achieved by the pumping of protons of H^+. Note that overall the energetics of (a) and (b) are identical and that some electron transport proteins systems function like (a) (see e.g. Figs 5.23, 5.24 and 5.26), whilst others have elements of both (a) and (b) (see Figs 5.14 and 5.16). 'Pure pumping of protons is not well documented for electron transport but bacteriorhodopsin (Chapter 6) and the ATP synthase (Chapter 7) correspond to this model.

transport being assured by co-ordinate conformational changes which make the redox site alternately accessible from either side of the membrane (Fig. 5.11). A conformational redox pump must co-ordinate a redox change, a conformational change, and a protonation change. The reversibly protonated site in this model need not necessarily be limited to a conventional redox centre – it is equally possible that a redox-induced conformational change alters the pK_a of an amino acid side chain. The closest approach to understanding such a scheme in molecular detail comes from bacteriorhodopsin (Chapter 6).

5.6 COMPLEX I (NADH–UQ OXIDOREDUCTASE)

Reviews and further reading   Yagi *et al*. 1998, Albracht and Hedderich 2000, Friedrich and Scheide 2000, Sazanov and Walker 2000

There are two reasons why NADH–UQ oxidoreductase (complex I) is the least well understood component of the mammalian mitochondrial electron transport chain. First, it is very large (about the same size, molecular weight 750 kDa, as the large subunit of a ribosome) and it has as many as 43 polypeptides. Second, the redox centres, apart from the flavin FMN, are iron–sulphur centres, which cannot be studied by optical spectroscopy, but instead require the more difficult technique of low temperature ESR (Section 5.4.1) for characterization and assessment of their redox state. What is certain about complex I is that it catalyses the transfer of two electrons from NADH to ubiquinone in a reaction that is associated with proton translocation across the membrane, and that it is inhibited by

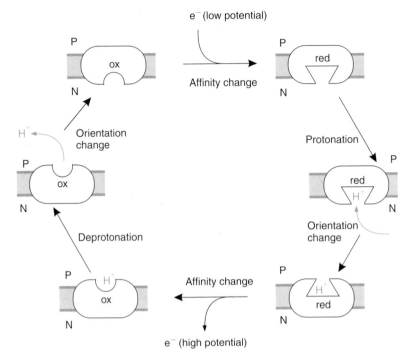

Figure 5.11   A model for a redox-driven proton pump.
A hypothetical model is shown for a redox pump with a stoichiometry of
$1H^+/e^-$. Reduction causes the proton binding site to adopt a high-affinity state,
sufficient to bind a proton from the N-phase. A spontaneous conformation change
now makes the binding site accessible to the P-phase (note: the protein itself does
not rotate in the membrane). Loss of the electron causes the protein to adopt a
low-affinity state, such that the protein releases the proton into the acidic environment of the P-phase. The cycle is completed by the binding site reorientating
itself towards the N-phase.

rotenone and piericidin A. Current evidence suggests that the proton translocation stoichiometry is $4H^+/2e^-$.

Acquisition of amino acid sequences, from the gene sequences in most cases, has advanced
our knowledge of this enzyme significantly. In particular, the finding that many bacteria have
a close counterpart to complex I has been a great help, because in these organisms the enzyme
appears to have 'only' 14 subunits (13 in *E. coli* where two proteins are fused as one). For reasons that are not understood, all the mitochondrial complexes I–IV have more subunits than
their counterparts in bacterial respiratory chains. Unfortunately, purification of intact bacterial forms of complex I has proved more difficult than for the mitochondrial enzyme, and thus
our knowledge of this enzyme is pieced together using sequence information from several
sources and biochemical information from the mitochondrial enzyme.

An added complication is that three different nomenclature systems are in use for the
enzyme; here we name the 14 subunits of the bacterial enzyme as NuoA to NuoN, where
Nuo stands for NADH–ubiquinone oxidoreductase and A to N is the order of the genes
within an operon (but see legend to Fig. 5.12 for details of other nomenclatures). Although

the experimental evidence is not completely definitive, it is generally accepted that only one copy of any subunit is present in the bacterial or mitochondrial enzymes. Analysis of the sequences of the Nuo proteins suggests that there are sufficient clusters of cysteine residues to provide eight Fe/S centres in complex I. It is possible to assign these on sequence considerations as 2Fe/2S or 4Fe/4S clusters as shown in Fig. 5.12. It is more difficult to assign these clusters to the ESR spectra obtained for the Fe/S centres of the enzyme, not least because only five signals can in general be clearly resolved. These signals are termed N1a, N1b, N2, N3 and N4. With the exception of N2, it has been possible to assign redox potentials in the range -300 to $-230\,mV$. Probably, therefore, these signals correspond to Fe/S centres that act in early steps of electron transfer from the reduced FMN (Fig. 5.12). The N2 signal has an $E_{m,7}$ in the range -20 to $-160\,mV$ (the exact value depending on the source of the complex), suggesting that it corresponds to the last centre that electrons pass through before transfer to ubiquinone.

Complex I is shown by electron microscopic studies to have an L-shape. Figure 5.12 shows approximately how the 14 subunits of the bacterial NADH dehydrogenase are thought to be arranged. This should be regarded tentatively as there is evidence that this shape is ionic strength dependent. The allocation of the subunits to the peripheral or membrane bilayer part of the enzyme is based both on sequence analysis (e.g. whether these subunits have predicted transmembrane α-helices) and identification of their counterparts in water-soluble or water-insoluble subcomplexes of the mitochondrial enzyme. There is some experimental evidence to suggest that NuoB and NuoI are located in a connecting region between the peripheral and membrane parts of the enzyme (Fig. 5.12). In the case of the mitochondrial enzyme, the additional 27 subunits are thought to be distributed between the peripheral sector ($\sim$16) and the membrane sector ($\sim$11). A curious feature of complex I is that the mitochondrial enzyme might possess extra functions, in the biogenesis of the protein, as an acyl carrier protein or as a hydroxysteroid reductase–isomerase.

The NuoE and NuoF counterparts in the mitochondrial enzyme are dissociable from the rest of the enzyme *in vitro* and catalyse the oxidation of NADH by ferricyanide. These two subunits, often called the flavoprotein fraction (FP, Fig. 5.12) contain Fe/S centres and the FMN molecule at the NADH active site. Further insight into these two subunits, together with NuoG, came when it was noted that regions of these proteins have considerable similarity to sequences within two subunits, α and γ, of a water-soluble H_2-NAD$^+$ oxidoreductase found in the bacterium *Ralstonia eutropha*. The α- and γ-subunits in this type of enzyme have FMN, Fe/S centres (two 4Fe/S and 2Fe/2S clusters) and possess NADH–ferricyanide–oxidoreductase activity, in common with the NuoE/NuoF segment of complex I. The *R. eutropha* α-subunit has considerable sequence similarity to sizeable stretches of both NuoE and NuoF, whilst its γ-subunit resembles NuoG (Fig. 5.12). These comparisons, and other features, strongly suggest that the NuoF subunit contains the site at which electrons are transferred to FMN from NADH.

NuoF and the α-chain of the *R. eutropha* enzyme contain a Cys-X-X-Cys-X-X-Cys motif diagnostic of a 4Fe/4S centre, almost certainly corresponding to signal N-3. The homologous regions of the γ-subunit of the *R. eutropha* enzyme and the NuoG protein contain cysteine ligands for a 2Fe/2S and two 4Fe/4S centres. Finally, homologous regions of the NuoE from several sources, including the bacterium *Paracoccus denitrificans* and the α-subunit of *R. eutropha* contain Cys residues, which contribute to binding a 2Fe/2S centre. The relationship between this type of soluble hydrogenase and subunits of complex I

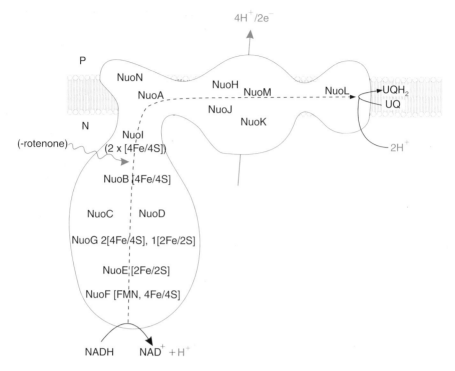

Figure 5.12   An outline structure for the mitochondrial (and bacterial) proton-translocating NADH–ubiquinone oxidoreductase (complex I).
The 14 subunits are named according to one nomenclature system (Nuo) used for the bacterial enzyme. Another bacterial system names the proteins as Nqo (NADH quinone oxidoreductase), whilst the mitochondrial enzyme is named with respect to the approximate molecular weights of the proteins and whether they are assigned to the flavoprotein fraction (FP), the so-called iron protein fraction (IP), stretches of characteristic amino acid sequence, or whether they are coded for on the mitochondrial genome (ND series); the latter are all predicted to be very hydrophobic proteins (and constitute what is sometimes called the hydrophobic fraction (HP) of the mitochondrial enzyme) that do not contain any known redox active groups. The equivalences are:

Bacteria		Mitochondria
NuoA	Nqo7	ND3
NuoB	Nqo6	PSST
NuoC	Nqo5	IP30K
NuoD	Nqo4	IP49K
NuoE	Nqo2	FP24K
NuoF	Nqo1	FP51K
NuoG	Nqo3	IP75K
NuoH	Nqo8	ND1
NuoI	Nqo9	TYKY
NuoJ	Nqo10	ND6
NuoK	Nqo11	ND4L
NuoL	Nqo12	ND5
NuoM	Nqo13	ND4
NuoN	Nqo14	ND2

has established the roles of the NuoF, E and G subunits, and their redox centres, in the early steps of the NADH–ubiquinone oxidoreductase reaction.

Further sequence comparisons suggest that Nuo A-D and H-N, are related to polypeptides of a different, membrane-bound, proton-pumping type of hydrogenase (see legend to Fig. 5.12). Sequence analyses are consistent with the proposal that the other three Fe/S centres are located at the 'connector' interface between the peripheral and the integral parts of the enzyme (Fig. 5.12). Surprisingly, no Fe/S centre appears to be within the bilayer. The Fe/S centre that was formerly assigned this role corresponds to signal N2, which has the highest redox potential and thus was assumed to be the component from which electrons are delivered to ubiquinone. This still could be the case, but with N2 provisionally assigned to NuoB, the transfer to UQ is likely to occur close to the connector region. There is also evidence that inhibitors such as rotenone bind to NuoB at an interface with NuoD, while other sequence comparison evidence also points to these two subunits being adjacent (Fig. 5.12). As we shall see subsequently (Sections 5.7, 5.12 and 5.13), a protein does not need extensive transmembrane helices in order to interact with ubiquinone.

What is even more puzzling is how proton translocation is achieved. There is speculation that the two 4Fe/4S clusters on NuoI, which are believed to be undetectable by ESR owing to spin coupling, play a key role. Many investigators suspect that the hydrophobic phase of the protein possesses additional uncharacterized redox groups. It seems certain that the hydrophobic domain, predicted to contain 54 transmembrane α-helices shared amongst the seven bacterial subunits (Fig. 5.12), must do something more than provide a conduit to deliver ubiquinol to the core of the membrane bilayer.

5.7 DELIVERING ELECTRONS TO UBIQUINONE WITHOUT PROTON TRANSLOCATION

In addition to complex I, at least three other enzymes feed electrons to UQ_{10} (Fig. 5.1). Succinate dehydrogenase or *complex II* transfers electrons from succinate as part of the

---

The 8 Fe/S clusters implicated as occurring in the enzyme are thought to be distributed as shown. As discussed in the text, it is not clear that distinct ESR signals have been characterized for all eight. One centre, N2, is distinguished from the rest because it has a significantly higher redox potential. This suggests that it acts as the last in a series of Fe/S centres and its assignment to NuoB is favoured. The location of N1a, N1b and N3 is not clear. At the time of writing it is suggested that the enzyme may contain a second molecule of FMN bound to the NuoB subunit. The enzyme is now regarded as being up from modules that are also found in other enzymes; NuoE, NuoF and NuoG are related to the soluble hydrogenase from *R. eutropha*; NuoBCDHIL to subunits in a membrane-bound, and in the case of NuoB and D also soluble, NiFe hydrogenases (see Friedrich and Scheide 2000; Albracht and Hedderich 2000). The narrow stalk between, and relative positions of, NuoL and NuoM is based on cryoelectron crystallography of complex I (Sazanov and Walker 2000). Roles for NuoH in ubiquinone binding and for Nuo LMN in proton translocation are based on sequence arguments. Note that it is possible that this structure is ionic strength dependent. At low ionic strengths the NuoEFG domain appears to bridge between the BCDI domain and AJKMN domains, giving a horseshoe shape.

TCA cycle. This is the only membrane-bound member of the cycle and its succinate-binding site faces the mitochondrial matrix. The second enzyme, also located on the matrix face of the membrane, is the ETF–ubiquinone oxidoreductase, while the third (not found in all types of mitochondria) is *s,n*-glycerophosphate dehydrogenase, which binds its glycerophosphate substrate at the outer surface of the membrane. All three are flavoproteins transferring electrons from substrate couples with mid-point potentials close to 0 mV. As would be expected on thermodynamic grounds, none is proton translocating. The feeding into the respiratory chain of electrons derived from the flavin-linked step in fatty acid oxidation is often overlooked but is functionally very significant in many mammalian cells.

5.7.1 Complex II (succinate dehydrogenase)

Reviews and further reading   Iverson *et al.* 1999, 2000, Lancaster *et al.* 1999, Ohnishi *et al.* 2000

As yet there is no structure available for succinate dehydrogenase (SDH), but it can be modelled with confidence from the recently determined structures of two bacterial fumarate reductases (Section 5.15), which catalyse the reverse reaction. Figure 5.13 shows that four polypeptides play a functional role in SDH. That furthest from the membrane bilayer contains the covalently bound FAD from which electrons pass sequentially into the membrane sector of the enzyme via three Fe/S centres located in the second peripheral subunit. The two integral membrane polypeptides contain one haem group sandwiched in between transmembrane helices. It is not understood exactly how the quinone is reduced, but the haem may transfer electrons to it from the 3Fe/4S centre. The quinone binding site must also receive protons from the N-(matrix) phase, balancing those released to the matrix upon oxidation of $FADH_2$ (Fig. 5.13). Thus the reduction of UQ by succinate is not associated with any charge movement across the membrane. The redox potential of the 4Fe/4S centre is much lower than those of the adjacent centres. As explained in Section 5.4, this is not a reason for excluding its role in the linear chain of centres (Fig. 5.13).

Complex II could be redesigned as a Δp consumer by taking the protons for quinone reduction from the P-phase. Although this would make no sense in mitochondria, in *Bacillus subtilis* a transfer of electrons occurs from succinate to menaquinone, which has a mid-point potential 114 mV more negative than ubiquinone. This is thermodynamically uphill and, to overcome the energetic barrier, this organism has a succinate dehydrogenase with two haems, one at each side of the membrane, and the quinone reduction site at the P-side. This succinate dehydrogenase consumes Δp as an electron has to move to the P-side from the catalytic centre at the N-side (Fig. 5.13b). It is clear that evolution has found a way to modify the bioenergetics of the succinate dehydrogenase reaction by switching between a one- and a two-haem succinate dehydrogenase.

5.7.2 Electron-transferring flavoprotein–ubiquinone oxidoreductase

The water-soluble ETF is located in the mitochondrial matrix, contains one molecule of FAD and accepts electrons from several dehydrogenases containing flavin. The latter include enzymes that catalyse one type of step in fatty acid oxidation or various steps in amino acid and choline catabolism. The resulting $FADH_2$ of ETF is oxidized by the

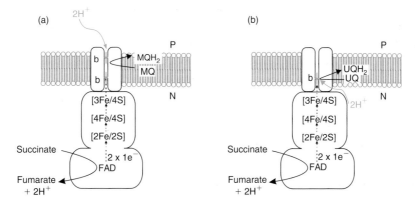

**Figure 5.13   Schematic models for the structural organizations of mito-
chondrial (a) and *B. subtilis* (b) succinate dehydrogenases based on the
crystal structures of the closely related fumarate reductases of *E. coli* and
W. succinogenes.**
The enzymes have four subunits, with that furthest from the membrane having a
covalently bound FAD at the active site. A second peripheral subunit contains
three Fe/S centres arranged approximately as shown and thus providing a route
for electrons into the membrane phase, which comprises two polypeptides, each
of which contributes three α-helices to the overall structure. For the mitochondrial enzyme (a) the site of ubiquinone reduction is believed to be on the N-side of
the membrane, possibly close to the haem group which is positioned on the basis
of the location of one of the two haem groups in the *W. succinogenes* fumarate
reductase. The haem is shown diagrammatically as sandwiched between helices of
the two membrane subunits, each of which has three helices. The Fe/S centres
are sometimes called S-1 [2Fe/2S], S-2 [4Fe/4S] and S-3 [3Fe/4S] with respective $E^{o\prime}$ values of 0 mV, -260 mV and 60 mV; the haem has $E^{o\prime} = -185$ mV.
Although the redox potentials do not become increasingly positive as electrons
move from succinate to ubiquinone, the positions of the redox groups and the
arguments in Section 5.4 and Fig. 5.7 implicate them all on the pathway for electron flow through the enzyme. For the *B. subtilis* enzyme (b) the site of
menaquinone reduction is believed to be at the P side of the membrane from
where protons are taken and two haem groups are present, presumed to be in similar positions as the haems in the *W. succinogenes* enzyme. The two haems have
$E^{o\prime}$ values of -95 mV, and 65 mV, being respectively located towards the P and
N sides of the membrane, overcoming the 160 mV difference between the haems.
Thus the membrane potential will act as a driving force for the movement of
electrons from the Fe/S centres to the site of menaquinone reduction at the P side.

ETF–ubiquinone oxidoreductase, which contains an FAD, an Fe/S centre and a ubiquinone
binding site. It is not certain whether the FAD or the Fe/S centre is the immediate electron
acceptor for ETF but the structure suggests that $FADH_2$ could transfer its electrons to
bound ubiquinone. The protein is mainly globular with 10 α-helices and 21 β-strands.

ETF–ubiquinone oxidoreductase does not contain any transmembrane helices and the association with the membrane can be attributed to a series of hydrophobic residues that contribute to an α-helix and a β-sheet, which are adjacent to the hydrophobic ubiquinone binding
pocket. The structure of this protein shows that quinones can be bound by proteins that are
largely globular (rather than transmembraneous), but provide a surface that can dip into the

membrane sufficiently for ubiquinone and ubiquinol to be able to exchange directly with the core of the lipid bilayer. Further examples of such proteins are now recognized in bacterial electron transfer chains (see Section 5.15).

5.7.3 *s,n*-Glycerophosphate dehydrogenase

This enzyme, which oxidizes *s,n*-glycerophosphate at the outer surface (P-side) of the inner mitochondrial membrane, contains FAD and at least one Fe/S centre. Its organization is presumably similar to that of ETF–ubiquinone oxidoreductase such that it can receive ubiquinone from, and deliver ubiquinol to, the hydrocarbon bilayer core of the membrane. The gene for this enzyme is found on genomes from those of man to yeast, but in higher eukaryotes its expression may differ significantly between cell types, making difficult generalizations about the role of *s,n*-glycerophosphate oxidation in cellular bioenergetics (Chapter 8).

5.8 UBIQUINONE AND COMPLEX III (bc_1 OR UQ–CYT c OXIDOREDUCTASE)

Reviews   Derry *et al.* 2000, Darrouzet *et al.* 2001

The transfer of electrons from ubiquinol to cyt c and the associated proton translocation is not a simple matter. The reaction is catalysed by complex III of the respiratory chain, which is also termed the cyt bc_1 complex, or more appropriately the ubiquinol–cytochrome c oxidoreductase. This complex is also found in many bacteria (Section 5.15) and is similar in many respects to the plastoquinol–plastocyanin oxidoreductase, or cyt bf complex, of thylakoids (Chapter 6).

The redox groups in cyt bc_1 comprise a 2Fe/2S centre, located on the Rieske protein, two b-type haems located on a single polypeptide, and the haem of cyt c_1. The Rieske protein 2Fe/2S cluster is attached to the polypeptide by chelation of one Fe to two cysteines and the other to two histidine residues. The polypeptide chain is folded as a globular structure incorporating the 2Fe/2S centre and extending into the aqueous layer beyond the bilayer on the P-face of the membrane and anchored via a hydrophobic N-terminal helix.

Cytochrome c_1 has a similar globular domain and hydrophobic anchor as the Rieske protein, except in this case it is the C-terminus that provides the anchor. The cyt b subunit has eight transmembrane α-helices. Four conserved histidine residues, two on each of two helices, are appropriately positioned to provide the axial ligands for the haems. One haem, with an $E_{m,7}$ of about $-100\,mV$ and thus known as b_L (sometimes called b_{566} because of its α-band absorption maximum) is located towards the P-(cytoplasmic) side of the mitochondrial membrane. The second haem, b_H, $E_{m,7}$ $+50\,mV$ (sometimes b_{560} because of its α-band at approximately 560 nm), is positioned towards the N-(matrix) side of the membrane.

The pathway of electron flow through the bc_1 complex is at first sight convoluted and we shall take some time to describe it in detail (Fig. 5.14). The discussion is for the mitochondrion; the bacterial bc_1 complexes are very similar.

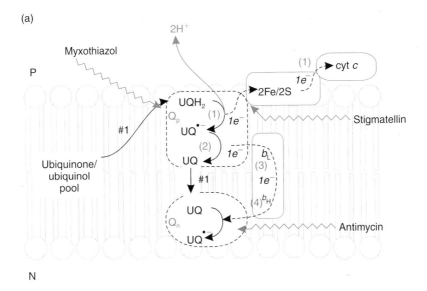

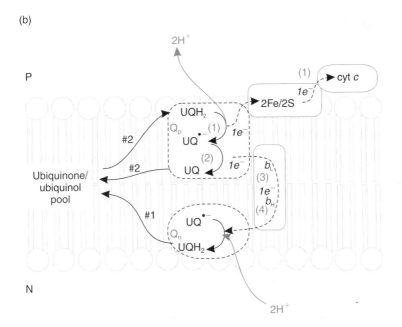

Figure 5.14   The Q-cycle in mitochondria.
(a) This illustrates the electron transfer events that follow the oxidation of a ubiquinol (#1) at the P-side of the inner mitochondrial membrane under conditions in which the quinone binding site at the N-side is initially either vacant or occupied by a ubiquinone molecule. There is evidence that an internal channel allows UQ to move from the Q_p to the Q_n site without equilibrating with the bulk pool (Section 5.8.6). (b) This illustrates the electron transfer events that follow the oxidation of a second ubiquinol (#2) at the P-side of the membrane when the Q_n is occupied by a ubisemiquinone radical. Note that the Q_p site has also been termed the Q_i or Q_c site (c indicating the cytoplasmic side of the membrane in bacterial cytochrome bc_1 complexes) in various systems, and the Q_p site is also known as the Q_o or Q_z site. The inhibitory sites of action of myxothiazol, stigmatellin and antimycin are also shown.

5.8.1 Stage 1: UQH$_2$ oxidation at Q$_p$

A pool of ubiquinone and ubiquinol exists in the inner mitochondrial membrane in large molar excess over the other components of the respiratory chain. The mid-point potential, $E_{m,7}$, for the UQH$_2$/UQ couple (in aqueous solution) is $+60$ mV (Fig. 5.9), while the actual $E_{h,7}$, which takes into account the size of the UQH$_2$/UQ ratio, is close to 0 mV. A molecule of UQH$_2$ from the pool diffuses to a binding site Q$_p$ (also termed Q$_o$ or Q$_z$) close to the P-face of the membrane and adjacent to the Rieske protein (Fig. 5.14). What then happens is that the oxidation of UQH$_2$ to UQ takes place in two stages:

(1) The first electron is transferred from UQH$_2$ to the Rieske protein (Fig. 5.14a), releasing two protons to the cytoplasm and leaving the free radical semiquinone anion species UQ$^{\bullet-}$ at the Q$_p$ site.
(2) The second electron is transferred to the b_L haem, which is also close to the P-face.

The $E_{m,7}$ for the UQH$_2$/UQ$^{\bullet-}$ couple is about $+280$ mV, close to that for the Rieske protein, and 220 mV more *positive* than the $E_{m,7}$ of the two electron oxidation ($+60$ mV, see above). This implies that the second stage of the oxidation will be energetically favourable as the semiquinone seeks to lose the second electron, and this is reflected in the E_m for the UQ$^{\bullet-}$/UQ couple, which at -160 mV is 220 mV more *negative* than that of the two-electron oxidation. The first one-electron oxidation step has thus generated the semiquinone anion, which is a very strong reductant. Under certain conditions it can be detected by ESR; it is also an intermediate of the bacterial photosynthetic reaction centre, which has a quinone binding site capable of stabilizing it (Section 6.2.2). As we shall see in Section 5.11, the unpaired electron on UQ$^{\bullet-}$ can in some circumstances be transferred directly to oxygen, thus generating the dangerous superoxide anion.

The electron received by the Rieske protein passes down the chain to cyt c_1, cyt c and cytochrome oxidase (Fig. 5.14a).

5.8.2 Stage 2: UQ reduction to UQ$^{\bullet-}$ at Q$_n$

The electron on b_L ($E_{m,7}$ -100 mV) now passes to the other haem, b_H ($E_{m,7}$ $+50$ mV). Note that these redox potentials were measured in the absence of a membrane potential, which would affect the distribution of electrons between them. At first inspection this looks as though the electron is losing 150 mV of redox potential; however, the two haems are on different sides of the hydrophobic core of the membrane across which is a membrane potential of some 150 mV (Fig. 5.14). Thus in experimental work the imposition of a membrane potential, positive on the P-side of the membrane, generated by ATP hydrolysis or by K$^+$-diffusion, causes electrons to move from b_H on the N-side of the hydrophobic core to b_L on the P-side.

The relative position of the two haems means that, in the normal operation of the complex, the electron retains its original energy on passing from b_L to b_H, since the drop to a more positive redox potential is compensated by the energetically unfavourable migration of the negatively charged electron from the P-side to the N-side of the membrane. In an uncoupled mitochondrion, energy would be dissipated at this step. This organization also implies that very high $\Delta\psi$ will retard electron transfer between the b-type haems, leading to a prolongation of the occupancy of the Q_p site by the ubisemiquinone and enhancing the chances of $O_2^{\bullet-}$ production (Section 5.11).

UQH$_2$ and UQ can in principle migrate freely from one side of the hydrophobic core to the other regardless of $\Delta\psi$, since these hydrophobic carriers are uncharged. A second quinone binding site, Q_n, in the close vicinity of b_H binds UQ and allows the transfer of the electron from the reduced b_H with the formation of the semiquinone anion UQ$^{\bullet-}$ (Fig. 5.14). At first glance this looks thermodynamically unlikely, since the $E_{m,7}$ for the UQ$^{\bullet-}$/UQ couple in free solution is -160 mV, while that for b_H is $+50$ mV. If, however, Q_n were to bind the semiquinone much more strongly than UQ, this would have the effect of shifting the $E_{m,7}$ to a more positive value, i.e. making the UQ more readily reducible. A ten-fold difference in the binding of the semiquinone relative to UQ shifts $E_{m,7}$ 60 mV more positive than if the reaction occured in free solution. A 300-fold stronger binding of the semiquinone would thus make the $E_{m,7}$ 150 mV more positive.

We have not cheated the first law of thermodynamics here, since the energy required for the addition of a second electron to generate unbound UQH$_2$ (see below) is proportionately *increased*, i.e. the E_m for the couple UQH$_{2\,\text{free}}$/UQ$^{\bullet-}_{\text{bound}}$ is made proportionately more negative. This is confirmed by actual measurements of the two $E_{m,7}$ values, using ESR to detect the semiquinone. We will come across this concept of driving an apparently unfavourable reaction by making a product very firmly/strongly bound again in Chapter 7 when we discuss the ATP synthase.

5.8.3 Stage 3: UQ$^{\bullet-}$ reduction to UQH$_2$ at Q$_n$

We now have a semiquinone firmly bound to Q_n. In the next part of the cycle (Fig. 5.14b), a second molecule of UQH$_2$ is oxidized at Q_p in a repeat of stage 1 – one electron passing to cyt c_1 and the other via b_L to b_H. This second electron now completes the reduction of UQ$^{\bullet-}$ to UQH$_2$, the two protons required for this being taken up from the matrix (Fig. 5.14b). The UQH$_2$ returns to the bulk pool and the cycle is completed.

Q_n and Q_p are not equivalent in this model: only Q_n stabilizes the semiquinone through strong binding – supported by its detection by ESR. At Q_p the redox potentials of the two steps are widely separated and the semiquinone has only a transient existence.

5.8.4 The thermodynamics of the Q-cycle

The overall reaction catalysed by the bc_1 complex involves the *net* oxidation of 1 UQH$_2$ to UQ (two UQH$_2$ oxidized in stage 1 and one UQ reduced in stage 3), the reduction of two cyt c_1, the release of 4H$^+$ at one side of the membrane and the uptake of 2H$^+$ from the other. In the model we have discussed, the major *charge* transfer across the membrane is the movement of the two electrons between the haems, which we placed on opposite sides of the hydrophobic barrier. In practice, Q_p is close to the cytoplasmic face, such that the release of protons at Q_p does not significantly contribute to the displacement of charge across the membrane, whereas Q_n is more deeply buried into the matrix side of the membrane, such that the entry of protons from the matrix contributes partially to the charge movement.

So far we have an elegant mechanism, but it may not be intuitively obvious why it can function as a proton *translocator*, i.e. removing protons from the matrix at low electrochemical potential and releasing them in the cytoplasm at a Δp some 200 mV higher.

To answer this, we shall consider two conditions, where Δp is present purely as a membrane potential (approximating to the condition in the respiratory chain) and where Δp is present purely as a ΔpH (as would occur in a thylakoid membrane, where a closely analogous cycle probably operates; Section 6.3), making the simplification that Q_n is close to the matrix face. In the first case the protons are present at equal concentrations on both sides of the membrane and the work that must be done is to push $2e^-$ from the cytoplasmic to the matrix side of the membrane against a high membrane potential. As stated above, this is energetically possible since the electrons are transferred from a negative (low) potential haem to a positive (high) potential haem. In the case of a pure ΔpH, the electrons would flood from b_L to b_H, since they have no $\Delta\psi$ to push against: this would drive UQH_2 oxidation at Q_p and UQ reduction at Q_n, enabling protons to be translocated against a high ΔpH. It is also worth recalling that the overall action of the bc_1 complex is oxidation of one ubiquinol by two molecules of cytochrome c, a significantly energetic downhill reaction.

One feature that often causes confusion is that, although the cyt bc_1 complex translocates two positive charges for each two electrons passing from ubiquinol to cytochrome c, $4H^+$ appear at the P-side when only $2H^+$ disappear from the N-side (Fig. 5.14). This apparent imbalance will be resolved when we discuss cytochrome c oxidase where $2H^+/2e^-$ appear at the P-side and $4H^+/2e^-$ are removed from the N-side.

5.8.5 Inhibitors of the Q-cycle

Antimycin, myxothiazol and stigmatellin are inhibitors of mitochondrial electron transport that act on the bc_1 complex. Antimycin acts at Q_n, preventing the formation of the relatively stable $UQ^{\bullet-}$. If, in the presence of this inhibitor, oxygen is added to an anaerobic suspension, a *reduction* of the b cytochromes occurs. This *oxidant-induced reduction* is consistent with the cycle because some Fe^{3+}-haems within cytochrome b could be reduced in the presence of antimycin (Fig. 5.14) by accepting an electron from the Q_p site as the other electron from UQH_2 passes to the Fe/S centre and onwards to cytochrome c.

Myxothiazol blocks events at Q_p, while stigmatellin inhibits electron transfer to the Rieske protein. It should be clear from inspection of Fig. 5.14 that oxidant-dependent reduction of the b cytochromes does not occur in the presence of these inhibitors.

5.8.6 The structure of complex III

Reviews and further reading   Yu *et al.* 1998, Zhang *et al.* 1998, 2000, Berry *et al.* 1999, 2000, Crofts *et al.* 1999a,b, Darrouzet *et al.* 2001

The functional information about complex III that is summarized in Sections 5.8.1–5.8.5 was all collected before any 3D structures were available. Subsequently, several crystal structures have been obtained. For the most part, these fully support the earlier biochemical investigations, but two features in particular substantially add to our understanding. One of the most important insights relates to how the electron transport pathway is bifurcated at the Q_p site; in other words, why does one electron transfer to the Rieske Fe/S centre, and thereafter to the cytochrome c_1, whilst the second passes to cytochrome b_L and then to cytochrome b_H (Fig. 5.14)?

Different structures of complex III have shown distinct spatial positions for the globular bulk of the Rieske protein (Plate B). In one structure it is close to the Q_p site, whilst in another it is docked on to cytochrome c_1. These changes in position can be correlated with the mechanism of the bc_1 complex. All the crystal structures show that the Fe/S centre of the Riekse protein is close to the surface of the polypeptide and that in the oxidized state of the centre this position is stabilized by electrostatic interactions with amino acid residues close to the Q_p centre. Reduction of the Fe/S centre lessens the interaction, with the consequence that the Rieske protein dissociates and moves to a new docking position on cytochrome c_1, which provides a socket into which the two exposed histidine ligands of the Fe/S centre insert like the prongs of an electrical plug (Plate B). When electron transfer to the haem of cytochrome c_1 occurs, the consequential oxidized state of the Fe/S centre has a diminished affinity for cytochrome c_1; the Rieske protein dissociates from and moves back towards the Q_p site (Fig. 5.15 and Plate B). Thus the globular domain of the Fe/S protein is mobile, being able to shuttle approximately 20 Å. This can occur on the submillisecond timescale and thus can match the required electron-transfer rates. There is significance to the shuttling distance being 20 Å as explained elsewhere (Section 5.4). The rate of electron transfer between two centres within proteins is critically dependent on the distance between the two centres. Once this exceeds 14 Å, the rate drops off dramatically. Consequently, when the Fe/S centre is bound to cytochrome c_1 it is too far from the Q_p site to accept an electron. Thus the second electron released upon ubiquinol oxidation at Q_p has, in effect, no alternative but to transfer to the b_L haem of the cytochrome b subunit.

The crystal structures of complex III show that it is a dimer in which the two monomeric units do *not* function independently. For example, the globular domain of the Rieske protein of one monomer interacts with the Q_p site and the cytochrome c_1 in the other (Fig. 5.15 and Plate B). The two monomers are packed together such that there are two separate cavities, the walls of which contain the quinone binding sites. This provides a second example of co-operation between monomers because, in any given one cavity, there is a Q_p site provided by one monomer and a Q_n site provided by the other.

What is the functional consequence of this organization? In principle, it means that the UQ molecule produced by oxidation of UQH_2 at a Q_p site can then diffuse to the Q_n site within the cavity without having to equilibrate with the bulk quinone pool. Such transfer of quinone or quinol within a cavity will lessen the number of required entries and exits of quinol/quinone from/to the bulk pool and thus may, in principle, contribute to catalytic efficiency.

The crystal structures of the cytochrome bc_1 complexes show the positions of as many as 11 subunits in each monomer. Eight of these have no catalytic role in the oxidation of ubiquinol. One of the small subunits targets another of the subunits to the mitochondrion from its site of synthesis in the cytoplasm while two of the others are peptidases that may catalyse removal of such targeting sequences.

5.9 CYTOCHROME *c* AND COMPLEX IV (CYTOCHROME *c* OXIDASE; FERROCYTOCHROME *c* : O_2 OXIDOREDUCTASE)

Reviews and further reading   Babcock 1999, Michel 1999, Abramson *et al.* 2000, 2001, Behr 2000, Proshlyakov *et al.* 2000, Zaslavsky and Gennis 2000, Gomes *et al.* 2001

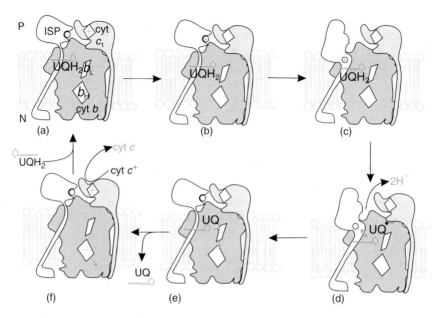

Figure 5.15   Movement of the globular domain of the Rieske iron–sulfur protein (ISP) during UQH$_2$ oxidation at the Q$_p$ site.

Only cyt b, cyt c_1 and the Rieske iron–sulfur protein (ISP) are shown in this model. Note that the mitochondrial complex III includes several other so-called core subunits which are peripheral to the membrane and face the N (matrix) phase. These are not required for electron transfer and are absent from the corresponding protein in bacteria. Complex III is a dimer within which the ISP of one monomer transfers electrons from the Q$_p$ site to the haem of cyt c_1 of the other monomer of the dimer as a result of its globular domain changing conformation and moving (see Plate 3). In (a) ISP is in its c_1 state with a reduced 2Fe/2S centre which is close to the surface of ISP, and held adjacent to cyt c_1 within which the oxidized haem is 14 Å or less from the Fe/S centre. The histidine ligands of the Fe/S centre are important for the interaction. This spatial arrangement allows electron transfer to the haem, leaving the Fe/S centre oxidized and with greater positive charge. This results (b) in a weakening of the interaction between ISP and cyt c_1, such that the Fe/S centre moves approximately 20 Å to be close to the Q$_p$ site; the b state is shown as adopted by ISP. Transfer of 1e$^-$ (c) from UQH$_2$ bound at the Q$_p$ site to the Fe/S centre lessens its affinity for this site and ubisemiquinone UQ$^{\bullet-}$ is formed at Q$_p$ as two protons are released to the P-phase (d). An electron from the UQ$^{\bullet-}$ is then transferred across the membrane through the b haems, and both UQ released and the Fe/S domain of ISP dissociate from the Q$_p$ site (e). In (f) an electron is transferred from the c_1 haem of cyt c, thus preparing the system for a further electron transfer event from ISP to cyt c_1, and binding of a new molecule of UQH$_2$, thus completing one passage through a cycle. In (f) an electron is shown on the b_H haem (cf. Fig. 5.14); this will be transferred to a ubiquinone bound nearby at the Q$_n$ site to generate a bound UQ$^{\bullet-}$ (cf. Fig. 5.14). This is not shown but it should be appreciated (cf. Fig. 5.14) that a second series of events (a) to (f) will result in formation at the Q$_n$ site of UQH$_2$ which will then dissociate, i.e. two circuits from (a) to (f) will oxidize two molecules of UQH$_2$ at Q$_p$, transfer two electrons to cyt c and reduce one molecule of UQH$_2$ at Q$_n$. Note that the Q$_n$ site that functions in concert with the Q$_p$ site shown the figure is provided largely by the cyt b subunit of the other monomer that is not shown here (see text). There is evidence for crystallography of an intermediate state (Plate B) for ISP between c_1 and b; this may be adopted in (b). (Adapted from Crofts *et al.* 1999.)

Complex III transfers electrons to cytochrome c, which is not isolated as a component of a complex, although it can bind stoichiometrically to cytochrome c oxidase. Cytochrome c is a peripheral protein located on the P-face of the inner mitochondrial membrane and may be readily solubilized from intact mitochondria. The mechanism of electron transfer to and from the haem of cyt c was discussed in Section 5.4.

The final step in the electron transport chain of mitochondria and certain species of respiratory bacteria operating under aerobic conditions is the sequential transfer of four electrons from the reduced cyt c pool to O_2 forming $2H_2O$ in a $4e^-$ reaction catalysed by a cytochrome c oxidase:

$$O_2 + 4e^- + 4H^+ \rightarrow 2H_2O \qquad [5.1]$$

The names cytochrome c oxidase and ferrocytochrome $c : O_2$ oxidoreductase refer to the catalysis of oxidation of cyt c by oxygen. Complex IV is an often-used alternative name, while earlier nomenclature, such as cytochrome oxidase and cytochrome $aa3$ oxidase, are also used.

The protons required for the reduction of oxygen are taken from the N-side of the membrane, whereas the electrons from the oxidation of ferrocytochrome c come from the P-phase. This means that the reduction of oxygen to water automatically generates a Δp. Note that protons used in the reduction of water are not translocated all the way across the membrane by this process; how far they move can be deduced from the structure (Fig. 5.16 and Section 5.9.1). This imbalance between proton uptake and release at the two sides of the membrane disappears when the overall oxidation of ubiquinol by oxygen is considered (Section 5.10). In addition, complex IV is a proton pump (Fig. 5.16).

Cytochrome oxidase poses several challenging problems including: (i) how the protons required for oxygen reduction are taken from the N-side of the membrane; (ii) the mechanism of oxygen reduction to water; and (iii) the mechanism by which the reduction of oxygen is coupled to the pumping of protons across the membrane. Understanding of all these issues has been advanced by the availability of 3D structures for the enzyme.

5.9.1 Structure of complex IV

References   Saraste 1999, Schultz and Chan 2001

High-resolution structures of the molecule, in both fully oxidized and fully reduced states, have been obtained from both a mitochondrial and a bacterial (*P. denitrificans*) source. The key catalytic functions are found in each case in subunits I and II (Plate C). Other subunits (as many as 11 for the mitochondrial enzyme) are not relevant to catalysis and are not considered here. The structures show that subunit II, in addition to two transmembrane helices, has a globular domain, folded as a beta barrel, which projects into the P-phase. This is the location of the copper A (Cu_A) centre, which has two copper atoms in a cluster with sulfur atoms. The binuclear copper centre undergoes a one-electron oxidation–reduction reaction, but why this is advantageous relative to a single copper atom (e.g. as in plastocyanin, Chapter 6) is not clear. A function of the Cu_A centre is to receive electrons, one at a time, from cytochrome c.

The two haem groups of complex IV, located approximately 15 Å below the P-surface of the bilayer (Fig. 5.16 and Plate C), are sandwiched between some of the 12 transmembrane

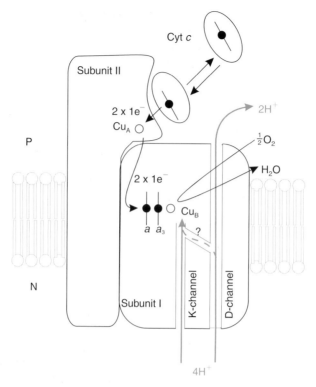

Figure 5.16   Schematic representation of subunits I and II of cytochrome *c* oxidase.

The crystal structure has established the relative positions of the *a* and a_3-type haems and the two copper centres (copper atoms represented by open spheres). An approximate deduced site of cytochrome *c* docking on to subunit II is shown; this site has to allow the haem of cytochrome *c* to come within at least 14 Å of the Cu_A site in order to facilitate sufficiently rapid electron transfer. The reaction is shown in terms of two electrons, and thus consumption of half an oxygen molecule, so as to facilitate comparison with the operation of the electron transport chain as a whole, which is traditionally analysed in terms of two electrons. Note that a cyt *c* molecule loses one electron upon oxidation and thus two molecules must dock in order to transfer two electrons into cytochrome oxidase. For each two electrons reaching an oxygen atom from cyt *c*, four positive charges are moved through the oxidase and thus across the membrane; two of these can be regarded as pumped all the way across the membrane but the other two charge movements result from the movement of two electrons from the P-side to meet two protons coming from the N-side. Currently, there is uncertainty as to pathways for proton movement from the N-side; the tentatively accepted contributions of the D- and K-channels (see text) are shown. All stoichiometries shown should be multiplied by two in order to account for reduction of one oxygen molecule (O_2).

α-helices of subunit I. Haem *a* is slightly closer to the Cu_A centre than the second haem, a_3. Haem *a* is the electron acceptor from Cu_A. Haem a_3 is within a few angstroms of haem *a*. The nomenclature for a_3 has distant origins in the scientific literature and, although not informative today, is retained for distinguishing the two *a*-type haems. The two haems are chemically identical but quite distinct spectroscopically, principally because one axial coordination position to the a_3 haem iron is not occupied by an amino acid side chain. This is the position where oxygen binds before its reduction to water. As known in some cases from crystal structures, a_3 is also the site of binding of several inhibitors, including cyanide, azide, nitric oxide and carbon monoxide.

Immediately adjacent to haem a_3 is a third copper atom, known as Cu_B; it has three histidine ligands, suggesting that a fourth co-ordination position may be occupied by a reaction product during some stages of the oxygen reduction reaction. The crystal structure surprisingly indicated that one of these histidine ligands was cross-linked through a covalent bond to a nearby tyrosine residue, a feature that has subsequently been confirmed by analysis of peptides.

Plausible channels, containing bound water molecules and located between some of the helices, can be identified. These could provide a pathway for conducting protons from the N-phase into the site of oxygen reduction, thus correlating with earlier data that this side is the origin of those protons. Two of these channels are known as D and K after a conserved aspartate and conserved lysine, which have their respective side chains projecting into these channels. There are no such obvious channels linking the haem a_3 to the P-side of the membrane, perhaps explaining the failure of protons from the P-phase to be recruited for the reduction of oxygen to water. Thus the structure does rationalize how the combination of electrons coming from cytochrome c in the P-phase plus protons from the N-phase with oxygen contributes to the generation of the protonmotive force (Fig. 5.16).

What the structure does not reveal is how protons can also be pumped across the membrane (Fig. 5.16). As explained earlier (Fig. 5.11), a proton pumping mechanism requires a conformational device to ensure that protons move unidirectionally across the membrane. One of the first structures of complex IV suggested that one of the three histidine ligands to the Cu_B centre might be able to adopt different spatial positions and protonation states; thus it was proposed to be a central component in a proton pumping mechanism. However, further crystal structures have not supported this so-called histidine cycle mechanism. Indeed, comparison of the fully oxidized and fully reduced structures of the bacterial enzyme have not shown any significant conformational changes, suggesting that complex IV functions essentially as a static molecule, in contrast to complex III (Section 5.8). On the other hand, it may be that the critical conformational changes only occur in oxidation states (Fig. 5.17) of the enzyme intermediate between the fully oxidized and fully reduced states.

In contrast, comparison of the structures of the mitochondrial enzyme in these two oxidation states has revealed a conformational change, associated with an aspartate residue, near the P-side of the membrane and some way distant from the cluster of the haem a, haem a_3 and Cu_B. This finding has generated a proposal for a proton pumping mechanism whereby the translocated protons move through a hydrogen-bonded system that is independent of the the D and K channels, and in which the conformational change required to ensure vectorial proton translocation is relayed from the site of oxygen reduction through conformational changes between the helices. A difficulty with this mechanism is that the key residue that undergoes the conformational change is not found in any of the proton pumping bacterial cytochrome c oxidases, nor indeed in all the mitochondrial enzymes. Furthermore, mutagenesis in the bacterial enzymes of counterparts to the residues implicated in the proton translocational pathway rule out any such pathway for the bacterial enzymes. Whilst the possibility of many mitochondrial cytochrome c oxidases using a different proton pumping mechanism than their bacterial counterparts cannot be

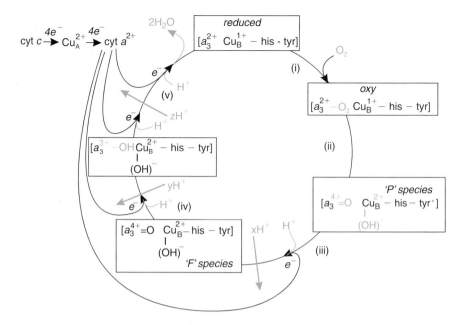

Figure 5.17   Simplified scheme for the reaction between cytochrome oxidase and oxygen its proton-pumping activity.

The fully reduced form of the enzyme is envisaged as binding oxygen to form an oxy form, which rapidly decays to give the P species in which the oxygen–oxygen bond has already broken to give bound O_2^- and hydroxide. This chemistry requires that four electrons are transferred to the original oxygen molecule; two come from the a_3 haem, which thus adopts the 4+ or ferryl oxidation state, one from the copper B, which thus becomes 2+, and the third from the tyrosine 244 side chain, which is crossed linked to histidine 240 that is a ligand to the Cu_B. The resulting tyrosine residue side chain is a radical (•), which is thought to be uncharged, implying that a proton has been lost. Formally, at least, this proton can be considered as having transferred to one of the original oxygen atoms to make a hydroxide. In the next step, delivery of an electron (from haem a) plus a proton, the latter arrival being rate limiting, restores the tyrosine radical to a normal tyrosine side chain state which generates the F state. A further electron, again supplied from haem a, together with another rate-limiting delivery of a proton, results in the a_3 haem being reduced to the +3 state and the bound oxygen acquiring a proton to become hydroxide. Further delivery of two protons and two electrons to the active site restores the haem and Cu_B, respectively, to their +2 and +1 oxidation states, and would allow each of the two oxygen atoms originally in oxygen to dissociate as water. Note, however, that the timing of water molecule release by the enzyme has not been measured. There is currently considerable controversy as to which steps in the reaction cycle are linked to proton pumping (i.e. the movement of protons addtional to those required for reduction of oxygen to water). Most investigators agree that the charge translocation mainly occurs at the three points shown, with the dispute being as to which of these steps is associated with movement of two charges right the way across the membrane, i.e. x, y or $z = 2$; the others being associated with movement of one charge, so as to account for the pumping of four protons right the way across the membrane for each oxygen molecule converted to water. Valence states and ligands in blue draw attention to changes between the different species in the cycle.

completely excluded, this notion runs counter to evolutionary and mechanistically unifying principles.

At the time of writing, most investigators of cytochrome oxidase favour a mechanism in which the route of proton pumping passes close to the haem a_3/Cu_B part of the protein. The temptation to imagine that the extended side chain of the a type haem is important for proton translocation must, however, be resisted as there are relatives of cytochrome c oxidase in bacteria that still pump protons despite having the standard haem b in both the haem binding sites. Nevertheless, there is increasing evidence that one or more propionates of the haem groups may play a role in the proton-pumping mechanism.

The fact that two putative proton channels, D and K, can be identified in cytochrome oxidase has led to attempts, by mutagenizing key residues, to determine whether one of these channels provides the protons required for reduction of oxygen while the second is concerned with the proton-pumping process. Initial studies did suggest such a division of function, but currently the situation is through to be more complex. For each eight protons needed when one O_2 is reduced (four for the water formation and four translocated), up to six are thought to travel via the D-channel with only one or two via the K-channel. Thus no clear distinction between pathways for chemical and translocated protons can be drawn.

5.9.2 Electron transfer and the reduction of oxygen

Complex IV must reduce one molecule of oxygen to two molecules of water whilst not releasing any reactive oxygen species, such as superoxide. One provisionally accepted mechanistic scheme is shown in Fig. 5.17, which shows a cycle of catalytic activity starting with the two centres (a_3 and Cu_B) at the catalytic site reduced. Note that this scheme does not involve in detail the other two centres, haem a and Cu_A.

In this scheme, oxygen binds to a_3 to give a transient oxy form (stage i); a previous suggestion that the oxygen atom might bridge between a_3 and Cu_B is now less favoured.

The O–O bond now breaks (stage ii), leaving a hydroxyl group, bound to the Cu_B, and a ferryl (i.e. $Fe^{4+}=O$) species at the haem. This requires the transfer of four electrons to the original oxygen molecule: one from Cu_B and two from a_3 (giving the +4 ferryl state), while the fourth is argued to come from the tyrosine side chain cross-linked to a histidine ligand of Cu_B. The resulting tyrosine radical is uncharged as it donates a proton to the oxygen atom that becomes hydroxide bound to Cu_B.

In stage iii, an electron is delivered to the tyrosine from haem a while a proton is taken up from the N-phase to reprotonate the tyrosine.

Stage iv requires that the ferryl ($Fe^{4+}=O$) species is reduced to a ferric hydroxide species (Fe^{3+}–OH), a process that requires one electron and one proton, the latter again originating from the N-phase. The electron is donated by haem a, which in the interim would have been re-reduced by transfer of an electron from Cu_A.

Completion of the cycle (stage v) requires two further electrons and two protons in order to release the two molecules of water and regenerate the reduced catalytic site. The complete cycle has removed four protons from the N-phase, which at the catalytic site have met four electrons which have originated in cyt c from the P-phase. Thus the P-phase has lost four negative charges and the N-phase has lost four positive charges; hence $\Delta\psi$ is generated (Fig. 5.16).

So far we have not considered how these steps are linked to the additional pumping of four protons per four electrons by the complex. This is currently a contentious area, but it is likely that protons are pumped at stages iii, iv and v; however, the stoichiometry at these individual steps is debated.

Any scheme has to satisfy thermodynamic constraints. In the present model, the Gibbs energy change resulting from the transfer of four electrons to an oxygen molecule is roughly balanced by that resulting from the formation of the highly oxidizing $a_3^{4+}=O$ and tyrosine radical species steps i and ii. Thus steps iii, iv and v are sufficiently exergonic to drive proton pumping. It is important that Fig. 5.17 does not involve the generation of any potentially toxic free-radical oxygen species.

Electrons enter the complex from cyt c on the P-side at an $E_{h,7}$ of about +290 mV for mitochondria in state 4, and are ultimately transferred to the $\frac{1}{2}O_2/H_2O$ couple with an $E_{h,7}$ in air-saturated medium at about +800 mV. However, because the site of proton uptake for O_2 reduction is the N-side (Fig. 5.16), the electrons effectively have to cross the membrane against a $\Delta\psi$ of some 180 mV. This reduces the available energy just as the reverse process in the bc_1 complex increased the energy. The redox span $\Delta E_{h,7}$ is therefore over 300 mV (i.e. +800 − 290 − 180 mV). Four electrons falling through this potential would be sufficient to translocate up to six protons across the membrane against a Δp of some 200 mV. However, unlike the remainder of the respiratory chain, complex IV is irreversible. The actual H^+/O stoichiometry for the proton-pumping activity alone is lower, $4H^+/4e^-$, reflecting this lack of reversibility. Recall again that the combination of this proton pumping and the meeting of protons and electrons within the protein (Fig. 5.16) results in an overall charge movement being $8q^+/4e^-$.

5.9.3 Control by nitric oxide of complex IV

Reviews   Brown 1995, Boveris *et al*. 2000, Moncada and Erusalimsky 2002

The oxygen binding site of complex IV oxidase also binds ligands such as CO, CN^- and N_3^-, which explains why these species are inhibitors of respiration. In addition to its role as a membrane-permeant second messenger, nitric oxide is also a reversible inhibitor of complex IV, competing with oxygen. Its slow reduction to the non-inhibitory nitrous oxide contributes to the reversibility of the inhibition. Plausibly physiological concentrations of NO decrease the apparent affinity of the complex for oxygen, and this may be of significance in the reduced oxygen environment of the intact cell. The restriction of respiratory chain activity may induce cell death under pathological conditions (Section 9.9).

5.10 OVERALL PROTON AND CHARGE MOVEMENTS CATALYSED BY THE RESPIRATORY CHAIN: CORRELATION WITH THE P/O RATIO

For every two electrons passing from ubiquinol to cytochrome c, complex III releases four protons at the P-side but takes up only two from the N-side (Fig. 5.14). In contrast, for every two electrons passing through complex IV, four protons are taken up from the N-side but only two released at the P-side (Fig. 5.16). The combined activity of complexes III and IV, therefore, removes the imbalance in proton consumption and release in the two phases. In addition to catalysing the reduction of oxygen to water, complex IV oxidase also acts as a proton pump with a stoichiometry of $2H^+/2e^-$ (Fig. 5.15). Thus the oxidase moves four

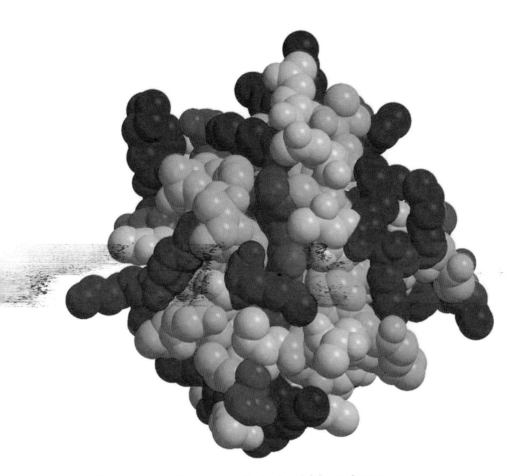

Plate A   Space filling model of mitochondrial cytochrome *c*.
The space filling model shows how only an edge of the haem (purple) is exposed.
The basic residues (arg, lys, his) are in blue; key lysine residues 13, and 86 and
87, which are involved in docking to cytochrome c_1 and cytochrome aa_3 oxidase,
are at the top, to the right and left, respectively. Acidic residues (e.g. asp) are in
red, neutral-polar residues (e.g. thr) in white and hydrophobics (e.g. val) in yellow.
The picture is based on PDB file 1HRC and was prepared by Dr Vilmos Fülöp.

P-side

N-side

P-side

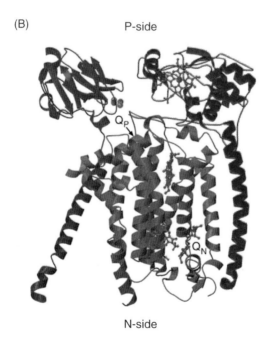

N-side

(B)

P-side

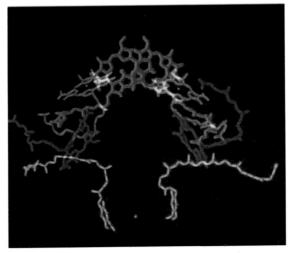

N-side

P-side

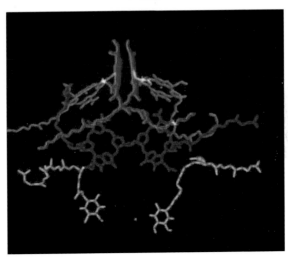

N-side

(B) The relative positions of the cofactors in the reaction centre (the two views are related to one another by a 90° rotation around the two-fold symmetry axis). Colour code: bacteriochlorophyll dimer, red; bacteriochlorophyll monomers, green; bacteriopheophytins, blue; ubiquinones (Q_A at right-hand side and Q_B at left-hand side) yellow; Fe atom (dot between two quinone sites) yellow. Reproduced with permission from Allen *et al.* (1987) *Proc. Natl Acad. Sci. USA* **84**, 5730–5734.

P-side

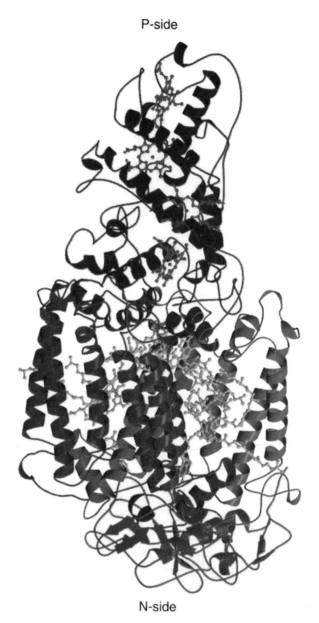

N-side

Plate G The three-dimensional structure of the *Rhodopseudomonas viridis* reaction centre.

The tetrahaem cytochrome *c* subunit is shown in blue with its haem groups in red; the L chain is grey, the M chain purple and the H chain green. The bacterio-cholorophylls and bacteriopheophytins are shown in yellow and the non-haem in red. The picture was drawn by Dr Vilmos Fülöp as for Plate B using PDB co-ordinates 1PRC.

P-side

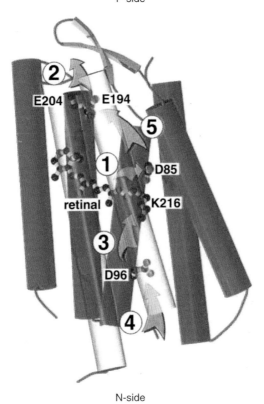

N-side

Plate H   The outline structure of bacteriorhodopsin showing the five sequential proton movements that can be correlated with steps in the photocycle (see Chapter 6, Section 6.5).
The one-letter amino acid code is used (D = aspartate, E = glutamate and K = lysine). The retinal is bound to the side chain of K216 via Schiff's base linkage. Reproduced with permission from Luecke (2000) *Biochim. Biophys. Acta* **1460**, 133–156.

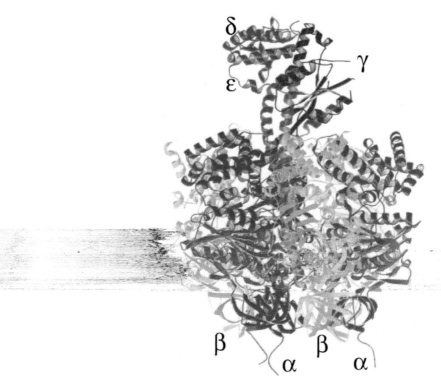

Plate I   The complete structures of all five types of polypeptide in mito-chondrial F₁ ATPase.

The α-chain is in red, the β-chain in yellow, γ in light and dark blue, δ green and ε magenta. Note how the α- and β-chains are arranged alternately around the central helical coiled coil γ-subunit. Drawn from PDB file 1E79. Reproduced with permission from Stock *et al.* (2000) *Curr. Opin. Struct. Biol.* **10**, 672–679.

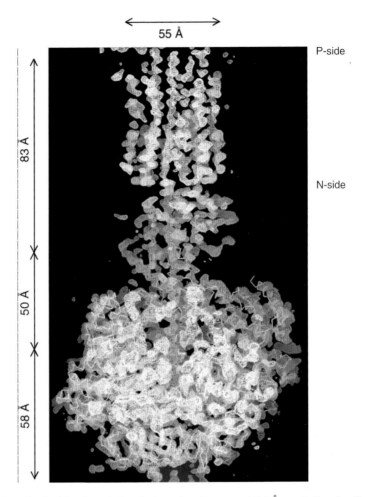

Plate J   A side view of the electron density map at 3.9 Å resolution for the yeast ATP synthase.

At the top is density for some of the 10c subunits, each of which forms two α-helices each of approximate length 50 Å. The ring of 10c subunits can be deduced to be organized such that the N-terminal helices are on the inside of the ring and the C-teminal helices on the outside. A plug of phospholipid probably seals the centre of the ring. The approximate P-side and N-side membrane bound-aries are indicated; it is probable that each c subunit projects out of the bilayer, at either side, by about 10 Å. N- and C-terminii are at the P-side, with the two helices joined by a loop at the N-side, which can thus make contact with F_1 subunits. It can be deduced that the glutamate (or, in some organisms, aspartate) that reacts with DCCD is located approximately half way across the bilayer on the outer C-terminal helices. Superimposed on the electron density is the main chain trace (C α atoms) for the α- (orange), β- (yellow) and γ- (green) chains of F_1. Drawn from PDB file 1QO1. Reproduced with permission from Stock *et al.* (1999) *Science* **286**, 1700–1705.

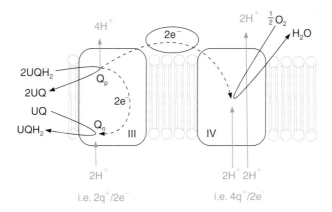

Figure 5.18   A schematic representation of the movement of protons and electrons by complexes III and IV.
See text.

charges across the membrane for each two electrons transferred to oxygen (Fig. 5.15), contrasting with the two charges per two electrons moved by complex III (Section 5.8 and Fig. 5.14). Thus, complexes III and IV have different charge per electron stoichiometries and therefore cannot be considered as thermodynamically equal or equivalent sites of proton translocation. As shown in Fig. 5.18, the overall proton (and charge) translocation stoichiometry for the transfer of $2e^-$ from ubiquinol to oxygen is $6/2e^-$. If the electrons originate from NADH, thus passing through complex I, which translocates four protons per two electrons, then the overall proton and charge stoichiometry is $10/2e^-$.

If the intact mitochondrion requires four protons (see Chapter 7) to be translocated for the generation and export of one ATP, then the P/O ratio for NADH oxidation would be $10/4 (=2.5)$ and for succinate $6/4 (=1.5)$. If the contributions of complexes III and IV are considered separately, then their $P/2e^-$ ratios are 0.5 and 1, respectively.

Since the complexes act in parallel with respect to the proton circuit (Fig. 4.2), each must generate the identical Δp. The lower charge translocation stoichiometry for complex III relative to complex IV thus accords with its smaller redox span, and hence lower ΔG available for proton translocation (Section 3.6.1). It is a frequently encountered misunderstanding to envisage that complex III makes 'less protonmotive force' than complex IV.

5.11 SUPEROXIDE PRODUCTION BY COMPLEXES I AND III

Reviews   Turrens 1997, Barja 1999

As discussed above, the $E_{m,7}$ values for the two-stage oxidation of UQH$_2$ at the Q$_p$ site of complex III via UQ$^{\bullet-}$ to UQ are, respectively, $+280\,mV$ and $-160\,mV$. This means that the UQ$^{\bullet-}$/UQ couple is highly reducing. Molecular oxygen appears to have access to the Q$_p$ site, since there is a small but finite chance that UQ$^{\bullet-}$ may donate its electron, not to b_L but to O$_2$, with the formation of the superoxide anion, O$_2^{\bullet-}$.

$$O_2 + 1e^- = O_2^{\bullet-} \qquad E_{m,7} = -160\,mV \qquad [5.2]$$

Superoxide is a free radical and capable of causing oxidative damage to the mitochondrion and its environment. Furthermore, the oxidation of $UQ^{\bullet-}$ bound at the Q_p site is dependent on the passage of the liberated electron across the membrane via the b cytochromes. Since this is opposed by the membrane potential, the actual $E_{h,7}$ of the $UQ^{\bullet-}/UQ$ couple will become more negative at high $\Delta\psi$. It would, therefore, be predicted that the production of $O_2^{\bullet-}$ would increase with $\Delta\psi$, and this is confirmed in studies with isolated mitochondria where there is a dramatic increase in the generation of reactive oxygen species, when $\Delta\psi$ increases when going from a typical respiratory state 3 to state 4. Estimates of the production of $O_2^{\bullet-}$ in respiring mitochondria vary from 0.1% to as much as 4% of total oxygen consumption, depending on the local oxygen concentration. It should be noted that, although complex IV catalyses four sequential $1e^-$ additions to molecular oxygen, the intermediates are tightly retained (Section 5.9) by the complex and no leakage of superoxide occurs.

The effects of the two characteristic complex III inhibitors, antimycin and myxothiazol, on $O_2^{\bullet-}$ generation are consistent with the above model. By preventing transfer of the first electron from UQH_2 to the Rieske protein, myxothiazol prevents formation of $UQ^{\bullet-}$ and inhibits $O_2^{\bullet-}$ generation. In contrast, antimycin acts at the Q_n site, preventing the transfer of electrons across the membrane via the b cytochromes, maximizing occupancy of the Q_p site by $UQ^{\bullet-}$ and hence increasing $O_2^{\bullet-}$ production.

Superoxide is generated at the outer, Q_p, quinone-binding site of complex III. The negative charge on the superoxide anion should firstly limit its membrane permeability and, secondly, oppose its entry into the matrix against the negative-inside membrane potential. It is therefore likely that the $O_2^{\bullet-}$ generated by complex III is released at the P-side into the intermembrane space. In cells, a cytoplasmic CuZn-superoxide dismutase, Sod1, detoxifies cytoplasmic $O_2^{\bullet-}$ by dismutation into hydrogen peroxide and water.

The mitochondrial matrix possesses a second, manganese-containing superoxide dismutase (Mn-SOD, or SOD2). Both isoforms carry out the reaction:

$$2O_2^{\bullet-} + 2H^+ = H_2O_2 + O_2 \qquad [5.3]$$

The two half-reactions are:

$$O_2^{\bullet-} + 2H^+ + 1e^- = H_2O_2 \qquad [5.4]$$

and

$$O_2^{\bullet-} = O_2 + 1e^- \qquad [5.5]$$

Since SOD2 is an essential enzyme (knockouts being embryonic lethal) this implies that $O_2^{\bullet-}$ can also be formed within the matrix. It is likely that complex I is the main source of matrix $O_2^{\bullet-}$. The relative importance of complexes I and III varies with the origin of the mitochondria, and the molecular details of its generation in complex I are less clear.

In isolated heart and brain mitochondria, $O_2^{\bullet-}$ production by complex I is upstream of the rotenone binding site, and this inhibitor increases $O_2^{\bullet-}$ production during electron transport through the complex from NAD^+-linked substrates. In contrast to complex III, complex I generation of $O_2^{\bullet-}$ is largely independent of $\Delta\psi_m$ during a state 4–state 3 transition. Conversely, with succinate as substrate, and in the absence of endogenous complex I substrates, complex I can still generate $O_2^{\bullet-}$ as a result of reversed electron transfer (Section 4.9). This is decreased by rotenone, thus locating the complex I $O_2^{\bullet-}$ generation site on the NADH side of the inhibitor binding site (Fig. 5.12).

5.12 OXIDATIVE STRESS

Reviews Vogel *et al.* 1999, Tanaka *et al.* 2000, Schafer and Buettner 2001

Mitochondria have developed defences to detoxify superoxide generated by the respiratory chain, the most important being the glutathione couple and superoxide dismutase 2 (Fig. 5.19). The three quantitatively most important redox couples in the mitochondrial matrix are GSSG/GSH (Section 3.3.5), $NADP^+/NADPH$ and $NAD^+/NADH$. In state 3 conditions their respective redox potentials are roughly $-240\,mV$, $-390\,mV$ and $-290\,mV$. The disequilibrium between the NADP and NAD pools is maintained both by the Δp-linked transhydrogenase (Section 5.13) and the NADP-linked isocitrate dehydrogenase.

Brain mitochondria have 5 nmol of total glutathione per milligram of protein and the pool is about 88% reduced in state 4. GSSG is reduced to GSH in the mitochondrial matrix

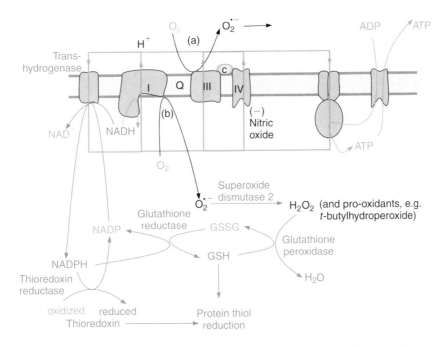

Figure 5.19 Mitochondrial 'oxidative stress': interactions between the GSSG/GSH couple and reactive oxygen species.
The Δp-driven mitochondrial transhydrogenase (together with NADP-linked isocitrate dehydrogenase) maintains a highly reduced NADP pool. NADPH reduces GSSG to GSH via glutathione reductase and reduces the protein thioredoxin. GSH and reduced thioredoxin maintain protein thiols in the reduced state. Using glutathione peroxidase, GSH detoxifies hydrogen peroxide generated by superoxide dismutase 2, and artificial pro-oxidants such as *t*-butylhydroperoxide. Superoxide, $O_2^{\bullet-}$ is the major reactive oxygen species (ROS) generated by the respiratory chain. Superoxide is generated at complex III and complex I (Section 5.11). The former site probably released $O_2^{\bullet-}$ into the intermembrane space, while that from complex I may be released into the matrix. In addition $O_2^{\bullet-}$ may be membrane permeant in its protonated form HO_2.

by the NADPH-linked enzyme glutathione reductase. The large thermodynamic disequilibrium between the GSSG/GSH and NADP$^+$/NADPH couples suggests that the reductase activity may exert a major control over the rate of GSH generation. Thioredoxin is a 12 kDa protein with a redox-active dithiol at its active site. Both thioredoxin and GSH help to maintain protein thiol groups in the reduced state. In addition GSH detoxifies H_2O_2 or artificial pro-oxidants such as t-butylhydroperoxide (t-BuOOH) via the enzyme glutathione peroxidase.

As discussed in Section 3.3.5, the absolute concentrations of GSH and GSSG, rather than merely their ratio, defines the redox potential of the GSSG/GSH couple. Even partial depletion of the pool greatly decreases the concentration of GSSG for a given redox potential, and this can pose kinetic problems for glutathione reductase, which uses GSSG as substrate (Fig. 3.4).

Oxidation of the GSSG/GSH couple, for example, by t-BuOOH, greatly sensitizes both isolated and *in situ* mitochondria (Fig. 5.19) to the permeability transition (Section 8.2.5), as does inhibition of the transhydrogenase by oxidation of the NAD$^+$/NADH pool by acetoacetate in liver mitochondria.

5.13 THE NICOTINAMIDE NUCLEOTIDE TRANSHYDROGENASE

Reviews   Jackson *et al.* 1999, Cotton *et al.* 2001

Although the mid-point potentials for the NAD$^+$/NADH and NADP$^+$/NADPH couples are the same (Table 3.2), the ratio NADPH/NADP$^+$ is much greater than NADH/NAD$^+$ in the mitochondrial matrix. One process (others include an NADP-linked isocitrate dehydrogenase) maintaining this disequilibrium is the protonmotive force-dependent transhydrogenase, which catalyses the following reaction, running in the direction from left to right:

$$NADP^+ + NADH + nH^+_{P\text{-phase}} = NADPH + NAD^+ + nH^+_{N\text{-phase}} \qquad [5.6]$$

where n is almost certainly 1. The observed mass-action ratio (Section 3.2) may exceed 500, which would thus be maintained by respiration-dependent Δp. The function of this transhydrogenase, which is also found in the cytoplasmic membranes of many bacterial species, has long been enigmatic. In mammalian mitochondria it is currently thought that provision of NADPH for reduction of glutathione (Section 5.12) is a significant role, along with the possibility that, together with the NAD$^+$- and NADP$^+$-linked isocitrate dehydrogenases, it plays a role in the fine tuning of the TCA cycle. In some parasites the transhydrogenase is argued to function as a protonmotive force generator, that is, the reaction runs from right to left in the above equation.

The mitochondrial transhydrogenase comprises a polypeptide of approximately 110 kDa. The enzyme has a 400-residue N-terminal globular component, which binds NAD(H), a central hydrophobic region predicted to contain 14 transmembrane helices, and a 200-residue C-terminal globular component, which binds NADP(H). Structures are available for both globular components, which in the intact enzyme are exposed to the mitochondrial matrix. The enzyme functions as a dimer. Direct hydride transfer from NADH on one component to NADP$^+$ on another requires the direct juxtaposition of binding sites for the two substrates. It is envisaged that, at any one time, this can occur only between one N-terminal/C-terminal component pairing, with the other pair within the dimer being in

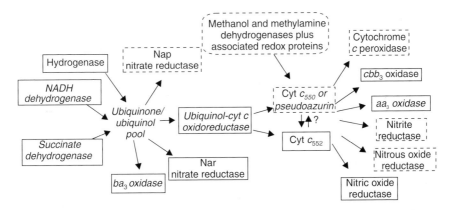

Figure 5.21 Organization of electron transport components in P. denitrificans.
Only the components in italics are thought to be constitutive. The other components are induced by appropriate growth conditions and are unlikely to be all present at once. NADH dehydrogenase, succinate dehydrogenase, ubiquinol cytochrome c oxidoreductase and aa_3 oxidase correspond to mitochondrial complexes I–IV. Continuous boxes indicate integral membrane components; dashed lines represent periplasmic components. Further details of methanol and methylamine oxidation is given in Fig. 5.23, and of nitrate respiration in Fig. 5.24.

overall $H^+/2e^-$ stoichiometry from UQH$_2$ to oxygen will be lower. Cytochrome ba_3 is very similar to the cytochrome bo_3 of *E. coli* (Section 5.15.2); both are members of the super-family of terminal oxidases known as the haem–copper oxidases. This family is defined by high sequence similarity within the largest subunit (subunit I, which has 12 transmembrane α-helices) and a binuclear active site consisting of a high-spin haem (a_3, b_3 or o_3) and a closely associated Cu$_B$.

The cbb_3 oxidase (Fig. 5.21) is also a member of this family with two *b*-type haem molecules, one designated b_3 to denote proximity to Cu$_B$, in place of the *a*-types. The Cu$_A$ domain of cyt aa_3 is replaced by a *c*-type cytochrome subunit. For as yet unknown reasons, these changes result in an oxidase with much higher affinity for oxygen than aa_3, whilst retaining (according to most investigators) the capacity for pumping 1H$^+$ per electron. Comparison of cbb_3 and aa_3 suggests that neither the farnesyl side-chain extension of the *a*-type haem nor its formyl group is an essential component in a proton translocation mechanism. The cbb_3 enzyme was first identified as an oxidase with very high affinity for oxygen that was required for symbiotic nitrogen fixation in *Rhizobial* species. The branched aerobic electron transport chain of *P. denitrificans* is a very common feature amongst the bacteria, but the reasons for it, the control of expression of the different components and the regulation of the distribution of electrons between the branches are not understood in detail.

Figure 5.21 also shows a hydrogenase, probably related to the membrane-bound hydrogenase that has other similarities to components of complex I (see Section 5.6). It can pass electrons directly to the UQ pool. Electrons can also be fed to UQ from succinate dehydrogenase as well as from a FAD-linked fatty acid oxidation step.

P. denitrificans can use final electron acceptors other than oxygen (Fig. 5.21). Amongst these is H_2O_2, a molecule commonly found in the soil environment of the bacteria. Reduction of H_2O_2 is catalysed by a periplasmic cyt *c* peroxidase (Fig. 5.21), which is a dihaem *c*-type cytochrome and thus differs from the enzyme with the same function that is in the intermembrane space in yeast mitochondria (Section 5.12). Anaerobic electron acceptors in *P. denitrificans* are described in Section 5.15.1(b), after we discuss how this organism is able to oxidize compounds that have only one carbon atom.

(a) Oxidation of compounds with one carbon atom

Reviews Goodwin and Anthony 1998, Anthony 2000

P. denitrificans is one of a restricted group of bacteria that can grow on methanol or methyl-amine as its sole carbon source. The dehydrogenases for these two compounds (Fig. 5.22) are found in the periplasm (P-phase). That for methanol contains pyrroloquinoline quinone (PQQ) as a cofactor, whilst methylamine dehydrogenase (for which a 3D structure has been determined), contains a novel type of redox centre, a tryptophyl-tryptophan involving a covalent bond between two tryptophan side chains. Electrons pass from the redox centre of trimethylamine dehydrogenase to a 1e⁻ carrier copper protein, amicyanin, which forms a sufficiently tight complex, mainly through apolar interactions, with the dehydrogenase to allow co-crystallization. From amicyanin, the electrons pass to *c*-type cytochromes, probably including cyt c_{550}, and then to oxygen via cyt aa_3 oxidase. Thus Δp is established in this short electron-transfer chain by the inward movement of electrons and outward pumping of protons through cyt aa_3, together with the release and uptake of protons at the two sides of the membrane associated with methylamine oxidation and oxygen reduction (Fig. 5.22).

In the case of methanol oxidation, the energetic considerations are similar, electrons being transferred from reduced PQQ, within methanol dehydrogenase, which is also a

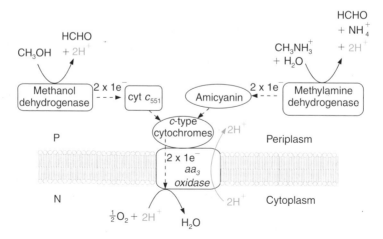

Figure 5.22 Schematic representation of periplasmic oxidation of methanol or methylamine in *P. denitrificans*.

protein of known structure, to c-type cytochromes (Fig. 5.22). Although the $E_{m,7}$ of the methanol–formaldehyde couple is similar to that for succinate–fumarate, the H^+/O stoichiometry for the oxidation of succinate (which involves both the bc_1 complex and cyt aa_3 oxidase) is greater than for methanol oxidation. Why a longer electron transport chain is not used for the oxidation of methanol or methylamine is not understood. The formaldehyde produced from oxidation of methanol or methylamine is oxidized by cytoplasmic (N-phase) enzymes to CO_2 with concomitant generation of NADH. In $P.$ $denitrificans$, the CO_2 thus produced is refixed into cell material. In other organisms that grow on methanol, some of the formaldehyde can be directly incorporated into cell material.

(b) Denitrification

The sequential reduction of NO_3^-, NO_2^-, NO and N_2O is catalysed by anaerobically grown $P.$ $denitrificans$ in the process known as denitrification (hence the name of the organism). The four reductases required to carry out this process receive electrons from the underlying electron transport system used in aerobic respiration (Figs 5.21 and 5.23).

The membrane-bound NO_3^- reductase (often called Nar) receives electrons from UQH_2 towards the P-side of the membrane. $2H^+/UQH_2$ are released to the periplasm and $2e^-$ pass inwards across the cytoplasmic membrane via two b-type haems to the site of nitrate reduction, which is at an Mo atom that is co-ordinated by a specific cofactor known as MGD, on the N-surface (Fig. 5.23). The $E_{m,7}$ for the NO_3^-/NO_2^- couple is $+420\,mV$. The inward movement of the electrons is equivalent to the transfer from cytoplasm to periplasm of two positive charges $(2q^+)$ per $2e^-$. Since UQ was originally reduced at the N-face of the membrane, the outward transfer of UQH_2 and the return to the N-phase of $2e^-$ can be considered as a good example of a redox loop mechanism (Fig. 5.10).

In common with many other organisms, $P.$ $denitrificans$ also possesses a periplasmic nitrate reductase (often called Nap). As with the membrane-bound enzyme, the nitrate is reduced at an Mo centre, which is co-ordinated by sulfur atoms provided by the two molecules of the MGD cofactor. The important bioenergetic distinction between the two types of nitrate reductase is that the periplasmic enzyme itself is not associated with proton translocation (Fig. 5.23). How exactly the electrons pass from UQH_2 within the core of the bilayer to Nap is not clear, but it is very probable that a protein known as NapC participates in this process. NapC is a tetra-haem c-type cytochrome but, as the haem groups are located in a globular domain, one must propose that this domain dips sufficiently into the bilayer to provide a binding site for ubiquinol. This would be analogous to the ETF–ubiquinone oxidoreductase that was discussed in Section 5.7.2. Explanations as to why two nitrate reductases can be present, as in $P.$ $denitrificans$, are complex and beyond the scope of this book.

Nitrite reductase is a soluble enzyme in the periplasm, contains both c- and d_1-type haem centres (hence usually called cytochrome cd_1) and can receive electrons from cyt bc_1 via either cyt c_{550} or a copper protein known as pseudoazurin, on the periplasmic face of the membrane. The crystal structure shows that the active site is the d_1 haem, unique to this type of enzyme and carrying several modifications to the usual haem ring, which is contained in a propeller-shaped structure made up from eight blades of four stranded β-sheet. The cytochrome c domain must therefore interact with either cyt c_{550} or pseudoazurin.

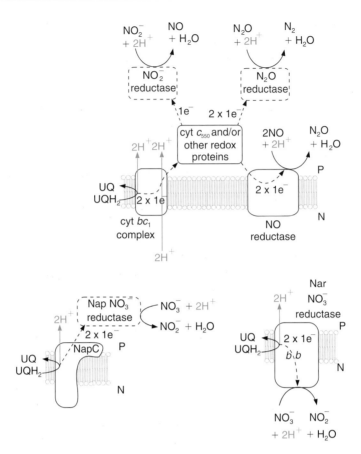

Figure 5.23   Electron transport pathways associated with nitrate reduction and the subsequent steps of denitrification in *P. denitrificans*.
There are two routes for electron flow from ubiquinol to nitrate. One, the Nap pathway, results in electron and proton release to the periplasm and hence no generation of Δp; the tetrahaem NapC protein is believed to provide the ubiquinol oxidation site. The other pathway uses the membrane-bound (Nar) reductase which has two *b*-type haems distributed across the membrane and a cytoplasmic-facing active site at which there is a molybdenum centre (containing a specialized pterin cofactor known as MGD). As explained in the text, this loop mechanism is associated with the same net positive charge translocation across the membrane as electron flow to the other three nitrogenous acceptors. The charge and proton translocation stoichiometry catalysed by the ubiquinol cytochrome *c* oxidoreductase is explained in Section 5.8 and Figs 5.14 and 5.18. Note that the reduction of nitrate by the membrane-bound nitrate reductase requires that negatively charged nitrate enters the cell despite the $\Delta\psi$ being negative inside. A nitrate–nitrite antiporter (cf. Chapter 8) is currently postulated to overcome this bioenergetic problem.

As these two proteins have very different structures, the concept of pseudospecificity has been invoked to explain how they can each dock onto cytochrome cd_1. The transfer of $2e^-$ through bc_1 to the periplasmic cyt c_{550}, coupled to the release to the periplasm of $4H^+/2e^-$ is, as in the case of the mitochondrial bc_1 complex, equivalent to $2q^+/2e^-$. Note that these nitrate and nitrite reductases, and the *E. coli* enzymes discussed below (Section 5.15.2), are distinct from the widespread enzymes with the same names that are responsible for the assimilation of nitrogen in bacteria and plants, and which are beyond the scope of this book.

Nitric oxide reductase is an integral membrane protein containing both *b*- and *c*-type haems, which from sequence analyses has unexpectedly proved to be closely related to the cbb_3 oxidase and thus to the haem–copper family of oxidases. A critical difference is that the copper at the active site is replaced by iron and that the enzyme does not translocate charge across the membrane. Thus as shown in Fig. 5.23, it is thought that both protons and electrons reach the active site from the same (periplasmic) side of the membrane and a proton-pumping activity is absent, consistent with the absence from nitric oxide reductases of key residues implicated in the proton-pumping activity of the oxidases. The unexpected discovery of a relationship between nitric oxide reductase and oxidases has led to the interesting proposal that nitrate respiration, and that of nitric oxide in particular, predated terrestrial oxygen respiration. Presumably nitrogen oxides must have been made by photochemical reactions between water and nitrogen. Electrons for NO reduction are supplied from cyt bc_1 via cyt c_{550}; thus, as in the case of nitrite reductase, the net outward charge transfer is $2q^+/2e^-$.

Finally nitrous oxide reductase – a copper-containing enzyme – is, like nitrite reductase, a soluble periplasmic enzyme. It contains the same Cu_A centre as in complex IV as well as a novel cluster of four copper atoms, bridged by sulfur, at the active site. Electrons reach this reductase from bc_1 via c_{550} and are associated with a $2q^+/2e^-$ outward charge transfer. It is notable that the $E_{m,7}$ of the N_2O/N_2 couple is even more positive than $\frac{1}{2}O_2/H_2O$, at $+1100\,mV$, although the concentration of N_2O *in vivo* may be so low that the actual E_h for the couple may be comparable to the $+800\,mV$ for the oxygen reaction.

Nitrite reductase, N_2O reductase and NO reductase thus all appear to serve as P-face electron sinks at the level of cyt *c* (Fig. 5.23), and as such all play integral roles in their respective pathways. Each is associated with $2q^+/2e^-$ stoichiometries for the spans from UQH_2, as indeed is nitrate reductase with its different pathway. Thus, despite the redox span from the UQH_2/UQ couple to NO_3^-/NO_2^- being far smaller than to N_2O/N_2, the charge transfer is the same (Fig. 5.23). It should be noted also that the onward flow of electrons from cytochrome c_{550} to oxygen via cyt aa_3 or cyt cbb_3, rather than to these reductases, leads to a higher $q^+/2e^-$ ratio. One complication in the scheme of Fig. 5.23 is that the role of cyt c_{550} in denitrification has been questioned by the finding that a mutant of *P. denitrificans* lacking cyt c_{550} is still able to denitrify. An explanation is that the copper protein pseudoazurin, similar to plastocyanin (Chapter 6), can substitute.

There is sometimes confusion as to whether a soluble periplasmic enzyme such as N_2O reductase can 'participate' in the generation of Δp. It should be clear that, although the activity of such enzymes *per se* does not contribute directly to Δp, their roles in transferring electrons to acceptors is necessary for the proton-translocating NADH dehydrogenase and bc_1 complex to function. On the other hand, comparison of Fig. 5.22 with Fig. 5.23 shows that transfer of electrons from the periplasmic methanol dehydrogenase to nitrous oxide reductase would not generate a Δp despite the large redox span. This illustrates the importance of considering not only redox spans but also the topology of electron flow in energy-transducing membranes.

5.15.2 *Escherichia coli*

Reviews and further reading   Unden and Bongaerts 1997, Bader *et al.* 1999, Iverson *et al.* 1999, 2000, Kobayashi and Ito 1999, Abramson *et al.* 2000, Dym *et al.* 2000

When *E. coli* is grown aerobically, many of the electron transport components are distinct from those found in either mammalian mitochondria or *P. denitrificans*. In particular, there are no detectable *c*-type cytochromes, and electron transport from UQH_2 to oxygen is insensitive to antimycin and myxothiazol, inhibitors of the bc_1 complex, which is absent from *E. coli*. Two oxidase enzymes, known as cyt bo_3 (but often called *bo*) and cyt *bd* can directly oxidize ubiquinol (Fig. 5.24). Cyt bo_3 is a member of the haem–copper super-family. A crystal structure (Plate D) of this bo_3 oxidase shows, as expected, that the spatial organization of the two haem groups and the adjacent copper is very similar to that in aa_3 oxidase, and that a ubiquinol binding site can be inferred, intercalated between helices of subunit I towards the P-side of the membrane. Thus the absent Cu_A site of cyt aa_3 is replaced by a quinol oxidation site. This cyt bo_3 is, in common with cytochrome aa_3,

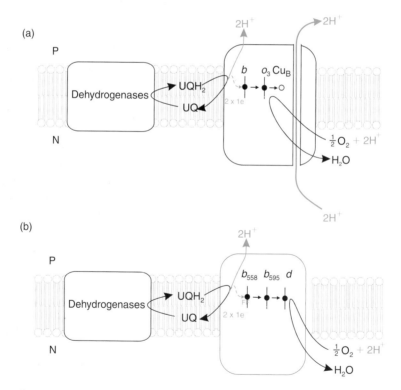

Figure 5.24   The *E. coli* aerobic electron transfer chain from ubiquinol to oxygen.
A crystal structure is available for the bo_3 oxidase (Plate D) and the ubiquinol oxidation has been modelled at approximately the location shown. The open cir-cle indicates the Cu_B centre. For cytochrome *bd* (b), a more approximate model based on biochemical data is presented. Note that cytochrome bo_3 (a) is related to cytochrome aa_3 and essentially performs the same function.

a proton pump with stoichiometry $2H^+/2e^-$ and overall charge movement of $4q^+/2e^-$ (cf. Section 5.9).

The cyt bd complex comprises two polypeptide chains and has two b-type haems, one called b_{558}, which is the electron acceptor from ubiquinol. The b_{595} is thought to form a haem pair with the distinctive porphyrin ring of the d-type haem. Strictly speaking, this last is a chlorin, owing to saturation of one of the pyrrole rings, but is markedly different from the d_1 haem in *P. denitrificans* nitrite reductase (Section 5.15.1). d haem is the site of oxygen reduction; there is no adjacent copper or non-haem iron. Subunit I probably folds into nine transmembrane helices and provides both axial ligands to the b_{558}. The other two haem groups are probably bound by both subunit I and the five helix subunit 2.

Cytochrome bd does not show any sequence similarity with the haem–copper family of oxidases; although perhaps surprising, this is consistent with the very different fold of the protein. There is no evidence that cyt bd is a proton pump and thus the stoichiometry of charge translocation is $2q^+/2e^-$, due purely to the inward movement of electrons from the site of ubiquinol oxidation, in other words a loop mechanism (Fig. 5.10, and cf. Fig. 4.9). Biochemical analysis and studies with mutants all suggest that all three haem groups are located towards the periplasmic side of the membrane. Cyt bd has a much higher affinity for oxygen than cyt bo_3 and is synthesized under conditions of low oxygen concentrations.

The lower stoichiometry of proton translocation achieved at low oxygen concentrations may be the price that has to be paid to attain a high catalytic rate of oxidase activity with no thermodynamic back pressure from the protonmotive force on the individual reaction steps of oxygen reduction.

Clearly *E. coli* has a truncated electron transport chain, in comparison with mitochondria and *P. denitificans*, with lower $q^+/2e^-$ and $H^+/2e^-$ ratios. If the H^+/ATP ratio for the ATP synthase is the same as for an organism where electron transfer is exclusively via bc_1 and aa_3, then it follows that the P/O ratio for $UQH_2 \rightarrow \frac{1}{2}O_2$ is also lower. Obviously the redox span for $UQH_2 \rightarrow \frac{1}{2}O_2$ is independent of the pathway. The electron transport system of *E. coli* illustrates that an organism may not always be seeking to maximize the stoichiometry of ATP production. Natural habitats of *E. coli* may be rich in potential substrates and the need to maximize ATP yield may not apply.

E. coli has two NADH dehydrogenases. One of these, the proton-translocating enzyme, is very similar to that of *P. denitrificans* (Section 5.15.1) and the mitochondrial complex I (Section 5.6). The second enzyme has a much simpler subunit composition and does not translocate protons.

Figure 5.25 shows that the *E. coli* respiratory chain can receive electrons from many electron donors. The oxidation of D-lactate is interesting because the dehydrogenase is a peripheral membrane protein of known structure, which has to dip into the membrane sufficiently to make contact with the head group of ubiquinone or menaquinone. As in the case of the mitochondrial ETF/UQ oxidoreductase (Section 5.7), transmembrane helices are not needed for the provision of a binding site for quinone.

E. coli is not restricted to aerobic growth. Under anaerobic conditions, part of the TCA cycle, 2-oxoglutarate dehydrogenase, ceases to function, in contrast to *P. denitrificans* and many other non-enteric bacteria. Pyruvate can be converted to formate or fumarate. Consequently, under anaerobic conditions, these products can respectively act as an electron donor and acceptor to the electron transport system that contains menaquinone rather

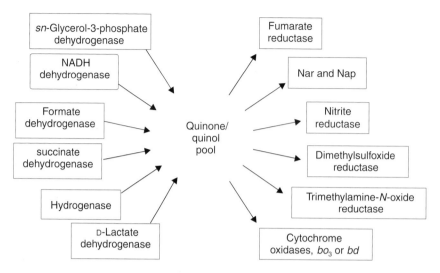

Figure 5.25   An overview of *E. coli* aerobic and anaerobic respiratory systems.
The components present depend on the growth conditions. Under anaerobic conditions, menaquinone replaces ubiquinone as the main quinone. Note that many of the enzymes (e.g. the nitrite reductase, the periplasmic nitrate reductase and the trimethylamine-N-oxide reductase) are located in, or have their active sites facing, the periplasm.

than ubiquinone under these conditions. Oxidation of formate to CO_2 and concomitant reduction of fumarate to succinate can generate Δp (Fig. 5.26).

Crystal structures of a membrane-bound formate dehydrogenase and fumarate reductase have now been obtained (Plate E). The formate binding site is in a globular domain exposed to the periplasm, which is connected to transmembrane helices between which two b-type haems are sandwiched, one towards the P-side and one towards the N-side. Protons are released to the periplasm, and electrons are transferred from the Mo/MGD centre in the active site via suitably positioned Fe/S centres to the haem at the N-side, where protons are taken up and quinol formed. Thus the enzyme acts as a Δp generator (Fig. 5.26). In contrast, fumarate reductase has its globular domain exposed to the N-side and, although it is related to succinate dehydrogenase, it has no haem groups. It is thought that quinol oxidation and proton release both occur at the N-side of the membrane and thus this reductase does not generate Δp.

A variety of anaerobic electron acceptors can be used by *E. coli* (Fig. 5.25). The expression of many of the necessary enzymes is dependent on the Fnr protein, which, under anaerobic conditions, acts as a transcriptional acitivator for many of the genes associated with anaerobic metabolism. NO_3^- is reduced to NO_2^- by two reductases that are very similar to those described above for *P. denitrificans*. However, NO_2^- is reduced to NH_4^+, rather than to NO, by a periplasmic nitrite reductase that contains five c-type haem groups. Dimethylsulfoxide and trimethylamine-*N*-oxide (both occur in natural environments, the latter especially in fish) can also serve as terminal electron acceptors via one or more reductases, which contain Mo at the active sites, bonded similarly as in nitrate reductases (Fig. 5.25).

It has recently emerged that the respiratory chain of *E. coli* can accept electrons from a post-translational process in the periplasm. Disulfide-bond formation in periplasmic

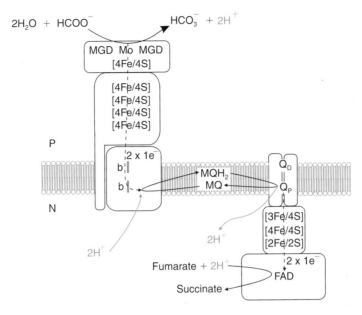

Figure 5.26   The structural basis of Δp generation as formate is oxidized by fumarate in _E. coli_.

The structures of both the formate dehydrogenase N (see also Plate E) and the fumarate reductase (Plate E) have been determined at high resolution. The active site of the formate dehydrogenase has an Mo atom coordinated by two molecules of the pterin cofactor MGD. Oxidation of formate at this site results in electron transfer via a series of Fe/S centres and two haems such that the electrons are taken from the P to the N side of the membrane where proton uptake occurs for formation of menaquinol. Δp is thus established by this reaction which represents the electron carrying arm of a redox loop (cf. Fig. 5.10). In contrast, the fumarate reductase does not transfer electrons across the membrane so that its functioning does not involve any generation or utilization of Δp. (Formate oxidation can also be linked to the Nar type of nitrate reductase (Fig. 5.23) which also contributes to Δp generation (see Plate E).) More details can be found in Jormakka _et al._ (2002).

proteins requires the oxidation of cysteine thiol groups. The primary oxidation event is catalysed by a periplasmic protein known as DsbA in which a disulfide bridge is reduced. The resulting two cysteines are in turn reoxidized by a disulfide in DsbB; the latter is reformed when DsbB is oxidized by the ubiquinone or menaquinone of the respiratory chain. The exact mechanism whereby two thiol groups in DsbB are oxidized back to a disulfide is, however, not known.

5.15.3 Relationship of _P. denitrificans_ and _E. coli_ electron transport proteins to those in other bacteria

Many of the features of the electron transport systems of _P. denitrificans_ and _E. coli_ are of general importance, as exemplified by the fact that some components, e.g. the membrane-bound and periplasmic nitrate reductases, are common to both despite the low overall similarity of these two organisms. The UQ/UQH_2 (or MQ/MQH_2) pools not only act as collectors of electrons from several sources, as in mitochondria, but also donate electrons to several alternative pathways, and thus serves as a 'cross-roads' for electron transfer (Fig. 5.25), compatible with quinone–quinol mobility within the bilayer. This feature is found in many other bacteria. It is not known how electrons distribute themselves between the different pathways; the simplest mechanism would be competition, which may also occur in plant mitochondria (Section 5.12). In some cases this explanation appears insufficient; thus nitrate reduction via the membrane-bound enzyme by *P. denitrificans* is blocked by the presence of oxygen. The locus of control may be a nitrate transport system (Fig. 5.23).

The c-type cytochromes, which in bacteria are far more varied and play a greater range of roles than in mitochondria, are also a common feature of bacterial electron transport. Such cytochromes are often water-soluble and almost invariably found in the periplasm or at least with a haem group exposed to the periplasm (e.g. cyt c_1). The presence in the periplasm of a large number of electron transfer proteins is also a general feature of redox reactions in Gram-negative bacteria. In Gram-positive bacteria, which do not have a periplasm, the c-type cytochromes appear to be more tightly associated with the cytoplasmic membrane and the range of metabolic activities associated with dehydrogenases and reductases at the P side is much more restricted than in Gram-negative organisms. It is notable that, in many organisms, periplasmic c-type cytochromes function at a junction point. NapC (Section 5.15.1) is an example of a large class of such proteins that are involved in electron transfer into and out of the periplasm.

Some of the electron transport chain components found in *P. denitrificans*, especially bc_1 and aa_3, are also relatively widely distributed, although these two cytochrome complexes are not always present together. Of particular interest is the widespread distribution of nitric oxide reductase, even in organisms that cannot denitrify. This may equip organisms to use nitric oxide that mammalian cells make as a defence against bacteria. As with cytochrome oxidase, there are variants of nitric oxide reductase; some can directly oxidize quinols, whilst others contain a Cu_A centre. The electron transfer components of *E. coli* are also found elsewhere. For example, a similar cytochrome *bd* oxidase with high affinity for oxygen terminates a ubiquinol oxidase system in *Azotobacter vinelandii* and *Klebsiella pneumoniae*. In these organisms a role of this oxidase is to maintain low oxygen concentrations in order to protect an oxygen-sensitive nitrogenase enzyme.

Overall, it is becoming clear that many of the components found in *P. denitrificans* and *E. coli* are also found in a range of other organisms, but in various combinations (see Section 5.15.4 for a recent example). It is as if the genes for these components have been transferred between organisms so as to provide the optimal apparatus for a particular growth environment. However, it should be appreciated that there are many other electron transport proteins found in the bacterial world which are distinct from the examples given here (see the Appendix). For example, in some denitrifying bacteria, nitrite reductase does not contain c and d_1 cytochromes but rather a copper protein – a high-resolution structure shows a trimeric protein with type I copper as in plastocyanin (Section 6.4) and type II copper as the active site.

Another important variation is that some bacteria either use or produce precipitated materials during respiration. In at least some cases it is currently envisaged that electron transfer must occur across the outer membrane. In the case of *Shewanella frigidimarina*, a multihaem c-type cytochrome on the exterior surface of the outer membrane is postulated to be involved in reduction of the soluble ferric iron to the insoluble ferrous form. A further distinct example is provided by sulfate reducing bacteria, which are rich in a variety of c-type cytochromes quite distinct from those in *P. denitrificans*, whilst we can expect that the archae will provide further variants of quinones and cytochromes. Nevertheless, the unifying themes can be expected to be: (a) quinones and c-type cytochromes acting as mobile components to connect enzymes that handle different electron donors and acceptors; and (b) spatial distribution of enzymes between the P- and N-sides of the membrane so as to provide organizations that will lead to the generation of Δp within the thermodynamic limits.

In the next section we illustrate how some of the electron transfer components identified in *P. denitrificans* and *E. coli* appear in organisms that have very different physiologies than these two models.

5.15.4 *Helicobacter pylori*

Review Kelly 1998

Helicobacter pylori is a bacterium that grows at very low oxygen concentrations and has attracted attention as it is a cause of several disease states, including gastric ulcers and possibly gastric cancer. It is an example of an organism where more knowledge of its electron transport system has been gained from the sequencing of its genome, now an increasingly rapid procedure, rather than from biochemical analyses. Most of the respiratory chain (Fig. 5.27) system could be readily identified from sequence similarities with known bacterial electron-transport components. Several features are notable, some of which are in common with *P. denitrificans*, others with *E. coli*. First, and unusually for a bacterial respiratory chain, there is only one oxidase and that, as might be expected from the

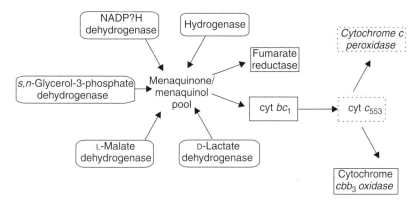

Figure 5.27 The electron transfer chain of *Helicobacter pylori* as deduced from the genome sequence.
The components were identified almost exclusively by assigning open reading frames in the genome sequence using sequence data bases. Cytochrome c_{553} is related to a c-type cytochrome found in other bacteria including *Campylobacter*. The other components shown in the figure have been introduced in the discussions of *P. denitrificans* and *E. coli*. Dotted lines indicate periplasmic location.

requirement for growth under low oxygen conditions, is the high affinity cbb_3-type (Section 5.15.1). Second, like *E coli*, the organism can use fumarate as an electron acceptor, although there are no reports of it growing anaerobically with this electron acceptor, while in common with *P. denitrificans* it has the respiratory chain enzyme that can reduce hydrogen peroxide to water. Third, the cytochrome bc_1 complex is evidently adapted to use menaquinol rather than ubiquinol, and fourth succinate dehydrogenase is absent. Thus the organism has, as it were, recruited a set of electron transport chain components that are found elsewhere in the bacterial world in order to provide a system that can satisfy its growth niche. How aerobic growth is possible without succinate dehydrogenase is beyond the scope of this book!

Not everything about the bioenergetics of *H. pylori* can be immediately deduced from the genome sequence. Thus, although an NADH dehydrogenase can be expected to be present, the critical NuoE and NuoF subunits (Fig. 5.12) of an NADH dehydrogenase are absent despite the presence of orthologues of other Nuo subunits. As NuoE and NuoF provide the catalytic site for NADH oxidation, and presumably the first electron-transfer steps within the enzyme (Fig. 5.12), it is difficult to understand the nature of the enzyme. However, an intriguing biochemical observation is that cytoplasmic membranes oxidize NADPH much more rapidly than NADH. The combination of data from the genome sequence and traditional biochemistry now directs future research into the possibility that this organism has a proton translocating NADPH dehydrogenase; similar approaches will be needed in many other contexts.

Finally, we note the presence of both a D-lactate menaquinone oxidoreductase (similar to the enzyme in *E. coli*) and an L-malate menaquinone oxidoreductase, emphasizing again the role of quinone as a junction point in electron transport systems. It should be clear that the bioenergetic interpretation of the genome sequence relies on the knowledge of bacterial electron transport systems gained previously by biochemical studies on a limited number of model organisms.

5.15.5 *Nitrobacter*

If an organism grows on a substrate with a relatively positive redox potential, it can be faced with the problem of how to generate NADH or NADPH for biosynthetic reactions. The example of *Nitrobacter* illustrates this aspect of electron transport.

Nitrobacter grows by oxidizing nitrite to nitrate ($E_{m,7}$ +420 mV) by a nitrite oxidoreductase, transferring electrons via a *c*-type cytochrome to a cyt aa_3 oxidase and reducing oxygen to water ($E_{m,7}$ +820 mV; Fig. 5.28). It is not immediately apparent, therefore, how the organism can reduce NAD^+ to NADH ($E_{m,7}$ −320 mV) for biosynthetic reactions. The solution comes from reversed electron transfer, which was introduced in a mitochondrial context in Section 4.9.

The short electron-transfer chain described above generates a Δp, which will, as in other bacterial genera, drive ATP synthesis and transport processes. *Nitrobacter* probably possesses a bc_1 complex (or an equivalent) and an NADH dehydrogenase complex, which, as in other electron transfer chains, are reversible. The Δp generated by $NO_2^- \rightarrow O_2$ is used to reverse both these complexes and proton re-entry drives a minority of the electrons originating from NO_2^- back through the complexes from *c*-type cytochrome to NADH (Fig. 5.28).

The mechanism of Δp generation has not been fully elucidated. Figure 5.28 presents a plausible, but not fully proven, scheme in which NO_2^- oxidation occurs on the N-face of

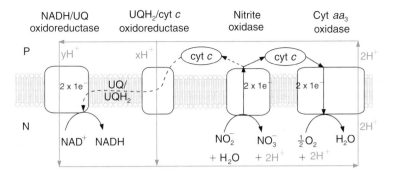

Figure 5.28   Protonmotive force generation and reversed electron transport in *Nitrobacter.*
As explained in the text, it is proposed that $\Delta\psi$ drives electrons energetically uphill from nitrite to the cytochrome c and that cytochrome aa_3 acts as proton pump. (---) indicates reversed electron flow. Not all details of this scheme have been fully substantiated, although the sequence of the cytochrome aa_3 shows the presence of all the amino acid residues implicated as important in proton pumping by this type of enzyme. The nitrite oxidase has sequence similarities with the membrane bound-type of nitrate reductase (Nar) (Fig. 5.19).

the membrane, transferring electrons to cyt c on the P-face. This may seem strange, since this charge movement *collapses* rather than generates Δp, but there is a thermodynamic reason. The $E_{m,7}$ for the cytochrome is $+270\,\text{mV}$, or $150\,\text{mV}$ more electronegative than for NO_2^-/NO_3^-. Δp, or rather the $170\,\text{mV}$ $\Delta\psi$, which is typical for bacterial cytoplasmic membranes, is needed to drive the reduction of the cytochrome. A related utilization of $\Delta\psi$ has been discussed in the context of the mitochondrial bc_1 complex (Section 5.8). From cyt c electrons pass to oxygen via the proton-pumping cytochrome aa_3 oxidase. Note that the electrons return to the N-face and that the net outward movement of positive charge is due to the proton pumping of the oxidase.

   Support for the above role of the membrane potential comes from studies with inverted membrane vesicles from *Nitrobacter*. Electron transfer from $NO_2^- \rightarrow O_2$ is slowed, rather than accelerated as in mitochondrial oxidations (Section 4.6), by conditions that decrease $\Delta\psi$ (e.g. the presence of protonophores or 'state 4/state 3' transition; Section 4.6). Such conditions result in a decreased reduction of cyt c and hence a decline in respiration. The energetics of *Nitrobacter* illustrate the beautiful economy of the chemiosmotic mechanism. Δp drives the initial step of substrate oxidation and reversed electron transport as well as more conventional processes such as ATP synthesis and substrate transport. The sequence of the *Nitrobacter* cytochrome aa_3 oxidase shows that key residues implicated in proton pumping (Section 5.9) are present.

5.15.6   *Thiobacillus ferrooxidans*

Reviews   Ingledew 1982, Blake *et al.* 1993

Nitrobacter is by no means the only example of an organism in which reversed electron transport is important. Another instance is *Thiobacillus ferrooxidans*, which oxidizes Fe^{2+}

to Fe^{3+} ($E_{m,7} = +780\,mV$). As with the oxidation of nitrite to nitrate, this reaction cannot directly reduce NAD^+ and thus a small proportion of the electrons derived from Fe^{2+} are transferred 'uphill' to NAD^+, whilst the remainder flow to oxygen with concomitant generation of Δp. The oxidation of Fe^{2+} occurs in the periplasm and electrons are transferred to an oxidase via a copper protein known as rusticyanin (Fig. 5.29). *T. ferrooxidans* typically grows at an external pH of 2, which is important for two reasons. First, the rate of uncatalysed oxidation of ferrous to ferric ion is slower than at pH 7 and, second, the reduction of oxygen to water, being a reaction that consumes protons, has a more positive E_h at the lower pH.

In assessing the bioenergetics of this organism it is important to keep in mind that the limits are set by the E_h values for the Fe^{2+}/Fe^{3+} and O_2/H_2O reactions at pH 2. With

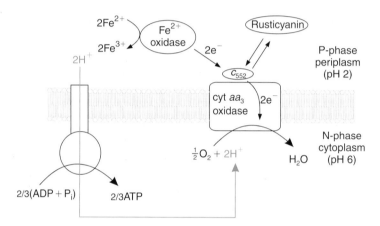

Figure 5.29   Electron transfer and ATP synthesis by *Thiobacillus ferrooxidans*.
Only at external pH values of approximately two does the organism grow by oxidizing Fe(II) at the expense of oxygen. The low pH means that the Fe(II) is more soluble and its non-catalysed oxidation to Fe(III) is much slower than at pH 7. Furthermore, the free-energy change for the overall oxidation of Fe(II) by oxygen is much greater at pH 2, owing to the pH-dependence of the redox potential of the oxygen–water reaction. Nevertheless, the overall energy available to the cell is still small. The oxidation of iron occurs in the periplasm or on the outer membrane and oxygen is reduced by a cytochrome aa_3 oxidase. It is thought that, as in all cytochrome oxidases, the protons required for reduction of oxygen are taken from the cytoplasm, but in contrast to other cytochrome aa_3 molecules there is reason to believe (see text) that there is no additional proton pumping for energetic reasons. Thus, the action of the protonmotive activity of the oxidase will contribute to making the electric potential more negative inside and also to consumption of cytoplasmic protons. In the absence of any electron transfer, the protonmotive force is zero; the large pH gradient is balanced by a membrane potential, positive inside the cells. The operation of cytochrome oxidase scarcely changes the pH but results in the membrane potential inside becoming less positive by approximately 180 mV, thus giving a net protonmotive force. This drives ATP synthesis (as shown) and reversed electron transport (not shown). The diagram shows a proton cyt unit which, if we assume that $3H^+$ are needed for each ATP mode (Chapter 7), means that 2/3ATP are made per $2e^-$ flowing to oxygen (i.e. $p/2e^- = 0.67$). The scheme does not show how the pH gradient is maintained.

reasonable estimates of the actual concentrations or partial pressure of the substrates, we can calculate that at pH 2 the redox span is approximately 300 mV. This means that, if transfer of one electron from Fe^{2+} to oxygen is associated with the movement of one charge across the membrane, then the maximum value of Δp could be 300 mV, whereas movement of two charges (Fig. 5.29) would restrict the Δp to 150 mV.

The low pH outside the cells has important consequences for the relative contributions of $\Delta \psi$ and ΔpH to the total Δp. During steady-state respiration, the cytoplasmic pH is estimated to be about 6, giving a ΔpH of 4 units, equivalent to 240 mV. If the total Δp were to be 300 mV, and thus much larger than found in other systems, $\Delta \psi$ would be 60 mV, positive inside. On the other hand, if Δp were to be 150 mV, then the $\Delta \psi$ would have to be 90 mV, negative outside. Currently, there is some uncertainty about the size of Δp and whether or not the terminal oxidase, now known from the genome sequence to be cytochrome aa_3, is the first example of this class of enzyme that does not pump protons (Fig. 5.29). Some key residues thought to be involved in proton pumping are absent from the sequence of the *T. ferrooxidans* oxidase. If it does turn out to have the 'standard' stoichiometry of proton pumping, then Δp will be restricted to approximately 150 mV. The implications for the stochiometry of ATP synthesis are explained in Fig. 5.29.

A final point of confusion concerning the bioenergetics of *T. ferrooxidans* concerns the magnitude of Δp in the absence of respiration. The pH difference can still be 4 units, apparently equivalent to a Δp of 240 mV, but there is no net Δp under these conditions. The $\Delta \psi$ is approximately 240 mV, positive inside the cells, and arises from an inwardly directed diffusion potential of protons. The onset of respiration and thus of outward positive charge movement effectively lessens the magnitude of this potential. It is a common mistake to imagine that the large pH gradient can be regarded as some type of gratis and bonus contribution to Δp.

5.15.7 The bioenergetics of methane synthesis by bacteria

Reviews and further reading   Chistoserdova *et al*. 1998, Thauer 1998, Gottschalk and Thauer 2001

Methanogenic bacteria are archaea, which obtain energy from several types of reaction in which methane is an end product. It was only established in the late 1980s that this methane formation is associated with electron transport driven H^+ or Na^+ translocation and that resultant ATP synthesis is by a chemiosmotic mechanism.

Two methanogenic organisms, *Methanosarcina barkeri* and *Methanosarcina mazei* strain *Gö*1, provided important clues about the bioenergetics of methanogenesis; both gain energy for growth from the reduction of either CH_3OH or CO_2 by H_2:

$$CH_3OH + H_2 \rightarrow CH_4 + H_2O \qquad [5.7]$$

or

$$CO_2 + 4H_2 \rightarrow CH_4 + 2H_2O \qquad [5.8]$$

The organisms can also grow on methanol alone, but discussion of this will be reserved until the fundamental electron pathways for reduction of CH_3OH and CO_2 have been described. A striking feature of methanogenesis is that it involves a number of water-soluble molecules that are rarely found outside methanogens. These include coenzyme M

Figure 5.30   Sequence and energetics of reactions involved in methane formation from CH_3OH or CO_2 plus H_2 in methanogenic bacteria.
As explained in the text, reaction step 1, in the direction as written, is endergonic, whereas steps 6 and 8 are significantly exergonic. These are the three reactions in which at least some components are membrane-bound and in which, therefore, coupling to ion translocation across the cytoplasmic membrane is possible. The reductive steps 4 and 5 are catalysed by water-soluble enzymes for which the reduced form of F_{420} is the electron donor. Re-reduction of F_{420} is catalysed by a water-soluble hydrogenase. Electrons for the reduction of *CoM-S-S-CoB* in step 8 are transferred from a membrane-bound hydrogenase, with some similarities to complex I, and in some methanogens, *b*-type cytochromes, to the reductase. This step is linked to proton translocation. Another membrane-bound hydrogenase, also with similarities to complex I, participates in step 1. X in step 1 is probably a polyferredoxin. Work to define the biochemistry of methane formation completely is continuing.

($HSCH_2CH_2SO_3^-$; *CoM*), coenzyme B (a molecule with a thiol group at the end of a chain of six CH_2 groups attached to a threonine phosphate; *CoB*) and F_{420}. The latter is a 5'-deazaflavin with a $E_{m,7}$ of $-370\,mV$, and is a structural and functional hybrid between nicotinamide and flavin coenzymes; it is a diffusible species in the cytoplasm of methanogenic bacteria.

There are a number of other unusual cofactors bound to the enzymes of methanogenesis (Fig. 5.30); some have now been discovered in bacteria that oxidize methanol.

(a) Reduction of $CH_3OH \rightarrow CH_4$ by H_2

Intact cells of *M. barkeri* can synthesize ATP as well as CH_4 in the presence of CH_3OH and H_2. ATP synthesis is chemiosmotic (rather than a result of substrate-level phosphorylation by a soluble enzyme system) by the following criteria:

(a) Protons are extruded.
(b) DCCD, presumed to be a specific inhibitor of an ATP synthase as in other organisms, inhibits ATP production, increases Δp and slows the rate of methane formation. Genome sequencing for a methanogen has shown that an ATP synthase of the type discussed in Chapter 7 is present, although certain distinctive features cause it to be classified as an A (archaeal)-type enzyme.
(c) Protonophores dissipate Δp but increase the rate of methane formation.

These observations parallel what would be observed in the analogous mitochondrial proton circuit (Chapter 4). An involvement of Na^+ can be eliminated, since this ion is not needed for methanogenesis from CH_3OH plus H_2, although Na^+ is required for growth of methanogens and some reactions of methanogenesis (see below).

Further understanding required preparation of functional inside-out membrane vesicles from *M. mazei Göl*. Addition of H_2 and CH_3SCoM to crude *M. mazeii Göl* vesicles resulted in ATP synthesis. (CH_3SCoM is formed in cells from CH_3OH and CoM in a reaction catalysed by methanol: CoM methyltransferase (step 9 in Fig. 5.30), an enzyme that has at its active site a haem-type ring (a corrinoid) with a Co, to which the methyl group is transiently attached, instead of Fe at the centre). The methyl group of CH_3SCoM was

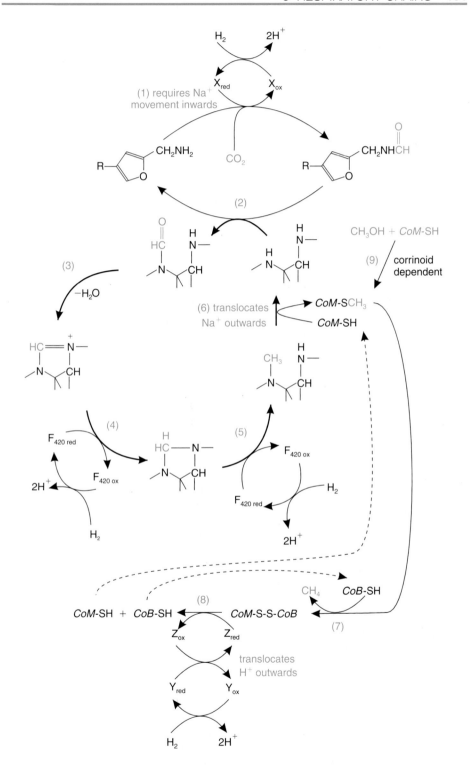

converted to methane through reaction with a second thiol-containing compound known as *CoB*-SH, and present in the preparation of vesicles, to give a heterodisulfide:

$$CH_3SCoM + CoB\text{-}SH \rightarrow CH_4 + CoM\text{-}S\text{-}S\text{-}CoB \qquad [5.9]$$

The enzyme catalysing this reaction is known as methyl coenzyme M reductase and was fortuitously present in the vesicle preparation. The crystal structure of this water-soluble enzyme reveals a deep channel, at the bottom of which is a cofactor known as F_{430}, which is a porphinoid; this is related to a haem group but contains a Ni atom rather than Fe at the centre. It is not known exactly how this unusual enzyme works, but the methyl group is widely believed to transfer from CH_3SCoM to the Ni ion. The *CoB*-SH also enters the channel and its oxidation to give the heterodisulfide is linked to the cleavage of the Ni-methyl bond and the release of methane.

The above reaction (step 7 in Fig. 5.30), catalysed by a water-soluble enzyme cannot directly drive ATP synthesis; rather it is the reduction of *CoM*-S-S-*CoB* back to the two separate thiol species that is catalysed by a membrane-bound enzyme system, which translocates protons across the cytoplasmic membrane. An electron source for this reduction is hydrogen, from which electrons are transferred via a membrane-bound hydrogenase with similarities to complex I that contributes to a proton-translocating electron transport chain (Fig. 5.30). In line with a chemiosmotic mechanism, the rate of step 8 (Fig. 5.30) catalysed by vesicles was accelerated by the onset of ATP synthesis or addition of protonophores.

(b) Reduction of $CO_2 \rightarrow CH_4$ by H_2

Growth of *M. barkeri* is also supported by the reduction of CO_2 by H_2. CO_2 is first taken up by covalent attachment to methanofuran (Fig. 5.30). After a first reduction step to a formylated derivative, using electrons derived from H_2 via a membrane-bound hydrogenase with some similarity to bacterial and mitochondrial NADH-ubiquinone oxidoreductases (Section 5.6), this formyl group is transferred to a pterin compound (step 2 in Fig. 5.30). After two further reductions in which electrons from H_2 are transferred, using water-soluble enzymes, via F_{420}, a methyl group is formed which can be transferred to CoM (see above and Fig. 5.21). The CH_3SCoM then reacts as described above to generate CH_4.

(c) Growth by disproportionation of CH_3OH

M. barkeri and *M. mazei Gö*1 can grow by disproportionation of methanol in the absence of H_2:

$$4CH_3OH \rightarrow 3CH_4 + CO_2 + 2H_2O \qquad [5.10]$$

The stoichiometry of this reaction shows that one molecule of methanol is used to provide the reductant required for methane formation from the other three molecules of methanol. ^{2}H-labelling shows that three hydrogens in each of the three methane molecules are derived from the methyl groups in CH_3OH via CH_3SCoM. The electrons released in the oxidation of the fourth methane are, of course, those needed for the reduction of three molecules of

CH_3SCoM to CH_4. The fourth molecule of CH_3SCoM is converted to CO_2 by the reverse of the reactions 1–6 shown in Fig. 5.30, and the electrons released used to drive the reductive reaction of methane formation (step 8 in Fig. 5.30).

(d) Growth on acetate

In many natural environments most methane is formed from acetate. This involves some sophisticated enzymology, whereby the methyl group of acetylCoA, formed from acetate in an ATP-consuming reaction, is separated by carbon–carbon bond cleavage and transferred to the pterin compound of methanogenesis. The resulting methyl derivative is thereafter processed by steps 6, 7 and 8 in Fig. 5.30. The electron source for step 8 comes from taking the carbonyl group from acetylCoA and oxidizing it to carbon dioxide. The electrons released by this oxidation are used to reduce protons to hydrogen using a membrane-bound hydrogenase, which couples the reaction to proton translocation. The hydrogen so obtained is then used as the electron donor for reduction of mixed disulfide (proton translocating step 8, Fig. 5.30). Thus there are two translocation steps involved in forming methane from acetate but the net stoichiometry of ATP production will be low, as one ATP is consumed for each acetate converted to acetylCoA in the first step. This is one of many examples in bacterial energetics where the net yield of ATP is small.

(e) The energetics of methanogenesis

Although Fig. 5.30 summarizes the likely electron transfer steps involved in the reduction of CO_2 to CH_4, it gives no information about the bioenergetics of the process except that we have already established that reduction of CoM-S-S-CoB is coupled to H^+ translocation and this would allow the cells to grow on CH_3OH plus hydrogen. Are there any energy-conserving steps in the more complex reduction of CO_2? Analysis of the thermodynamics of the individual steps suggests that the methyl transfer step from pterin to CoM (step 6 in Fig. 5.30) is exergonic. Reduction of added formaldehyde (which attaches spontaneously to the pterin so as to enter the sequence after step 4) resulted in the generation of an electrochemical gradient although Na^+, rather than H^+, was translocated out of the cells. This is consistent with the enzyme catalysing step 6 (Fig. 5.31) being membrane bound. The exact role of the Na^+ gradient thought to be generated by step 6 (Fig. 5.30) is unclear, but it should be noted that methanogenic bacteria generally require Na^+ for growth. The proton and sodium electrochemical gradients appear to be interchangeable via a Na^+/H^+ exchanger.

At physiological concentrations of H_2, which are very considerably below the standard state value of 1 atm, the first reaction (step 1 in Fig. 5.30) is significantly endergonic. It is, therefore, very probable that this step is driven by the inward movement of Na^+ down the electrochemical gradient set up by reactions 6 and 8.

Whereas $\Delta G^{o\prime}$ for the reduction of CO_2 by H_2 is $-131\,kJ\,mol^{-1}$ per mol CH_4 produced, the actual ΔG is likely to be closer to $-30\,kJ\,mol^{-1}$. Since ΔG_p, the free energy for ATP synthesis, is likely to be about $+50\,kJ\,mol^{-1}$, this means that a theoretical maximum of about 0.6 ATP can be generated per mol of CH_4 formed from CO_2. This is a further example of the chemiosmotic mechanism allowing non-integral stoichiometries (see Chapter 4).

5.15.8 *Propionigenium modestum*

Review   Dimroth *et al.* 2001

We next turn to an example of bacterial energy transduction that does not involve electron transport but which, in common with electron transport-dependent energy transduction, involves the co-operation of two ion pumps, and so is appropriately discussed in this chapter. *P. modestum* is an anaerobic bacterium that ferments succinate to propionate by a short reaction sequence:

$$\text{Succinate} \rightarrow \text{succinyl CoA} \rightarrow \text{methylmalonyl CoA} \rightarrow \text{propionyl CoA} \rightarrow \text{propionate}$$
$$[5.11]$$

The decarboxylation of methylmalonyl CoA has a ΔG of about $-27\,\text{kJ mol}^{-1}$, close to that for the overall fermentation. The decarboxylase is a membrane-bound, biotin-dependent enzyme that pumps 2Na^+ out of the cell for each CO_2 released. The Na^+ electrochemical gradient thereby set up could be up to $12\text{–}15\,\text{kJ mol}^{-1}$ at equilibrium (since two ions are pumped) and is known to drive ATP synthesis through a Na^+-dependent $F_1.F_o$ ATP synthase that is discussed further in Chapter 7. Since a typical ΔG_p in a bacterial cell might be $45\text{–}50\,\text{kJ mol}^{-1}$, it is likely that three, or more likely four, Na^+ ions might be required per ATP synthesized for energetic reasons.

The energetics of this organism reinforce the significance of non-integral coupling stoichiometries in bacterial energetics; this organism could not by definition exist if one ATP had to be formed by a soluble enzyme system for each molecule of succinate fermented! This is also not the only example of bacterial energy conservation being linked to a decarboxylation reaction. A further example is given in Chapter 8.

Spinach chloroplasts are subjected to an acid bath (Fig. 4.15) in order to generate ATP in the dark, while the 'Z' scheme of non-cyclic electron transfer generates O_2 and transfers electrons to a negative E_m component

6 PHOTOSYNTHETIC GENERATORS OF PROTONMOTIVE FORCE

6.1 INTRODUCTION

A photosynthetic organism captures light energy in order to drive the otherwise endergonic synthesis of molecules needed for the growth and maintenance of the organism. A central feature of photosynthesis is the conversion of light energy into redox energy, meaning that photon capture causes a component to change its redox potential from being relatively electropositive to being highly electronegative. The electrons released from this component are utilized to generate a Δp, flowing either through a cyclic pathway back to re-reduce the original component, or in a non-cyclic pathway to reduce additional electron acceptors (ultimately $NADP^+$ in the case of photosynthesis catalysed by thylakoids in chloroplasts). In this latter case, a continual electron supply to the photon-sensitive component is required (obtained from H_2O in the thylakoid example).

The production of ATP by photosynthetic energy-transducing membranes involves a proton circuit, which is closely analogous to that already described for mitochondria and respiratory bacteria. Thus a Δp in the region of 200 mV across a proton-impermeable membrane is used to drive a proton-translocating ATPase in the direction of ATP synthesis. In the case of photosynthetic bacteria, Δp may also drive other endergonic processes (Fig. 1.8) – including reversed electron transport to generate NADH (see below). The ATPase (or ATP synthase) is identical to the mitochondrial enzyme except in detail (Chapter 7). The distinction between the respiratory and photosynthetic systems is in the nature of the primary generator of Δp, yet even here a number of familiar components recur, including cytochromes, quinones and Fe/S centres. Photosynthetic activity in *Halobacteria*, see Section 6.5, is distinctive: photon capture leads to a direct generation of Δp in the absence of electron transfer.

The two features that are unique to photosynthetic systems are the antennae, responsible for the trapping of photons, and the reaction centres, to which the energy from light is directed. A component in the reaction centre becomes electronically excited as a result of the

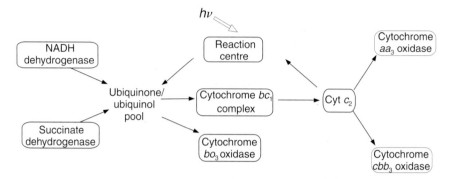

Figure 6.1   Light-driven cyclic electron transport system and its relationship to respiratory electron transport in *R. sphaeroides*.
This is a simplified version. Deletion of the gene for cytochrome c_2 does not prevent cyclic electron transport because an alternative *c*-type cytochrome can act as substitute. Electron transport in the closely related organism *Rhodobacter capsulatus* is similar except that cytochrome aa_3 is absent. Other aspects of electron transport in these two organisms, including anaerobic respiratory pathways, are given in Ferguson *et al.* (1987).

absorption of a photon. An electron can be released from this excited state at a potential which is up to 1 V more negative than the potential of donors to the reaction centre. Thus the electron lost from the reaction centre is replaced by an electron at much more positive potential this regenerates the ground state of the component in the reaction centre that underwent excitation. In this way light energy is directly transduced into redox potential energy. This sequence of photochemical events is often denoted by the shorthand

$$P* \xrightarrow{\;-e^-\;} P^+ \xrightarrow{\;+e^-\;} P \tag{6.1}$$

where P indicates a pigment in the reaction centre, * indicates an electronically excited state and P^+ the cation form of the ground state of the pigment. The components that accept the electron (e^-) from P* and donate it to P^+ are described later.

In the case of the representative purple photosynthetic bacterium, *Rhodobacter sphaeroides*, the electron released from the reaction centre feeds into a bulk pool of ubiquinol from which it passes via a proton translocating cytochrome bc_1 complex (Chapter 5) to a cyt c_2 (closely related to mitochondrial cyt *c* and *P. denitrificans* cyt c_{550} (Chapter 5)). Cyt c_2 in turn acts as the donor of electrons to the reaction centre completing the cyclic electron flow (Fig. 6.1).

Whilst a cyclic electron transfer pathway is also present in the thylakoids of chloroplasts (Section 6.4), the key difference from photosynthetic bacteria is the non-cyclic pathway in which electrons are extracted from water, pass through a reaction centre, a proton translocating electron transfer chain (which again has similarities to the cytochrome bc_1 complex), a second reaction centre and are ultimately donated to $NADP^+$, at a redox potential 1.1 V more negative than the $\frac{1}{2}O_2/H_2O$ couple (Section 6.4). The chloroplast thus not only accomplishes an 'uphill' electron transfer but at the same time generates the Δp for ATP synthesis. The ATP and NADPH are used in the Calvin cycle, the dark reactions of photosynthesis in which CO_2 is fixed.

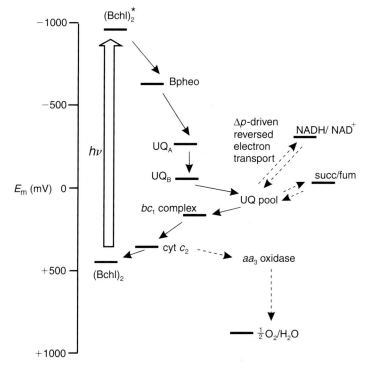

Figure 6.2   Pathways of electron transfer in *R. sphaeroides* in relation to the redox potentials of the components.
As explained in the text (Bchl)$_2$, Bchl, Bpheo, UQ$_A$ and UQ$_B$ are redox centres in the reaction centre.

6.2 THE LIGHT REACTION OF PHOTOSYNTHESIS IN *RHODOBACTER SPHAEROIDES* AND RELATED ORGANISMS

The heavily pigmented membranes of photosynthetic organisms act as antennae, absorbing light and funnelling the resultant energy to the reaction centres. In the case of *R. sphaeroides*, the photochemically active pigment has an absorption maximum at 870 nm and thus is known as P$_{870}$. The equivalent energy of a 870 nm photon amounts to 1.42 eV (Section 3.5); thus the energy transfer process is highly effective in the sense that 70% of the energy captured by the reaction centre is conserved in the resulting redox change of approximately 1 V (Fig. 6.2).

6.2.1 Antennae

Review   Cogdell *et al.* 1999

The photochemical activity of reaction centres depends upon the delivery of light at a specific wavelength (870 nm for the commonly studied reaction centre of *R. sphaeroides*). Energy of this wavelength can be obtained by direct absorption of incident light with this wavelength and also by transfer, through mechanisms to be described below, from components of the reaction centre that absorb at shorter wavelengths. However, even in bright sunlight, an individual pigment molecule will be hit by an incident photon only about once a second. Much higher rates of energy arrival at the reaction centre are required to match the turnover capacity of the centre, which is $>100\,s^{-1}$. Some collecting process is evidently required.

Furthermore, most of the incident photons will have the wrong wavelengths to be efficiently absorbed by the pigments of the reaction centre. Effective light absorption over a wide range of wavelengths shorter than 870 nm is achieved by an assembly of polypeptides with attached pigment molecules. These polypeptides, which bind bacteriochlorophyll and carotenoid pigments, are known as light-harvesting, or antenna, complexes and surround the reaction centre. That such antenna are not strictly necessary for photosynthesis is established by the existence of bacterial mutants lacking them but which will nevertheless grow photosynthetically, albeit only in bright light.

Use of antennae to speed up the rate of photochemistry in the reaction centres is clearly more effective in biosynthetic terms than inserting very many copies of the reaction centre into the membrane to achieve a high rate of overall photochemistry. Thus, over 99% of the bacteriochlorophyll molecules in a photosynthetic membrane are involved, together with carotenoid molecules, in absorbing light at shorter wavelengths than 870 nm (*R. sphaeroides*) and transferring it down an energy gradient to the lower energy absorption band at 870 nm. The transfer can occur by one of two mechanisms. The first, known as resonance energy transfer, is intermolecular, depending on an overlap between the fluorescence emission spectrum of a donor molecule and the excitation spectrum of an acceptor. Factors that affect the efficiency of such transfer include the relative orientation of donor and acceptor as well as the distance between the donor and acceptor (an inverse sixth power relationship). Energy transfer by this mechanism (which is *not* an emission followed by reabsorption of light) occurs over a mean distance of 20 Å in about $10^{-12}\,s$.

At intermolecular separations of less than about 15 Å direct interactions between molecular orbitals can occur, such that excitation energy is effectively shared between two molecules in a process known as delocalized exciton coupling and involving electron exchange. This process occurs at faster rates than resonance energy transfer, which is thus not significant at small intermolecular separations.

In an organism such as *R. sphaeroides*, the light-harvesting or antennae chlorophylls are associated with two light-harvesting complexes known as LH 1 and LH 2. The former is a protein complex that is closely associated with the reaction centre, whilst LH 2 is located further away from the reaction centres but sufficiently close to LH 1 to permit energy transfer to it (Fig. 6.3). Each of the light-harvesting complexes has two types of polypeptide chains, known as α and β (which differ somewhat in the two complexes). LH 1 contains 32 molecules of bacteriochlorophyll (870) and 16 carotenoid molecules, while the LH 2 of known structure (see below) has 27 molecules of bacteriochlorophyll and nine carotenoid molecules. The amino acid sequences of the polypeptides, which contain approximately 50 amino acids, indicate that each will form a single transmembrane α-helix in both LH 1 and LH 2.

X-ray diffraction analysis of a light-harvesting complex (LH 2 type) from *Rhodopseudomonas acidophila* at 2.5 Å resolution showed that the polypeptides are arranged as an $\alpha_9\beta_9$ complex organized as two concentric circles with the nine β-chains on the outside. Nine of the 27 bacteriochlorophylls (known as B800 because that is the wavelength of maximum absorbance) are intercalated between the β-chains and lie more or less parallel to the plane of the membrane at about 27 Å below the periplasmic surface of the

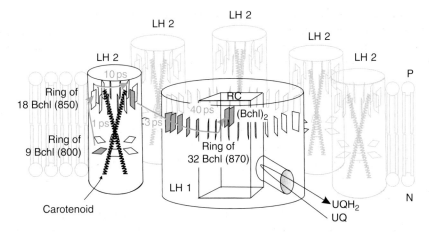

Figure 6.3   **The probable organization of the two light-harvesting complexes, LH 1 and LH 2 (antennae), and the reaction centre in a purple non-sulfur photosynthetic bacterium such as *R. sphaeroides.***
For LH 2 each cylinder represents an $\alpha_9\beta_9$ polypeptide unit, whereas for LH 1 the $\alpha_{16}\beta_{16}$ assembly may form an incomplete cylinder so as to allow access of ubiquinone–ubiquinol to/from the reaction centre (RC), which is envisaged as being largely surrounded by LH 1. The advantage of having many of the chlorophyll molecules of the antennae system at the same depth in the membrane as the special pair of chlorophylls in the reaction centre (towards the periplasmic side – see later) is that it ensures that a minimum distance over which the energy has to migrate (see text). The rings of bacteriochlorophylls B850 (LH 2) and B870 (LH 1) are coplanar, thus minimizing the distance over which the excitation energy has to transfer en route to the reaction centre. Several LH 2 complexes will surround one LH 1. (Note that the inside of the $\alpha_9\beta_9$ assembly is filled with phospholipids in the membrane and thus is not a pore.) For the purposes of illustration, the migration of excitation energy is shown as starting from a Bchl (800) molecule on LH 2, transferring to a ring of Bchl (850) within the same LH 2, where it may remain for up to 10 ps before migration to an LH 1, from where on the 40 ps timescale it transfers into the reaction centre special pair of chlorophylls (see Section 6.2). Other possibilities for excitation energy transfer include initial capture by the carotenoids of LH 2 and migration between several LH 2 molecules before reaching LH 1.

protein. The radius of the outer ring of β-subunits is 34 Å, leaving ample room for the chlorophyll rings, which are approximately 20 Å from each other, to be included. In contrast, the ring of α-subunits is only 18 Å in radius and thus there is insufficient room for bacteriochlorophyll molecules between the helices. Consequently, the other 18 bacteriochlorophylls (B850) lie parallel to the transmembrane helices and thus perpendicular to the plane of the membrane (Fig. 6.3).

The differences in absorbance maxima for the chemically identical bacteriochlorophyll molecules arise from the different environments provided by the protein scaffold. The B850 chlorophylls are only 1 nm below the periplasmic surface of the protein and approximately 9 Å from each other. Thus the majority of the bacteriochlorophylls in LH 2 are located towards the periplasmic side of the membrane, which, as we shall see later (Section 6.2.2), is the location of the chlorophylls that absorb light in the reaction centre to initiate photochemistry.

Curiously, another LH 2 from a related bacterium has an $\alpha_8\beta_8$ structure with rings of 8 and 16 chlorophylls, but is otherwise similarly designed to that from *R. acidophilia*. LH 2 complexes have carotenoid molecules, nine for the $\alpha_9\beta_9$ structure, which extend from the periplasmic to the cytoplasmic side of the protein, making close (van der Waals) contacts with the bacteriochlorophylls.

It is currently thought that the LH 1 complex is also either circular, or approximately circular, and that it surrounds the reaction centre. The subunit stoichiometry of $\alpha_{16}\beta_{16}$ would provide sufficient space inside a circular structure to accommodate the reaction centre (Fig. 6.3). However, a full circle of LH 1 would not leave any obvious way for ubiquinone–ubiquinol to exchange into and out of the reaction centre (Section 6.2.2). Thus it is likely that LH 1 might be an incomplete cylinder, leaving a path for diffusion of ubiquinone into and ubiquinol out of the reaction centre. A protein known as PufX may be integrated into LH 1 so as to provide a 'passageway' for quinone. The bacteriochlorophyll in LH 1 has its maximum absorbance at approximately 870 nm, the difference from LH 2 again arising from the environment within the protein.

From the moment of absorption of light by a component in LH 2, it takes approximately 100 ps for the excitation energy to reach the reaction centre. The resonance transfer mechanism accounts for the transfer from bacteriochlorophyll in LH 2 to LH 1 and from the latter to the reaction centre (Fig. 6.3). In contrast, transfer within a light-harvesting complex between closely adjacent bacteriochlorophyll molecules, or from carotenoid pigments to bacteriochlorophyll, is always by the delocalized exciton coupling mechanism. The excited state lifetime of carotenoids is too short to permit resonance energy transfer; this restriction means that at least part of a carotenoid molecule was predicted to be little further than the van der Waals distance from a bacteriochlorophyll; the structure has confirmed this. Very rapid transfer of energy between pigments and onwards to the reaction centre is essential if loss of energy by fluorescence or conversion to heat is to be avoided.

Structural information suggests that a typical series of light-harvesting events might be as follows. Incident light of wavelength shorter than 800 nm, provided it does not correspond to any wavelengths that chlorophylls absorb at, is absorbed by carotenoids of LH 2. Within 100 fs the energy (note that light harvesting is concerned with energy and *not*, a common misconception, electron transfer) will be transferred to either group of bacteriochlorophylls, i.e. those absorbing at either 800 nm or 850 nm. Energy absorbed by the 800 nm component will then be transferred downhill within 1 ps to the 850 nm chlorophylls. Energy absorbed by the B850 molecules is effectively mobile on the femtosecond timescale within their LH 2 ring (i.e. it can be thought of as hopping around a storage ring). Such hopping can continue for up to 1 ns before the energy is dissipated, for instance as fluorescence. However, usually the energy is transferred after approximately 3 ps from LH 2 to the chlorophylls of LH 1, which are the same depth in the membrane. The circular nature of LH 1 and LH 2 means that no particular defined orientation of the two complexes will be needed as each B850 in LH 2 is equivalent in a topological sense and thus energy transfer from any one of them is equally probable. It is estimated that the donor on LH 2 and a LH 1 acceptor bacteriochlorophyll come within 3 nm for this process, a distance that is compatible with effective energy transfer via resonance energy transfer.

Once absorbed by LH 1, the energy can again migrate around a ring of bacteriochlorophyll molecules, this time B870, before, on the 30–40 ps timescale, migrating to bacteriochlorophylls of the reaction centre (Fig. 6.3), in particular the special pair (Section 6.2.2) which is at the same depth in the membrane as the ring of bacteriochlorophylls in LH 1 (Fig. 6.3). As the reaction centre is not circular, this transfer will not occur with equal probability from all parts of the LH 1, but it is thought that many of the chlorophylls of the LH 1 will be within approximately 40 Å of the special pair, thus again facilitating the resonance energy transfer process.

The energy transfer process is very efficient; it is estimated that as much as 90% of the absorbed photons are delivered to the reaction centres of photosynthetic bacteria. The property of both bacteriochlorophylls and carotenoids of absorbing light at a range of wavelengths under 800 nm provides the capability to absorb this energy.

6.2.2 The bacterial photosynthetic reaction centre

Reviews   Van Brederode and Jones 2000, Heathcote *et al.* 2002

The first two membrane proteins for which high-resolution structures were obtained by X-ray diffraction analysis were both bacterial photosynthetic reaction centres. Although that from *Rhodopseudomonas viridis* was the first and seminal structural determination, we shall mainly discuss what has been learned about photosynthesis from study of the *R. sphaeroides* centre, since this has been studied much more extensively at the functional level.

Purified reaction centres from *R. sphaeroides* comprise three polypeptide chains, H, L and M, together with four molecules of bacteriochlorophyll (Bchl), two molecules of bacteriopheophytin (Bpheo), two molecules of UQ and one molecule of non-haem iron. It turns out that spectroscopic and biochemical studies on isolated reaction centres correlate in a very satisfying manner with the structure. First we shall review the key findings from the functional studies.

(a) P_{870} to Bpheo

Spectroscopic studies of reaction centres revealed that illumination caused a loss of absorbance (bleaching) at 870 nm consistent with the loss of an electron from a component absorbing at this wavelength. This was supported by the finding that ferricyanide ($E_{m,7}$ for $Fe(CN)_6^{3-}/Fe(CN)_6^{4-}$ +420 mV) caused a similar bleaching in the dark. The component absorbing at 870 nm was termed P_{870} and it was proposed that the absorption of a quantum led (within about a femtosec) to a transient excited state, P_{870}^*, in which an electron was raised to an higher energy level – increasing the ease with which the electron can be lost. This is the same as saying that the $E_{m,7}$ for P_{870}^+/P_{870}^* is very negative relative to P_{870}^+/P_{870}.

The electron would then be transferred to an acceptor to generate the bleached (oxidized) product, P_{870}^+. It should be noted that the P_{870}^+/P_{870} redox couple has a rather *positive* $E_{m,7}$ (about +500 mV) so that it can act as an electron acceptor from the cyclic electron transport system (Fig. 6.2).

Spectroscopic experiments further indicated that P_{870} was a unique Bchl dimer. The oxidized state P_{870}^+ had an ESR spectrum with a linewidth that was consistent with an unpaired electron delocalized over both Bchl rings. Note that, in contrast to a haem group, the electron is lost from the tetrapyrrole rings of the Bchl dimer; unlike Fe^{2+}, Mg^{2+} cannot give up an electron. The crystal structure of the reaction centre is consistent with this model, with two closely juxtaposed Bchl molecules (Plates F and G).

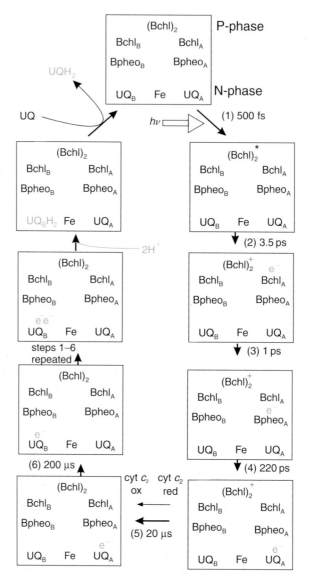

Figure 6.4   Two electron gating and time course of electron movement through the bacterial photosynthetic reaction centre from *R. sphaeroides*.
The electron is shown as migrating down only one branch (A – see text) of the reaction centre. Transfer of the electron from $(Bchl)_2$ to Bpheo via the monomeric Bchl has been controversial but is now widely accepted.

Rapid excitation of reaction centres with picosecond laser pulses, combined with rapid recording of visible absorption spectra, at first suggested that the immediate acceptor of the electron lost from P_{870}^+ was a Bpheo molecule, which is a chlorophyll derivative where the Mg^{2+} is replaced by two protons. The transfer of an electron from P_{870}^* to Bpheo can be detected in less than 10 ps (Fig. 6.4), and the resulting $(Bchl)_2^+ \ldots (Bpheo)^-$ biradical (often termed P^F) has a characteristic spectrum. Studies of isolated Bpheo in non-polar solvents suggest that its E_m in the reaction centre is about $-550\,mV$, or more than 1 V more negative than P_{870} in its unexcited state (Fig. 6.2). At low temperatures the spectra of the two Bpheo molecules could be resolved and this permitted demonstration that only one Bpheo normally accepted as electron

The two additional molecules of Bchl were originally thought to be inactive and were termed the voyeur chlorophylls. However, subsequent to the elucidation of the crystal structure (Plates F and G), which showed that they flanked P_{870}, additional rapid spectroscopic measurements in conjunction with femtosec flash excitation studies have indicated that one of the voyeur chlorophylls is an intermediate in the passage of electrons from the special pair to Bpheo. Figure 6.4 shows a current view of the timescale of electron transfer in these initial stages.

(b) Bpheo to UQ

The biradical P_F (i.e. P_{870}^+.Bpheo$^-$) is highly unstable, and within 200 ps, the electron is transferred from Bpheo$^-$ to UQ. There are two UQ binding sites, designated A and B. The addition of a single electron to the UQ at site A results in the formation of the free-radical semiquinone anion, $UQ^{\bullet-}$, which we have previously met in the context of the bc_1 complex (Section 5.8). The effective $E_{m,7}$ of the $UQ^{\bullet-}/UQ$ couple is about $-180\,mV$. The electron is further transferred to the second bound quinone at site B. The timescale for these electron transfers is very rapid (Fig. 6.4). We now have a UQ at A and a $UQ^{\bullet-}$ at B. The latter must be stabilized by its binding site within the protein because there is usually a strong tendency for the radical ion to disproportionate into $UQ + UQH_2$.

(c) Transfer of the second electron and release of UQH_2

Reduced-cytochrome c_2 is the electron donor responsible for reducing P_{870}^+ *in situ*. Since the $E_{m,7}$ of the P_{870}^+/P_{870} couple ($+500\,mV$) is more positive than that of the cyt c_2 couple ($+340\,mV$), the reduction is thermodynamically favoured (Fig. 6.2). A second photon now causes a second electron to pass from P_{870} to the quinone binding site B, once more via a transient $UQ^{\bullet-}$ at A. The UQ at site A thus switches between the oxidized and anionic semiquinone forms and never becomes fully reduced (Fig. 6.4). In contrast, that at B now becomes fully reduced as UQ^{2-}, following which two protons are taken up to give UQH_2 (Fig. 6.4). The UQH_2 is then released to the bulk UQH_2/UQ pool. The two bound UQ molecules thus act as a *two-electron gate*, transducing the one-electron photochemical event into a 2e$^-$ transfer. The protonations of the bound UQ_B play an essential role in the generation of Δp and the cyclic electron transfer is completed by a pathway from the bulk UQH_2 back to cyt c_2; this will be discussed in Section 6.3.

(d) Structural correlations

When the pathway of electron transfer deduced by spectroscopy is correlated with the structure of the reaction centre (Plates F and G), we see that transfer of electrons from the $(Bchl)_2$ dimer via Bpheo to UQ_A and then UQ_B fits with the spatial distribution of the groups. However, two obvious puzzles present themselves. Firstly, there appear to be two branches down which the electron might flow, since the redox carriers are arranged in the reaction centre with close to twofold symmetry. Yet all the evidence points to only the right-hand branch (as depicted in Plate F and sometimes called the A branch because spatially it lies above the Q_A site) being significantly active. The reason why the electrons flow more slowly down the left-hand branch is not known for certain but, of the factors affecting rates of electron transfer (Section 5.4), distance can be excluded. A plausible explanation, which is supported by some studies with reaction centres containing site-directed mutations, is that the energies of the $P^+(Bchl)^-$, and to a lesser extent of P^+Bpheo^-, states are significantly higher for the left hand or B branch compared with the A branch. The initial transfer of an electron from the P to Bchl may be energetically uphill for the B branch, whereas it is downhill for the A branch. Thus the precise environment of the Bchl molecules, and to a lesser extent of the Bpheo molecules, appears to be the determining factor. Second, no role has so far been found for the non-haem iron, which the structure shows lies between the two quinone sites; the iron atom can be removed without unduly affecting the performance of the reaction centre.

The two quinone binding sites have distinct properties. It is clearly important that both sites can stabilize the anionic form of UQ, but at the same time the B site must be suitable for the protonation events. Thus the B site is more polar and there is a pathway that can be discerned for UQ from the bulk phase to enter the B site with the head group coming in first. The B site is not in direct contact with the aqueous phase suggesting that side groups of the protein may be responsible for transferring protons from the bulk phase. A glutamate residue (212) on the L-chain lies between this site and the aqueous phase. Site-directed mutagenesis of this amino acid drastically attenuates the rate of protonation without affecting the rate of electron transfer to UQ_B, thus implicating the carboxylate side chain in proton transfer. Other adjacent acidic residues e.g. aspartate-213, plus chains of bound water molecules seen in very high resolution structures are also important features that aid the transfer of protons from the aqueous phase to the Q_B site.

X-ray diffraction data of purified reaction centres does not give information on the orientation of the complex in the intact membrane. However, cyt c_2 is located in the periplasm, in common with many other bacterial c-type cytochromes, while the H-subunit was susceptible to proteolytic digestion and recognition by antibodies only in inside-out membrane vesicles (i.e. chromatophores, see Chapter 1). Thus the reaction centre is orientated with the special Bchl pair towards the outside (periplasmic) surface of the bacterial cytoplasmic membrane, and with the two UQ binding sites towards the cytoplasm.

The L- and M-chains each have five transmembrane α-helices, while the single α-helix of the H-subunit also spans the membrane (Plate F). Electron transfer takes place through the redox groups bound to the L-subunit. These α-helices contain predominantly hydrophobic amino acids and appear to provide a rigid scaffold for the redox groups. The importance of minimizing relative molecular motion of these groups is illustrated by the finding that the rates of some of the electron transfer steps from the Bchl special pair to the the Q_A site increase with decreasing temperature.

(e) Charge movements

With the orientation shown in Plate F, light will cause an inward movement of negative charge from cyt c_2 to UQ_B, where the translocated electrons meet with protons coming from the cytoplasm. Thus the net effect of both of these charge movements is to transfer negative charge into the cell (i.e. from P-phase to N-phase), contributing to the generation of a Δp (positive and relatively acidic outside).

The inward movement of the electron through the reaction centre had already been detected using the carotenoid band shift (see Chapter 4) as an indicator of membrane potential. Using chromatophores, three phases of development of a membrane potential followed a short saturating flash of light. The first corresponded kinetically to the transfer of the electron from P_{870} to UQ_B; the second to the reduction of P_{870}^+ by cytochrome c_2, whilst the third, and much the slowest, phase corresponded to the return of the electrons from UQH_2 to cyt c_2, and could be blocked by antimycin or myxothiazol. It should be noted that the transfer of electrons between the UQ_A and UQ_B sites is parallel to the membrane and does not contribute to the establishment of a membrane potential.

The contribution of each electron transfer within the reaction centre to the development of membrane potential is related to the distance moved by the electron perpendicular to the membrane and to the dielectric constant of the surrounding membrane. Thus the movement from cyt c_2 to the special pair would make less contribution than the transfer from Bpheo to Q_A through a more hydrophobic (i.e. low dielectric constant) environment. Finally, the uptake of the protons into the Q_B site contributes about 10% to the overall charge separation across the membrane.

Despite the satisfying correlation between structure and function, it should be appreciated that not everything is known about the functioning of the reaction centre. The necessity for the special pair as a central component of a reaction centre is not so clear now that it is known that both types of plant photosystem (Section 6.4) have different organizations of the pairs of chlorophylls that are equivalent to the special pair in bacteria. The origin of the close to twofold symmetry and the reasons why only one branch is photochemically active may relate to the evolution from a primitive reaction centre in which both branches are active, as indeed is thought to be the case for the photosystem I of green plants (Section 6.4.3).

An essential feature of a reaction centre is that it is not reversible, i.e. the electron must not be allowed to return from $UQ^{\bullet -}$ at site A to reduce P_{870}^+, even though the semiquinone is thermodynamically capable to do this by virtue of the rather negative $E_{m,7}$ of its redox couple with UQ (Fig. 5.6). Reversal occurs 10^4 times more slowly than the forward reaction. This is the reason for the almost perfect quantum yield, i.e. one photon results in the creation of one low-potential electron. The cost of this irreversibility is the loss of redox potential as the electron passes from P_{870}^* to the quinones and the dissipation of 30% of the absorbed energy (Section 6.1).

The factors that prevent such reversal from $UQ_A^{\bullet -}$, or from any of the other components in the reaction centre, to the cationic state of the special pair are probably several. First, in accordance with the arguments in Section 5.4, the distance from the Q_A site to the special pair is sufficiently large (~ 30 Å) that the back reaction would, at room temperature, be expected to proceed at no more than 10 per second, significantly slower than the estimated rate of reduction of P^+ by cytochrome c_2 (Fig. 6.4). Second, the reaction Q_A to Bph is energetically uphill to the extent that it will not be significant, either kinetically or thermodynamically, even though the separation (10 Å) of the two centres is consistent with rapid electron transfer.

A third factor is needed to explain why the electron does not return from Bph^- to P^+. Here the distance is short, and there is a very large thermodynamic driving force. The reaction is precluded because the theory of electron transfer shows (Section 5.4) that, counter to simple intuition, the rate of electron transfer declines dramatically when the driving force exceeds a certain value. This is the so-called *inverted region* that is found when a plot of rate against driving force is made on a theoretical basis with the distance between two electron-transfer centres kept constant.

6.2.3 The *R. viridis* reaction centre

The *R. viridis* reaction centre differs in one major respect from that in *R. sphaeroides* by having an additional polypeptide subunit, which contains four *c*-type haems (Plate G). The haem nearest the special pair of bacteriochlorophylls (designated P_{960} in this organism because the absorption maximum and exact structure of the BChl are different than in *R. sphaeroides*) is the immediate donor to the reaction centre. The electron is thus transferred over a distance of approximately 20 Å (see Section 5.4) from this haem ($E_{m,7} = +370\,\text{mV}$). The other haems have redox potentials of $+10\,\text{mV}$, $+300\,\text{mV}$ and $-60\,\text{mV}$, listed in order of increasing distance from the special pair. It is not completely certain which of the haems accepts electrons from the cyt c_2, but as explained in Section 5.4, it is likely that the electron passes via the lower potential haems on a route that starts with cyt c_2 donating an electron to the haem furthest from the reaction centre. There is no satisfactory explanation as to why this tetrahaem *c*-type cytochrome is dispensable in *R. sphaeroides* and certain other organisms (e.g. *Rhodobacter capsulatus*). The presence of this cytochrome allowed, if it and the menaquinone at the Q_A site (see below) were pre-reduced, a form of the reaction centre containing $(\text{Bchl})_2$ and Bpheo^- to accumulate, because $(\text{Bchl})_2^+$ was eventually reduced by the cytochrome and electrons could not pass from Bpheo^- to the quinone. ESR studies gave important evidence that the electron resided on Bpheo.

The other difference in *R. viridis* is that the Q_A site is occupied by menaquinone rather than UQ. The Q_B site was found to be empty in the crystals of the reaction centre, consistent with the ability of UQ at this site to dissociate and equilibrate with the bulk pool.

6.3 THE GENERATION BY ILLUMINATION OR RESPIRATION OF Δ*p* IN PHOTOSYNTHETIC BACTERIA

We have already seen from the structure of the reaction centre that absorption of light causes the movement of negative charge into the cell and that the optical spectral changes in carotenoids (Chapter 4) can be used to follow this and other charge separations (Section 6.2.2). The slowest of the three phases of development of the carotenoid shift that are observed following exposure of chromatophores to very short saturating flashes of light is blocked by inhibitors of the cyt bc_1 complex (Section 6.2.2) and therefore must correspond to the movement of charge across the membrane by this complex. Such carotenoid measurements, together with measurements of light-dependent proton uptake by chromatophores, provided important evidence in favour of the Q-cycle mechanism described for the cytochrome bc_1 complex in Chapter 5.

Since electrons are retained within the cyclic pathway while protons are taken up and released, only the latter need be considered when calculating the overall charge movements per cycle. Since the reaction centre takes one proton per electron from the cytoplasm while the bc_1 complex takes up one proton from the cytoplasm but releases two protons to the periplasm, two protons are translocated for each electron handled. An implication of this stoichiometry is that, if the H^+/ATP ratio is 3 (Chapters 4 and 7), then the $\text{ATP}/2e^-$ ratio will be 4/3.

Cyclic electron transfer in *R. sphaeroides* and related organisms generates Δ*p* but does not produce reducing equivalents for biosynthesis. Provision of such reducing equivalents, i.e. NAD(P)H, often requires Δ*p* to drive reversed electron transport as well as ATP synthesis.

Thus if, for example, the organism is growing on H_2 and CO_2, then electrons from H_2 will be fed via hydrogenase into the cyclic electron transport system at the level of ubiquinone and driven by reversed electron transfer through a rotenone-sensitive NADH dehydrogenase (probably analogous to complex I, Section 5.6) to give NADH. Subsequent formation of NADPH, presumably via the transhydrogenase reaction (Section 5.13), then provides the reductant for CO_2 fixation. On the other hand, if the organism is growing on malate, which has approximately the same oxidation state as the average cell material, electrons will not be fed into the cyclic system and thus reversed electron transport will not be significant. Thus the extent and the location at the electrons from a growth substrate feed into the cyclic electron transport system are dependent on the substrate.

Sulfide and succinate are further examples of substrates in which electrons are fed in at ubiquinone and there are also electron donors in some organisms that donate to the c-type cytochromes. It is crucial that the cyclic electron transport system does not become over-reduced; if every component were to be reduced, then cyclic electron transport could not occur; the mechanism whereby over-reduction is avoided is not fully understood.

Some photosynthetic bacteria have nitrogenase, which catalyses the reductive fixing of nitrogen to ammonia. The nitrogenase requires a reductant with a more negative potential than NAD(P)H. For a photosynthetic bacterium growing on malate, it is not clear how this reductant is generated but there is increasing evidence that the cytoplasmic membrane contains a novel enzyme (Rnf complex) that utilizes the proton motive force to drive electrons uphill from ubiquinol or NADH to generate the required reductant, which is probably an iron–sulfur protein.

R. sphaeroides, in common with many other photosynthetic bacteria, can grow aerobically in the dark. Oxygen represses the synthesis of bacterochlorophyll and carotenoids, and so the reaction centre is absent. However, the b and c cytochromes are retained, and three terminal oxidases can be induced, which in the case of *R. sphaeroides* include a cytochrome aa_3 oxidase (Fig. 6.1), which is very similar to the mitochondrial complex IV (Section 5.9). By using some constitutitve cytochromes, the bacterium can, therefore, switch very economically between anaerobic photosynthetic growth and aerobic growth in the dark by assembling respiratory and photosynthetic chains with common components (Fig. 6.1).

Cyt c_2 has long been regarded as an essential component in the cyclic electron transport system. However, recent gene deletion experiments with both *R. sphaeroides* and *R. capsulatus* have shown that cyt c_2 is dispensible for photosynthesis because an alternative c-type cytochrome can substitute. The implications of this finding are not fully understood, but it is still thought that, in wild-type cells, cyt c_2 plays a major role in cyclic electron transport. Organisms such as *R. sphaeroides* can also use certain anaerobic electron acceptors, including at least some of the oxides of nitrogen. Their electron transport chains are thus even more versatile than Fig. 6.1 suggests because they can also incorporate some of the components shown for *P. denitrificans* in Section 5.15.1.

6.3.1 Photosynthesis in green-sulfur bacteria and heliobacteria

Reviews   White 2001, Heathcote *et al.* 2002

It will shortly become apparent that the reaction centre in *R. sphaeroides* is closely related to photosystem II in green plants. The other photosystem in green plants has a close relative in green-sulfur bacteria and heliobacteria. No structure has been obtained for this type of bacterial reaction centre, but sequencing and definition of cofactor content of purified reaction centres has clearly established the similarity. There is one important difference in that only in the bacterial protein are the two major subunits of the reaction centre identical; we will return to the significance of this later (Section 6.4.3).

Possession of the photosystem I type of reaction centre has important bioenergetic consequences. As Fig. 6.5 shows, the reaction centre does not reduce ubiquinone, but rather an iron–sulfur protein known generically as ferredoxin. This molecule has a more negative redox potential than ubiquinol and thus the output of the reaction centre can be used to reduce NAD^+ without the need for the protonmotive force-driven reversed electron flow that occurs in *R. sphaeroides*.

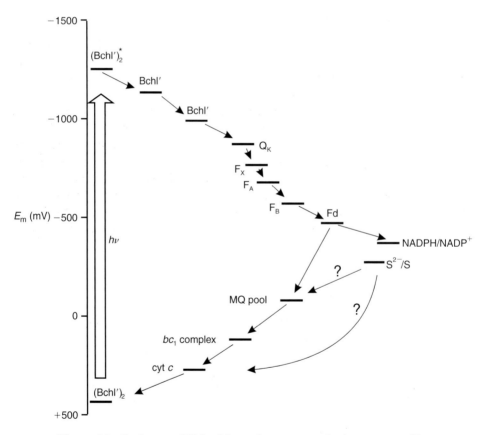

Figure 6.5   Pathways of light-driven electron transfer in a green-sulfur bacterium in relation to the redox potentials of the components.
Bchl' designates that the bacteriochlorophyll in these organisms may be modified relative to that in purple bacteria (Fig. 6.2). By analogy with photosystem 1 (Fig. 6.11) Q_K is phylloquinone, and F_X, F_A and F_B are Fe/S centres; together with the Bchl' molecules these make up the cofactors of the reaction centre. MQ is menaquinone and Fd is ferredoxin. '?' indicates that the entry point into the electron transport system of electrons derived from sulfide oxidation is uncertain.

An alternative fate for the reduced ferredoxin is as a reductant for menaquinone (presumably catalysed by ferredoxin–menaquinone oxidoreductase), which thus permits cyclic electron transport and thus generation of Δp (Fig. 6.5). If electrons from a growth substrate such as sulfide are fed in at the level of cytochrome c, it is apparent that the photochemical reaction is sufficient to transfer them to NAD^+; no protonmotive force-driven step is needed. Such linear flux of electrons from sulfide to NAD^+ must, of course, be accompanied by some concomitant cyclic electron transfer to generate Δp. Finally, the light-harvesting apparatus in green-sulfur bacteria is different from that in *R. sphaeroides*; however, this topic is outside the scope of this book.

6.4 THE ELECTRON-TRANSFER AND LIGHT-CAPTURE PATHWAY IN GREEN PLANTS AND ALGAE

Review   Soriano *et al.* 1999

Photosynthetic electron transfer in chloroplasts has two features not found in purple bacteria: (a) it can be non-cyclic, resulting in a stoichiometric oxidation of H_2O and reduction of $NADP^+$; and (b) two independent light reactions act in series to encompass the redox span from $H_2O/\frac{1}{2}O_2$ to $NADP^+/NADPH$ (Fig. 6.6).

The presence of two reaction centres was indicated by a classical observation known as the *red drop*. Algae illuminated with light in the range from 400 to 680 nm very effectively evolved oxygen. However, if light with a wavelength greater than 680 nm was used, then the efficiency fell very sharply and light >690 nm was essentially ineffective. This in itself merely showed that there was a component which required light <690 nm, either through direct illumination or by energy transfer from shorter wavelengths. What was more striking was that the oxygen evolution produced by relatively weak, non-saturating light at 650 nm could be increased by simultaneous illumination at 700 nm (the enhancement effect).

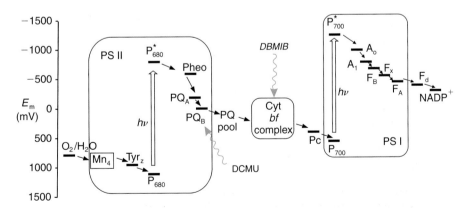

Figure 6.6   Two light-driven reactions, catalysed by PSI and PSII, operate in series in thylakoid membranes to drive electrons from water to $NADP^+$. The Z-scheme representation is shown with excited state redox potentials where appropriate. The sites of action of DCMU and DBMIB are shown schematically. PQ is paltoquinone and Pc plastocyanin. Details of the components within the two photosystems (PSI and PSII) are given in Figs 6.8 and 6.10.

The interpretation was that two photosystems were involved, one of which had an absorbance centred at 700 nm, and could be supplied either by direct illumination with 700 nm light or by energy transfer from pigments in the antennae that absorbed the 650 nm light. The second photosystem, with an absorbance maximum at 680 nm, could be excited by wavelengths up to that value, but not by the lower energy 700 nm light. In a development of this experiment, it was shown that, if the two light beams were applied separately, there was a transient release of oxygen when the shorter wavelength was applied after the 700 nm light.

The two photosystems are now known to be arranged in series (Fig. 6.6). The system that requires 680 nm light for excitation is known as photosystem II (PSII) and acts to abstract electrons from water and raise them to a sufficiently negative potential to reduce plastoquinone (PQ) to plastoquinol (PQH_2); the latter are respectively very similar to UQ and UQH_2 in structure (Fig. 5.6) and function. PQH_2 acts as the reductant for the cytochrome bf complex (sometimes called b_6f but the '6' is redundant), which is in turn functionally very similar to the cytochrome bc_1 complexes of respiratory chains. However, whereas the globular domain of cytochrome c_1 is similar to the structure of cytochrome c, cytochrome f is a mainly β-sheet protein with the N atom of the amino terminus of the polypeptide acting, uniquely, as one of the axial ligands to the haem iron. Nevertheless, cytochrome f is a c-type cytochrome because the haem is attached to the protein through the characteristic C-X-X-C-H motif. There is currently no clue as to the evolutionary origin of cytochrome f (f refers to *frons*, the Latin for 'foliage' or 'leaf', thus disguising that it is a c-type cytochrome). The cytochrome b polypeptide of the cytochrome bf complex has four transmembrane helices, which provide four side chains of conserved histidine residues for ligation of the two haems that are sandwiched between the helices. As with the cytochrome bc_1 complex, one of these haems is nearer to the P-side of the membrane and the other to the N-side. The cytochrome bf complex also contains a Rieske-type high-potential Fe/S protein.

Whereas the mitochondrial bc_1 complex acts as an electron donor to the peripheral cytochrome c, the bf complex passes electrons to plastocyanin, which is a peripheral protein located at the lumenal side of the thylakoid membrane. The redox centre in plastocyanin is a Cu ion, which undergoes $1e^-$ oxidation–reduction reactions between its +1 and +2 oxidation states. The environment of the Cu within the protein is such that its $E_{m,7}$ is +370 mV, very different from what it would be in aqueous solution. This, together with the characteristic ESR and blue absorbance spectra, is diagnostic of a type I Cu centre. These are also found in azurin and pseudoazurin – common components of bacterial electron transfer chains. X-ray diffraction studies show the Cu to have a highly distorted tetrahedral co-ordination geometry with ligands from the sulfurs of cysteine and methioinine, as well as two histidine side chains. An important point to note is that the $E_{m,7}$ span from PQH_2 to plastocyanin is very similar to UQH_2 to cytochrome c in respiratory systems.

Plastocyanin acts as the electron donor to photosystem I (PSI), which is excited by 700 nm light. PSI raises the energy of the electrons obtained from plastocyanin sufficiently to reduce ferredoxin ($E_m = -530$ mV) in a one electron reaction. Finally, reduced ferredoxin (Fd) reduces $NADP^+$ to NADPH via an enzyme known as ferredoxin-$NADP^+$ oxidoreductase.

Figure 6.6 shows that the redox potential of the electron, as it is driven from water to NADPH, follows the shape of a distorted letter N with the uprights significantly displaced. This scheme has become known as the Z-scheme because it was once presented in a format displaced by 90° from the present convention in which redox potential is shown on the vertical axis.

6.4.1 Plant and algal antenna systems

Reviews and further reading   Kuhlbrandt *et al.* 1994, Bibby *et al.* 2001, Boekema *et al.* 2001

As in bacterial photosynthesis, light-harvesting or antennae complexes are required in the thylakoid membrane even though the two photosystems, and especially photosystem I (see later), have much antennae chlorophyll associated with them. There are distinct polypeptide complexes, associated with the two photosystems I and II, with non-covalently attached chlorophylls (two types a and ab) and carotenoid, known as LHC I and LHC II. The latter is mainly made up of a polypeptide, which binds approximately half the total chlorophyll that is found in a green plant. The polypeptide has a molecular weight of approximately 25 kDa and has associated with it seven molecules of chlorophyll a, five molecules of chlorophyll b and two carotenoid molecules.

The structure of this complex has been obtained at 3.4 Å resolution by the electron diffraction/microscopy mode of analysis first developed for the analysis of two-dimensional crystals of bacteriorhodopsin. The polypeptide chain forms three transmembrane α-helices with two of these being titled significantly away from an angle of 90° to the membrane plane and being significantly longer ($\sim$35 Å) than typical transmembrane helices (Fig. 6.7). These two helices, together with the two molecules of carotenoid, cross over in the centre of the membrane. Between them the three helices provide a scaffold for the binding of the chlorophyll molecules, which are distributed towards both sides of the membrane. In the centre of the complex, chlorophyll a molecules are in close contact with chlorophyll b so as to permit rapid energy transfer, whilst the nearby carotenoids are available to quench any toxic singlet oxygen that might be formed.

Much of the energy transfer within the LHC II will be via the resonance energy transfer mechanism, although some of the chlorophylls within a monomeric unit may be sufficiently juxtaposed and appropriately oriented to involve delocalized exciton-coupling (Section 6.2.1). LHC II occurs as a trimeric unit and this may have implications for its action in capturing energy from light. LHC II is not structurally related to any other membrane protein, which prompts the question as to why it has not adopted the advantageous circular structure seen in bacterial light-harvesting proteins (Section 6.2.1). There is no obvious answer to this question and indeed a variety of light-harvesting protein organizations have been observed amongst photosynthetic organisms. However, it is notable that the LHC II interacts with photosystems, which themselves have chlorophylls other than those involved in photochemistry. Hence energy might be delivered into the plant photosynthetic reaction centres at a variety of depths in the membrane, in contrast to the bacterial reaction centre in which energy transfer at the P-side of the membrane is optimal. Furthermore, the bacterial light-harvesting centres are positioned to receive incident light from the P-side of the membrane and thus there is little obvious design advantage in placing chlorophylls towards the N-side. In contrast, the topology of the thylakoid is such that the incident light will strike the membrane on the N-side. Nevertheless, it has recently been shown that, when the cyanobacterium Synechocystis PCC 6803 is grown under conditions of iron limitation, then its photosystem I is surrounded by a ring of polypeptides, known as CP43′, which is related to the CP43 component that binds antennae chlorophylls in photosytem II (see Section 6.4.2). This ring seems to compensate for lower levels of phycobilisomes (the usual type of light-harvesting apparatus in this type of algae) and photosystem I in response to iron deficiency.

P-phase

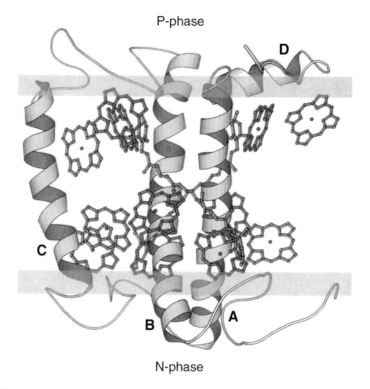

N-phase

Figure 6.7   The structure of a light-harvesting complex from thylakoids, showing the disposition of the chlorophylls.
Side view of the thylakoid LHC II monomer. The three transmembrane helices are labelled A, B and C. Helix C is perpendicular to the membrane plane and fully embedded in the bilayer, helices A and B are longer, are inclined at 30° and protude into the stroma (N-phase). D is a short amphiphilic helix at the lumenal membrane surface. Many of the 12 chlorophylls are oriented almost perpendicular to the membrane surface. Phytyl side chains have been omitted. Two molecules of carotenoid can be seen in the centre of the protein; these act as an internal cross-brace, linking loops of polypeptide on opposite surfaces of the membrane. A threonine residue towards the N-terminus at the stroma side (N-phase) is a site of reversible phosphorylation (see Section 6.4.5 and Fig. 6.13). Modified from Kuhlbrandt *et al.* (1994) with permission.

6.4.2 Photosystem II

Reviews   Rutherford and Faller 2001, Zouni *et al.* 2001, Heathcote *et al.* 2002

The green plant photosystems have been more difficult to purify and characterize than the single reaction centre of the purple bacteria. One difficulty is that less than 1% of the pigments are involved in the photoreaction, with the remainder acting as antennae. Recently, a crystal structure has been obtained for the photosystem II from the thermophilic cyanobacterium *Synechococcus elongatus*. This shows, as anticipated from sequence analyses together with earlier structural work with plant photosystem II using electron microscopy and cofactor analysis, that this photosystem has many similarities with the purple bacterial

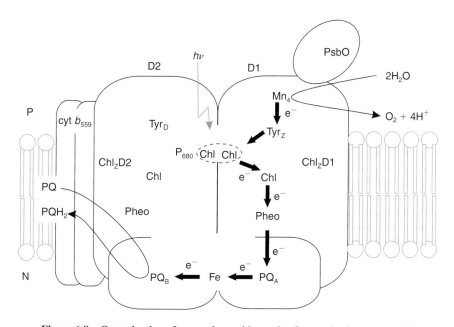

Figure 6.8 Organization of core polypeptides and cofactors in photosystem II.
The diagram is based on the 3.8 Å crystal structure of the PSII from the thermophilic cyanobacterium *Synechococcus elongatus*. By analogy with the bacterial
reaction centre it is thought that there is only one active branch from the P_{680}
centre (assumed to be largely contributed by one of the two clustered chlorophylls) to plastoquinone at Q_A. The roles of monomeric Chl and pheophytin
(Ph) are deduced from the structure and by analogy with the bacterial reaction
centre. Only the core D1 and D2 subunits, the PsbO subunit, which appears to
cap the water-splitting manganese cluster, and the two single transmembrane
helices that each provide one histidine ligand to the haem of cyt b_{559} are shown.
Note that the haem of cyt b_{559} is close to the stromal (N) side of the protein. Chl_z
(D1 and D2), each being around 20 Å from the photochemically active chlorophylls, may play a role in energy transfer from antenna chlorophylls, which are
located on two subunits, CP43 (12 chlorophylls) and CP47 (14 chlorophylls)
(not shown here), which each have six transmembrane helices and are located
external to D1 and D2. Without the intermediate role of Chl_z (D1 and D2) the
distances from the cholorphylls on CP43 and CP47 to the P_{680} would probably be
too long for efficient energy transfer. The *S. elongatus* PSII also has a bound
c-type cytochrome subunit located on the lumenal side; as the function of this
polypeptide is not known, and it does not occur in plant PSII, it is also omitted
from this figure. (Based on Zouni *et al.* 2001, *Nature* **409**, 739–743.)

photosynthetic reaction centre. Features in common include an approximate twofold
symmetry relationship with the molecule, with the two major polypeptides D1 and D2 each
having five transmembrane helices, which provide binding sites for a voyeur chlorophyll,
pheophytin and plastoquinone at A and B sites (Fig. 6.8). It is presumed that only one branch
is photochemically active. Amongst the most important differences are the presence of a
manganese cluster on the D1 polypeptide and a greater spatial separation between the two
chlorophylls, one of which forms the radical cation P^+ species, that are clustered together
on the twofold axis (Fig. 6.8).

The manganese cluster relates to the fundamental difference in the action of PSII relative to the bacterial counterpart; in the latter, the chlorophyll radical cation P^+ is reduced by cytochrome c_2, but in PSII the electrons originate from water which is thought to be oxidized at the manganese cluster. It was assumed that the P^+ species formed in PSII is associated with the two adjacent chlorophylls (Fig. 6.8), but it is now clear that these cannot be considered as a special pair in the way that they behave in the bacterial system. Thus in PSII the positive charge appears to be largely associated with one of these two chlorophylls; the significance of this difference from the bacterial system is not clear.

The water-splitting reaction is, together with its opposite number, the terminal oxidase of the respiratory chain, one of the most intriguing reactions in bioenergetics. In air the E_h for the $O_2/2H_2O$ couple is $+810\,mV$ (Section 5.9). Thus to abstract electrons from water requires a redox centre, which is even more electropositive than this value and capable of reacting spontaneously with water. The water-splitting centre contains four manganese atoms, which are seen in the crystal structure to be arranged in the form of a Y-shape. Unfortunately, the resolution of this structure is tantalizingly just insufficient to be able to identify the amino acid side chains and other ligands to the four manganese atoms. Four sequential events have been shown to be required for abstraction of $4e^-$ from two molecules of water to yield O_2 and release $4H^+$ into the lumen of the thylakoid. A possible mechanism, involving the four oxidation states (traditionally called S states) commonly described for the water splitting reaction is given in Fig. 6.9.

The electrons from the water-splitting reaction are not transferred to the P_{680} directly, but rather via a specific tyrosine residue side chain. In the crystal structure this side chain on the D1 polypeptide could be located at a position $7\,\text{Å}$ from its closest approach to the Mn cluster and around $12\,\text{Å}$ from the photochemically active chlorophyll. These distances are sufficiently short so as to ensure rapid electron transfer. Loss of an electron to P_{680} from this tyrosine generates a radical that is a neutral species since a proton is also lost, probably to a neighbouring histidine residue. The tyrosine residue, named Z^+ before its molecular identification, partly through study of the properties of a PSII in which the critical tyrosine had been changed to phenylalanine by site-directed mutagenesis, in turn regains an electron from the Mn cluster. There is a tyrosine residue in the equivalent position on polypeptide D2 but, although it can form a radical, its distance from the Mn cluster is thought to preclude any role for it in electron transfer events.

The organization of the cofactors shown by the crystal structure is clearly consistent with the pathway of electron flow from P to quinone being very similar to that in the bacterial reaction centre (Plates F and G). The Q_B site is believed to be the site of action of the inhibitor 3-(3,4-dichlorophenyl)-1,1-dimethylurea (DCMU), which blocks the activity of PSII. The similarity to the bacterial reaction centre is strengthened by a mutant of the latter, which is sensitive to DCMU and for which the EPR spectrum of bound $UQ^{\bullet-}$ has similarities with that of $PQ^{\bullet-}$ bound to PSII.

The *S. elongatus* PSII structure reveals one molecule of a *b*-type cytochrome known as cytochrome b_{559}, although PSII from other sources have been firmly thought, probably incorrectly, to contain two of these molecules, which are not found in the bacterial reaction centre. Cytochrome b_{559} has two transmembrane α-helices provided by two different, but short polypeptides (Fig. 6.8). The haem group is sandwiched between these, with each helix providing one histidine axial ligand to the Fe.

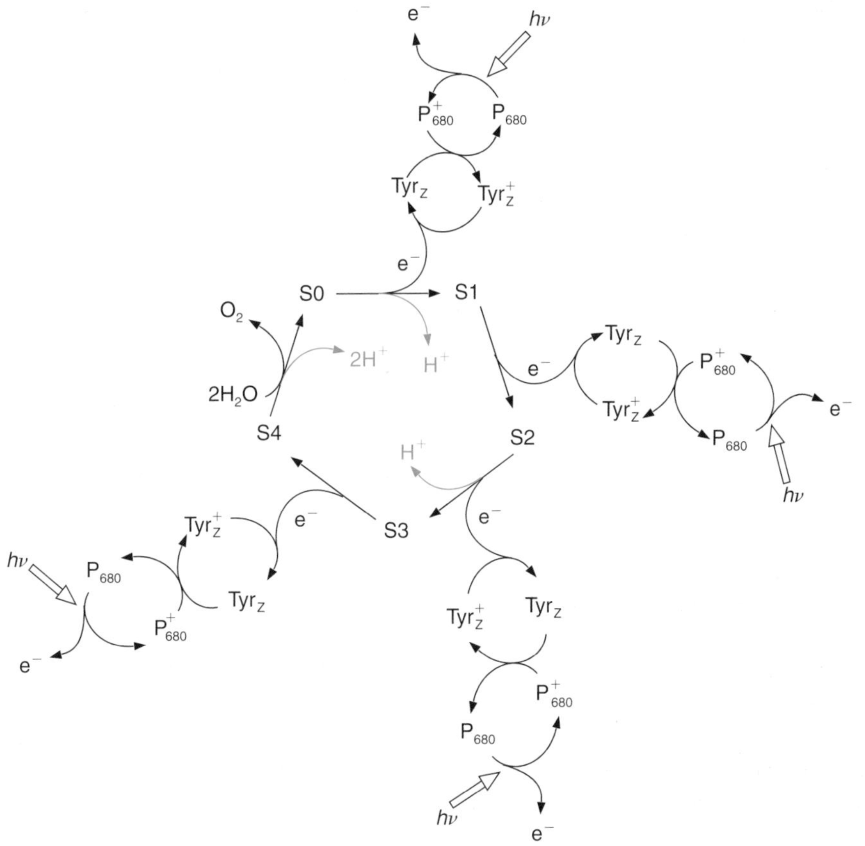

Figure 6.9   The water-splitting reaction of photosystem II.
Four quanta are required to abstract $4e^-$ from two H_2O. In the dark, the water-splitting centre is in state S1. The steps at which H_2O binds are not known, but the H^+ and O_2 release steps are thought to be as shown. The S4 state has four positive charges due to the transfer of $4e^-$ into the photosystem. The water-splitting centre contains Mn, but it is not clear how many electrons originate from the Mn rather than an amino acid side chain. Ca^{2+} is required for the transition from S3 to S4. Removal of Ca^{2+} traps the S3 state, in which ESR indicates that an electron has been lost from a histidine adjacent to a Mn atom.

The function of this cytochrome b_{559} is still not known. A common view has been that it provides a path for electron flow from PQH_2 at the Q_B site back to the P^+. Such a back reaction is what the bacterial reaction centre and PSII are designed to avoid, as it would dissipate the energy captured from light as heat. The apparently wasteful proposed pathway (which could not generate Δp) may be needed under conditions of high light and temperature when electron flow from the water-splitting centre to P^+ might be inadequate to reduce the latter sufficiently rapidly. P^+ is such a highly oxidizing species that it would cause damage to the components of the thylakoid membrane if it persists for a significant period. Indeed, D1 is damaged during normal conditions of illumination with the result that it is one of the fastest turning-over polypeptides known in biology. However, the structure of PSII shows that the haem of cytochrome b_{559} is too far from any other cofactor to have the proposed role in electron transfer back to P^+ unless a redox active β-carotene, which extends from b_{559} via Chl_Z to the region of the P centre, participates in the proposed protective electron transfer pathway.

6.4.3 Photosystem I

Reviews and further reading   Guergova *et al.* 2001, Jordan *et al.* 2001, Heathcote *et al.* 2002

The understanding of photosystem I has been considerably enhanced by the acquisition of a high (2.5 Å) resolution crystal structure for the protein isolated from the same thermophilic cyanobacterium (*S. elongatus*) used for the PSII structural studies. The key points to emerge from the structure are shown in Fig. 6.10. It is very complex, with each monomeric unit within an overall trimeric assembly comprising 12 polypeptides, bearing 90 molecules of chlorophyll and 22 carotenoids, in addition to the six chlorophylls, two phylloquinones and three Fe/S centres that make up the reaction centre itself. As in PSII and the purple bacterial reaction centre, there is again a twofold axis with each set of cofactors duplicated on each side of the molecule. In contrast to the other systems, it currently appears that both branches are active, although one of them probably operates at a faster rate than the other.

Note that an important difference between PSI and either the bacterial reaction centre or PSII is that single electron delivery is required at the N-side of the membrane; there is no requirement for one side to deliver two electrons sequentially to distinct quinone binding sites. In common with PSII, two closely neighbouring chlorophyll molecules provide the centre at which the P^+ species is assumed to form, but they are not organized in the same juxtaposed special pair arrangement as seen for the bacterial reaction centre. Furthermore, the unpaired electron of the P^+ species is associated with only one of the two chlorophyll molecules (Fig. 6.10) and thus functionally the two chlorophylls, which are not chemically identical, may not act as a unit in the manner of the special pair that operates in the bacterial reaction centre.

Currently, the advantage of a functional pair of chlorophylls in the bacterial reaction centre is not clear, given that the feature is agreed to be absent in PSII and its occurrence in PSI is debated. The electron lost from the chlorophyll on the PsaB subunit (Fig. 6.10), to give the P^+ species, passes to the A_0 chlorophylls on either side of the twofold axis (Fig. 6.10). How it gets there is not certain, but the chlorophylls that lie between the 'pair' of chlorophylls at the P_{700} centre and the two A_0 chlorophylls seem likely to be involved, just as the analogous molecules are in the bacterial reaction centre.

Kinetic studies have shown at least one of the A_0 chlorophylls to be reached by an electron after about 10 ps. After about a further 100 ps, the electron has reached a Q_K (also called A1) centre, which is a phylloquinone (also known as vitamin K_1) (Fig. 6.10). From there the electron migrates to the cluster of three Fe/S centres, which are located at the N-side of the complex. These in turn can reduce the water-soluble ferredoxin iron, which has an $E_{m,7}$ of -530 mV and so is extremely electronegative. The latter is the electron donor for several reactions including FNR-catalysed reduction of NADP, mainly for the Calvin cycle, and reduction of nitrite to ammonia for assimilation of nitrogen. The carotenoid band shift (Section 6.3), which was first detected in chloroplasts, indicates that the electron is transferred across the membrane in less than 20 ns.

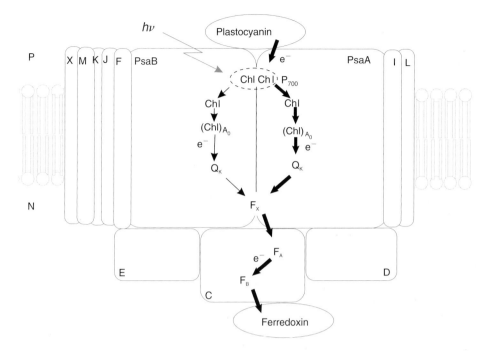

Figure 6.10   Structural organization of photosystem I.
The model is based on the crystal structure of the protein from the cyanobacterium *Synechococcus elongatus*. There are 12 polypeptides, named in the Psa
series. PsaA and PsaB are the largest, each forming 11 transmembrane α-helices;
they have significant sequence similarity to each other. Note that PsaA and PsaB
are much larger than their core counterparts D1 and D2 in PSII (Fig. 6.8). These
two polypeptides bind 79, and the other eight transmembrane α-helical subunits
shown (they each have from one or two helices) 11 (i.e. there are in total 90
chlorophyll molecules that play a role in light-harvesting). The chlorophyll
molecules involved in light-driven electron transfer through the photosystem can
be divided into three groups of two. One of these is P_{700}, loss of an electron from
which results in a cation radical being formed on the chlorophyll bound to the
PsaB subunit. The two chlorophylls in P_{700} are not chemically identical; that on
PsaA is an epimer of the normal chlorophyll a. It is currently thought that the
electron can migrate down both sides of the reaction centre but at unequal rates.
It is not known whether the electron migrates to the A_0 chlorophylls via the
chlorophylls that are adjacent to the P_{700}. Plastocyanin docks on to two short
helices, one provided by each of PsaA and PsaB, which run along the surface of
the membrane and thus connect two transmembrane α-helices. Note that the diagram is only roughly to scale. Approximate distances are: P to Chl, 12 Å; Chl to
$(Chl)_{A_0}$, 8 Å; A_0 to Qk, 9 Å; Qk to Fx, 14 Å; Fx to Fa, 15 Å; Fa to Fb, 12 Å.
A small number of additional subunits (e.g. PsaH) are present in plant PSI. Note
that PSI is a trimer but only a monomer is shown.

At the lumenal side of the PSI the P^+ is reduced by plastocyanin, whose copper centre is
though to dock within 14 Å, thus permitting direct electron transfer on an adequate
timescale. The redox potential of the P^+ species in PSI is estimated as $+450\,mV$ and is thus
appropriate to accept electrons from plastocyanin (E_h about $+250\,mV$).

An important difference between the core of PSI, and both PSII and the bacterial reaction centre is that PSI is larger. Thus each of the two main subunits in PSI have eleven helices rather than five and, in addition, there are several other transmembrane helices provided by the other subunits in PSI (Fig. 6.9). Although PSII does include more light-harvesting pigments relative to the bacterial reaction centre, it is clear that an important distinctive feature of PSI is that it comprises both a reaction centre and a set of light-harvesting antennae. However, the subunits CP43 and CP47 of PSII each with six helices, plus D1 and D2, are together similar to the PsaA and PsaB subunits of PSI. As noted earlier (Section 6.4.1) in a cyanobacterium, PSI can be surrounded by a ring of CP43′ subunits.

As discussed earlier (Section 6.3.1), green-sulfur bacteria have a reaction centre with considerable resemblance to that of PSI with the intriguing difference that PsaA and PsaB are replaced by two copies of a single core subunit. The bacterial analogue is a true homodimer, which can be expected to have a twofold axis. This strongly suggests that electron transfer may occur at equal rates down both sides of the bacterial reaction centre. This supports the use of both branches in PSI.

An intriguing issue concerning photosynthesis is exactly how the reaction centres evolved and when the two classes (I and II) of reaction centre diverged from one another.

6.4.4 Δ*p* generation by the Z-scheme

A carotenoid shift response indicates that both the photosystems translocate charge across the membrane (Figs 6.8 and 6.9). Evidence for the orientation of PSII comes from the observation that the protons liberated in the cleavage of H_2O are initially released into the lumen, indicating that oxidation of water occurs on the P-side of the membrane. Also, a radical anion form of plastoquinone bound to the reaction centre must be located close to N-side of the membrane, since it can be made accessible to impermeant electron acceptors such as ferricyanide after brief trypsin treatment. Ferredoxin and ferredoxin-$NADP^+$-reductase are accessible to added antibodies, whereas plastocyanin is not. These observations all suggest that PSI is oriented across the membrane as shown in Fig. 6.10.

The translocation of each electron from water to $NADP^+$ through the two photosystems is equivalent to the translocation of two positive charges into the lumen (Fig. 6.11). In addition to this, proton translocation is normally associated with the *bf* complex. If this were to function analogously to the closely related bc_1 complexes of mitochondria and purple bacteria, it would be predicted that four protons would appear in the lumen for each pair of electrons flowing from Q_B to plastocyanin. At high light intensities some experimental observations have suggested that only $2H^+$ per $2e^-$ are released, in which case the complex would not contribute any movement of charge across the membrane (Fig. 6.11). If this is true, then the *bf* complex must function differently from the bc_1 complex. However, most investigators now believe that the cytochrome *bf* complex follows the Q-cycle mechanism under all usual conditions. A definite difference from the bc_1 complex is insensitivity to both antimycin and myxothiazol. 2,5-Dibromo-3-methyl-6-isopropylbenzoquinone (DBMIB) is an inhibitor of the cytochrome *bf* complex, which acts at the Q_p site (i.e. on the lumenal side of the membrane) and is thus equivalent to the locus of action of myxothiazol on the cyt bc_1 complex (Section 5.8.1).

The *bf* complex has the two *b*-type haems characteristic of both the mitochondrial and purple bacterial bc_1 complexes. They differ in $E_{m,7}$, similarly to their counterparts in the

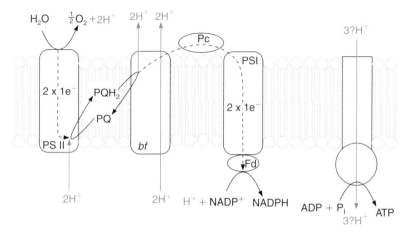

Figure 6.11   Stoichiometry of charge translocation and generation of Δp associated with electron transfer from water to $NADP^+$.
The movement of the electrons through the two photosystems moves negative charge from the P- to the N-phase. The other passage of electrons across the membrane, from plastoquinol to plastocyanin will release four protons to the P-phase, and cause uptake of two from the N-phase, thus moving two positive charges per two electrons into the P-phase (this assumes that the Q-cycle mechanism applies to the cytochrome bf complex, see text). Thus, six positive charges reach the P-phase for each $2e^-$ passing from H_2O to $NADP^+$, 6 H^+ are released into the P-phase but only 5 H^+ are taken from the N-phase. This is because reduction of $NADP^+$ requires $2e^-$ but 1 H^+. $P/2e^-$ ratios depend on the H^+/ATP ratio, shown here as 3 (but see text and Chapter 7 for uncertainty over this stoichiometry).

cyt bc_1 complexes, where the haem on the N-side of the membrane has an $E_{m,7}$ some 150 mV more positive than the other.

Figure 6.11 shows the overall proton movements occurring in non-cyclic electron transport in the thylakoid. The overall stoichiometry is $6H^+/2e^-$ (unless one believes that the Q-cycle does not operate for the bf complex when the ratio is 4) delivered to the lumen. If n protons must be translocated through the thylakoid ATP synthase to synthesize 1 ATP, then $6/n$ ATP can be synthesized for each NADPH synthesized.

The value of n is a matter of uncertainty (Chapter 7) but its value should be considered in the context that the Calvin cycle uses 1.5 ATP per NADPH. Thus if $n = 4$, there is an exact matching of the synthesis of 1 NADPH and 1.5 ATP for each two electrons transferred from water to ferredoxin. However, as will be discussed in Chapter 7, there are reasons to suppose that $n = 4.6$ in which case the yield of ATP would not be sufficient to match the requirements of the Calvin cycle.

6.4.5 Cyclic electron transport and balancing the two photosystems

Reviews   Bendall and Monasse 1995, Allen and Forsberg 2001, Haldrup *et al.* 2001

The main fate of the NADPH produced by non-cyclic electron flow is for the Calvin cycle, which fixes CO_2 in an overall process that requires 3 ATP for each 2 NADPH. Thus, as discussed above, there may be a shortfall of ATP. The situation is different in C4 plants where between 5 and 6 ATP molecules are consumed per CO_2 fixed. One mechanism in which

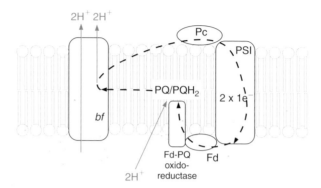

Figure 6.12   Physiological cyclic electron transport in thylakoids – a plausible scheme.

oxygen acts as an electron acceptor to make good this shortfall is described in Section 6.4.6, but an alternative strategy is cyclic electron transport, which occurs when electrons are able to return from ferredoxin to PQ (Fig. 6.12). Cyclic electron transport can occur (but like all cyclic processes is difficult to observe in the steady state) when thylakoids are illuminated with 700 nm light such that only PSI is active. Under these conditions, ATP can still be synthesized. Furthermore, there are cells in which only PSI appears to be active and cyclic electron transport would seem obligatory. These include heterocysts of cyanobacteria, which fix nitrogen and thus require an anaerobic environment, and the bundle-sheath cells of some C_4 plants. However, measurement of quantum efficiencies of PSI and PSII in leaves at ambient CO_2 suggests that cyclic phosphorylation is in general a minor contributor, but may be important during an induction phase when leaves are illuminated.

Although the idea of cyclic electron flow has been accepted for some time, the pathway is surprisingly not yet understood. A plausible scheme is shown in Fig. 6.12, involving a ferredoxin/PQ oxidoreductase. Such an enzyme and its constituent polypeptides have not been firmly identified (at the time of writing some researchers prefer a scheme in which electrons are passed from ferredoxin to the cyt *bf* complex). It is possible that there is involvement of polypeptides with sequence similarities to components of mitochondrial NADH dehydrogenase (complex I) that are predicted from gene sequencing to occur in thylakoids. Recall that mitochondrial complex I is also a quinone reductase (there is no convincing evidence, or rational role, for an NADH dehydrogenase in thylakoids of green plants). Whatever the uncertainty surrounding the molecular components involved in transfer of electrons in cyclic electron transport from ferredoxin to PQ, it is known that antimycin inhibits the process. The locus of this inhibition is not the cyt *bf* complex (see Section 6.4.4). It should also be recalled that cyclic electron flow in the thylakoid has a close relative in the cyclic electron transfer system of green-sulfur bacteria (Section 6.4); in these bacteria the general view is that ferredoxin transfers electrons to ubiquinone and not to the cytochrome bc_1 complex. This argument supports the scheme shown in Fig. 6.12.

Figures 6.11 and 6.12 together show that, assuming a Q-cycle mechanism for the *bf* complex, each turn of the cycle would result in four protons translocated per $2e^-$. The latter stoichiometry is the same as for cyclic electron transport in green (or purple) bacteria.

What regulates the relative activities of the two photosystems and thus of cyclic and non-cyclic electron transport? Some clue may come from the arrangement of the photosystems

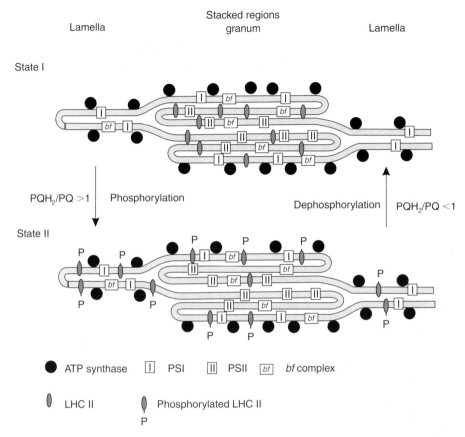

Figure 6.13   Phosphorylation of light-harvesting complexes may affect their distribution between stacked (rich in PSII) and unstacked (rich in PSI) regions of the thylakoid membrane.

in the thylakoid membrane. PSII can be found in the stacked regions of the thylakoids, whereas PSI, which has to deal with the large ferredoxin substrate, is restricted to the unstacked regions (Figs 1.6 and 6.13). Since the light-harvesting complexes, carotenoids and chlorophylls, transfer energy to the photosystems by resonance energy transfer (Section 6.2.1), the effectiveness of which decreases as the sixth power of the distance, the separation between light-harvesting complexes and photosystems will be critical. Light-harvesting complexes can be phosphorylated, and this is thought to cause them to be excluded from the stacked regions, thus decreasing the energy transferred to PSII. The extent of phosphorylation increases as the ratio PQH_2/PQ increases (implying that PSII is becoming more active than PSI), directing the light-harvesting complexes towards PSI in the non-stacked regions and restoring the balance.

Whereas at one time an electrostatic repulsion of LHC II from the stacked regions was thought to be the main consequence of phosphorylation, there is now evidence that the phosphorylated form of LHC II has a specific affinity for the H-subunit of PSI (although this is not present in *S. elongatus*; Fig. 6.10) and docks to it. If the phosphorylated form of LHC II has a lower affinity for PSII than the dephosphorylated form, the overall effect will be to displace an equilibrium in favour of the mobile LHC II binding to PSI.

The condition to which chloroplasts revert in the dark, in which LHC II is predominantly associated with photosystem II, is often known as state I, whilst state II refers to the situation in which at least some of the LHC II is considered to have migrated to the stroma lamellae region enriched in photosystem I. The terminology of states I and II is not to be confused with the description of mitochondrial respiratory states (Chapter 4). In the case of the thylakoid, state I was originally defined as the condition in which PSI was overexcited (in which case the thylakoid showed a relatively high fluorescence) and state II the condition where PSII received excess excitation and fluorescence was relatively low. Changes of state are thus induced by absorption of excess excitational energy by one of the two photosystems; the changes occur reversibly over several minutes. State I correlates with direct preferential excitation of PSI, whilst state II is where incident would preferentially excite PSII. Thus transition to state II involves redirection of absorbed energy to PSI. As cyclic electron transport generates ATP but not NADPH, it is often proposed that an ATP requirement of a photosynthetic cell may be involved in controlling the transition to state II.

6.4.6 Other electron donors and acceptors

The non-physiological electron acceptor ferricyanide allows the Hill reaction, a light-dependent oxygen evolution in the absence of $NADP^+$, to be observed. The $E_{m,7}$ of the $Fe(CN)_6^{4-}/Fe(CN)_6^{3-}$ couple, $+420\,mV$, is sufficiently positive to accept electrons from reduced plastocyanin. However, since plastocyanin is on the lumenal side of the ferricyanide-impermeable membrane, the couple can only accept electrons from a donor on the stromal side of the PSI complex. Thus oxygen evolution with ferricyanide as acceptor requires the operation of both photosystems. However, if in addition to ferricyanide, a benzo-quinone is also present, electrons can be transferred across the membrane from bulk phase plastoquinol (or possibly the Q_B site) to the external ferricyanide without the involvement of PSI (Fig. 6.14).

Operation of PSI in isolation can be achieved by donating electrons from ascorbate via DCPIP to plastocyanin. Illumination will then drive electrons through PSI to $NADP^+$; alternatively $NADP^+$ can be replaced by the oxidized form of the non-physiological acceptor methyl viologen (Fig. 6.14). As would be predicted, either of these reactions can occur in the presence of the PSII inhibitor, DCMU, and can be driven by light of $700\,nm$, which, as explained earlier (Section 6.4), is incompetent to activate PSII. Figure 6.14 also shows how another non-physiological electron carrier, PMS, can allow cyclic electron transfer around PSI to occur. Because the reduced form of PMS carries two electrons plus a proton, this form of cyclic electron transport generates a Δp and hence ATP synthesis can be observed.

Thylakoids also catalyse a process known as pseudocyclic electron transport and phosphorylation. Reduced ferredoxin (Fd) can react with oxygen to give superoxide:

$$4Fd_{red} + 4O_2 \rightarrow 4Fd_{ox} + 4O_2^{\bullet-} \qquad [6.2]$$

The subsequent activity of superoxide dismutase and added catalase in an *in vitro* experiment results in the formation of $3O_2$:

$$4O_2^{\bullet-} + 4H^+ \rightarrow 2H_2O_2 + 2O_2 \qquad \text{(superoxide dismutase)} \qquad [6.3]$$

$$2H_2O_2 \rightarrow 2H_2O + O_2 \qquad \text{(catalase)} \qquad [6.4]$$

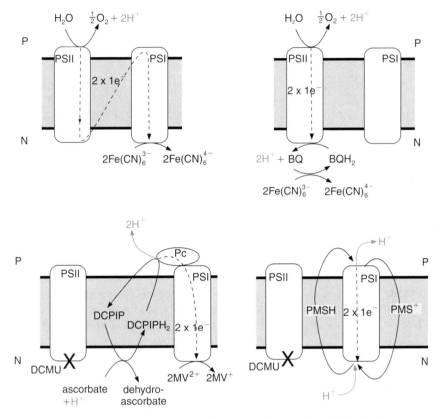

Figure 6.14 Use of redox mediators to dissect the thylakoid electron transport system.
Reaction shown at top left-hand side is the classic Hill reaction.

such that the overall reaction becomes:

$$4Fd_{red} + 4H^+ + O_2 \rightarrow 4Fd_{ox} + 2H_2O \qquad [6.5]$$

If these reactions are compared with that for the normal electron flow through both photosystems from H_2O to Fd:

$$4Fd_{ox} + 2H_2O \rightarrow 4Fd_{red} + 4H^+ + O_2 \qquad [6.6]$$

It is apparent that the cycling of Fd between these two reactions results in no net oxygen consumption or formation, allowing a pseudocyclic electron flow to proceed through both photosystems, with the generation of a Δp and ATP synthesis.

The intact chloroplast does not contain catalase and any H_2O_2 is reduced by ascorbate rather than generating oxygen. Thus when Fd is oxidized *in vivo* by O_2, one molecule of O_2 will be consumed for each $4e^-$ originating from water. This overall reaction will lead to ATP synthesis without the formation of NADPH. This is believed to occur in the intact cell as an alternative to cyclic electon transport to make good any shortfall in ATP synthesis (Section 6.4.5).

In the absence of ferredoxin a similar reaction can be observed if methyl viologen is added, since this dye also reacts with oxygen to give superoxide:

$$MV^+ + O_2 \rightarrow MV^{2+} + O_2^{\bullet-} \hspace{3cm} [6.7]$$

If, however, superoxide dismutase and/or catalase are absent or inhibited, then net O_2 consumption will be observed because formation of $4MV^+$ produces one O_2, whilst oxidation of $4MV^+$ consumes $4O_2$. This is an example of the Mehler reaction. A variant is when DCMU is present to inhibit PSII and electrons are donated to PSI from ascorbate via DCPIP and plastocyanin. A light-dependent oxygen consumption is observed driven by the operation of PSI alone.

6.4.7 The proton circuit

The steady-state Δp in thylakoids is, as discussed earlier (Section 4.2.4), present almost exclusively as a ΔpH owing to the permeability of the thylakoid membrane to Mg^{2+} and Cl^-. One important consequence of this is that electron transport can be uncoupled from ATP synthesis by ammonium ions or other weak bases, which enter as neutral species, increasing the internal pH as they protonate. Additional Cl^- uptake occurs in response to further proton pumping with the result that a massive accumulation of NH_4Cl occurs and the thylakoids burst.

Both the photosystems and the *bf* complex contribute to the net translocation of protons across the membrane. In the steady state, ΔpH can exceed 3 pH units, estimated from the accumulation of radiolabelled amines or the quenching of 9-aminoacridine fluorescence. The transient $\Delta\psi$ decays too rapidly to be measured by radiolabelled anion distribution, but can be followed from the decay of the carotenoid shift following single and flash-activated turnover of the photosystems. The timescale of the electron transfer reactions under these conditions is much shorter than for ion movements.

The chloroplast ATP synthase (Chapter 7) is essentially the sole consumer of Δp in the thylakoid. ATP-dependent H^+-uptake can be observed in the dark following activation of the latent enzyme by an imposed Δp. It is important that, in the dark, this ATP synthase does not wastefully hydrolyse ATP. However, it has recently been argued that some steady-state membrane potential must exist in the thylakoid in order to drive the ATP synthase (Chapter 7).

6.5 BACTERIORHODOPSIN AND HALORHODOPSIN

Reviews and further reading   Haupts *et al.* 1999, Beja *et al.* 2000, Lanyi 2000, Lanyi and Luecke 2001, Rouhani *et al.* 2001

Bacteriorhodopsin is a protein with seven transmembrane α-helices (named A–G) connected by short loops such that little of the protein protrudes from a membrane bilayer (Plate H). Covalently attached via its aldehyde group (i.e. via a Schiff's base) to the side chain of a lysine on one of the helices, G, is a retinal molecule, which is approximately oriented transversely relative to the membrane. Bacteriorhodopsin was first discovered in the archaebacterium *Halobacterium salinarum* (formerly *halobium*) where it acts as an ancillary generator of Δp by capturing energy from light and using it to pump protons outward across the cytoplasmic membrane; thus there is augmentation of the Δp generated by aerobic respiration under conditions of limiting oxygen.

The DNA sequence from an unnamed and uncultivated marine bacterium recently indicated that an analogue of bacteriorhodopsin might be present in a eubacterium; this has been confirmed by expression of the gene in *E. coli*, supplementation with the pigment retinal and incorporation into phospholipid vesicles to show proton pumping. The exact physiological role in the marine organism, which is common, is presently unclear but it is now recognized that bacteriorhodopsin is probably widely distributed amongst marine eubacteria.

We are closer to a molecular description for the proton-pumping mechanism for bacteriorhodopsin than for any other pump, but, as we shall see, active transport is a subtle process, depending on a carefully orchestrated series of sequential events within the protein. Unfortunately, active transport in this case, and perhaps in all transporters, is not going to be explained at the molecular level by a simple two- or three-state model of the system.

There are several reasons why bacteriorhodopsin has revealed more about an active transport mechanism than any other protein. First, the events that follow the absorption of light have been characterized in some detail by the methods of spectroscopy, in particular fast recordings of the time dependence of the visible absorption spectrum of the retinal. This has enabled the formulation of a photocycle involving at least seven species (Fig. 6.15). Second, high-resolution structures have been obtained for the protein not only in its dark state but also in states that are believed to correspond to discrete intermediates within the photocycle. Third, expression systems have allowed extensive study of molecules carrying specific mutations.

6.5.1 Structure of bacteriorhodopsin and sequence of proton transfer events

A current description of the action of bacteriorhodopsin is as follows. In the dark bR state, the N atom of the Schiff's base (Figs 6.15 and 6.16) is protonated (established by resonance Raman spectroscopy) and the carbon–carbon double bonds of the retinal are all in the *trans* geometry. The *extracytoplasmic* channel between the Schiff's base N atom and the aqueous phase on the P-side of the protein contains the side chains of several key amino acid residues and water molecules (Fig. 6.16). The opposite intracytoplasmic channel appears to be both narrower and lacking any well-defined water molecules in the dark-adapted bR state (Figs 6.15 and 6.16). A so-called π bulge distorts helix G; this means that instead of the normal H-bonding pattern within the α-helix, two peptide bond carbonyls are H-bonded to water molecules.

Mutagenesis has strongly implicated two aspartate residues, D85 in the extracytoplasmic and D96 in the intracytoplasmic channels as key participants in the pumping of protons. Fourier transform infrared spectroscopy (FTIR) has indicated, using mutants in which each of the two aspartates is separately replaced by asparagine, that in this dark state their respective pK_a values are about 2.5 and 10, respectively. Thus the side chain of D85 is deprotonated, whilst that of D96 is protonated. Other kinds of experiment indicate that the pK_a of the Schiff's base is around 12.

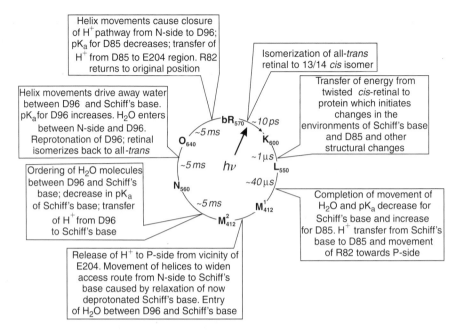

Figure 6.15   The photocycle of bacteriorhodopsin with details of the proton movements and some of the structural changes that accompany the transition between each state.

Note that for simplicity the J state, a precursor to K that forms on a 550 fs timescale, is not shown. It is also likely that there are other substates within the M intermediate. M1/412 and M2/412 are sometimes called the early and late M states, respectively. The numbers associated with each state are the absorption maxima for each species; note that these differ slightly from one investigator to another. More details of some of the photo-cycle intermediates and of the pK changes of some side chains are given in Fig. 6.16 and in the text.

Illumination of the dark bR state of the molecule causes, demonstrated by many experimental approaches, the following proton transfer events to occur in strict sequence (Plate H):

(a) A proton migrates from the Schiff's base to the side chain of D85.

(b) A proton is lost from the extracellular channel to the medium. The exact residue releasing this proton is unclear as glutamates E194 and E204 plus a water molecule appear to act as a unit.

These events raise the following questions: (i) what has caused a change in pK values underlying step (a); (ii) what molecular events cause the protonation of aspartate 85 to be linked to the release of the proton from the surface of the protein; and (iii) can we be sure that all the changes in protonation are due to migrating protons rather than to movement of hydroxyl in the opposite direction?

(c) The next step is a reprotonation of the Schiff's base by transfer of a proton from D96. This must require some change in pK in these groups and at the same time a pathway must exist for the proton transfer. A problem here is that the structure of the dark bR protein shows that these two groups are separated by 10 Å without any obvious intervening water molecules or side chains that could act as waystations for the proton (Fig. 6.16).

It is important to realize that, whilst an electron can migrate from one centre to another on a timescale of milliseconds over distances of up to 14 Å (Chapter 5), a proton can only move unaided between 1 and 2 Å on a comparable timescale. A further key point here is to understand why the proton in step (a) does not simply return from D85 to reprotonate the Schiff's base.

(d) Next, a proton is obtained from the N-phase to reprotonate D96.

(e) Proton translocation is completed by transfer of a proton from D85 to the glutamate/ water cluster that was the donor to the extracellular aqueous phase in step (b). This means that the now-unprotonated D85 can regain its favourable electrostatic interactions with the Schiff's base nitrogen atom.

The next issue is how to correlate these proton transfer steps with the photocycle. The Schiff's base nitrogen atom loses its proton [step (a)] coincident with the transition between the spectroscopically identified L and M1 states. Step (b) proton release to the medium, correlates with the M1 to M2 transition. Step (c), the reprotonation of the Schiff's base, detected by FTIR, corresponds to the transition between the M2 and N states. Step (d) corresponds to the N to O transition, and O to Br corresponds to step (e). It should be noted that the protonation states of the Schiff's base do not correlate with the geometric isomerization of the retinal.

The bR state is all *trans* in the dark, but immediately following the absorption of light, the K state, which does not involve proton translocation, is formed, with the formation of the isomerized *cis* 13/14 bond, which persists until the O state is reached. The different absorption spectra through the cycle clearly show that the retinal is in distinct environments and/or conformations throughout the cycle, but exactly how each of these relate to the function of the protein is not clear. In a general sense, the photocycle and associated proton pumping is driven by thermal relaxations of the retinal back from the *cis* state to the original all-*trans* conformation. These relaxations will involve transfer of energy from the retinal to the polypeptide chain, thus driving conformational movements (see below).

6.5.2 Relating structural changes to the mechanism of proton pumping by bacteriorhodopsin

The above description of the activity of bacteriorhodopsin leaves many questions unanswered, some of which are raised above. It is clear that these questions will only be answered by structural information for bacteriorhodopsin at various stages in the photocycle. There are two approaches to meeting this requirement. The first is to study the structure, under illuminated conditions, of mutant proteins, which are blocked at certain points of the photocycle. In this way, it is hoped that virtually all the molecules in a crystal will acquire the same structure, which will be representative of a true structural intermediate in the photocycle, i.e. it is assumed that the mutation has not induced the appearance of an irrelevant structure. Two such illuminated mutant proteins in particular have been examined by X-ray crystallography; in one (D96N), aspartate 96 is changed to asparagine and, in the other (E204Q), glutamate 204 is changed to glutamine. The first of these mutants is highly defective at reprotonating the Schiff's base nitrogen [step (c)] and the second is unable to release a proton properly to the extracellular medium [step (b)]. Illumination of both proteins indicates that the M state of the photocycle can be reached, but the structures are not the same.

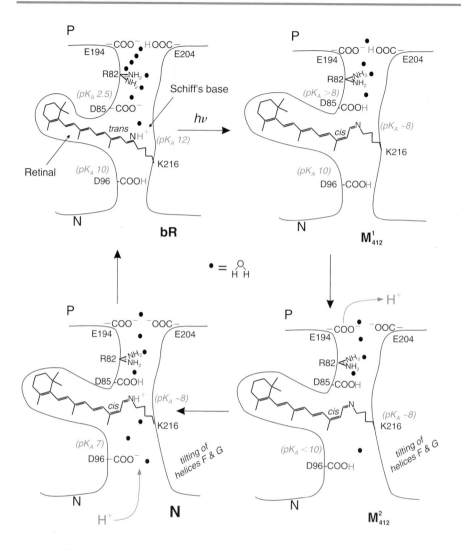

Figure 6.16   Representation of some of the conformational changes in retinal and the protein, together with proton movements in the bacteriorhodopsin photo-cycle.

Structural information for M^2_{412} and N states comes from X-ray crystallographic or electron diffraction analysis of illuminated mutant proteins that are thought to accumulate in these two states. The diagram for the M^1_{412} is an extrapolation from the dark bR and M^2_{412} states. Note that there is not a continuous movement of one proton from the N-side of the membrane to the P-side. The first movement is from the protonated Schiff's base of the retinal to aspartate 85 and the second from the P-side surface of the protein to the external aqueous medium. The exact origin of this proton is not known for certain; it is shown here as the side chain of glutamate 204, but it may be shared between this side chain, that of glutamate 194 and several water molecules. Note the movement of arginine 82 towards the P side (●). The third step in proton transfer is from aspartate 96 to the Schiff's base, the fourth is the reprotonation of aspartate 96 by a proton originating in the cytoplasm (N-side) and the final step is proton transfer from aspartate 85 to the glutamates 194/204-water cluster at the P-surface. These changes clearly involve the

Given that the reprotonation of the Schiff's base is believed to occur after the release of the proton to the outside (P-side), it is argued that the D96N mutant gives insight into the structure of a slightly later M2 state than that for the E204Q. The D96N structure was determined first and showed that significant movement of the residues around the Schiff's base and D85 had occurred together with loss of at least one bound water molecule from this region (Fig. 6.16). Overall, the effect is that the environments of the Schiff's base nitrogen and D85 have changed in order to lower the pK of the former and raise that of the latter, explaining the reversal of the relative protonation states. The structure of the illuminated D96N protein also shows that arginine 82 (R82) has moved towards the P-face of the protein, implying that this might be responsible for promoting the dissociation of the proton from the surface. This view is confirmed by the structure of the E204Q mutant, which shows the same movement of R82, even though the proton release to the medium is blocked. Thus the R82 movement must be a contributory cause of the proton release to the P phase rather than a consequence of it.

On the cytoplasmic side, the two illuminated mutant structures show quite distinct differences from the dark structure. However, in each case, there is evidence for a developing conformational change through movement of the F and G helices, and the beginning of the appearance of water molecules between D96 and the Schiff's base. This is envisaged as resulting from the relaxation of the deprotonated retinal, including the relay of movement of its 13-methyl group to a tryptophan residue. This, in turn, causes displacement of the peptide carbonyl group of the unusual π bulge in helix G. The result is that the intracellular channel has widened and the nitrogen of the Schiff's base can now be regarded as having switched, so that it is coupled to the cytoplasmic rather than the P-side of the membrane. Thus the transition from M1 to M2 can be seen as crucial in the switching or gating mechanism that controls proton access to/from this nitrogen atom; indeed, rotation of the C13–C14 double bond has at this stage turned the nitrogen atom of the Schiff's base away from D85 and 0.7 Å towards the cytoplasmic side of the protein. The proposal that the cytoplasmic (N-side) channel opens is supported by bacteriorhodopsin carrying the mutation phenylalanine-219 leucine (F219L); this accumulates in the N state upon illumination, arguably because the small side chain at residue 219 permits the longer lifetime of the N intermediate.

Electron crystallographic studies have shown that this N state is characterized by a large conformational change towards the cytoplasmic surface such that movements of helices E, F and G have opened up access of D96 to the inflow of water molecules. This would have

rearrangement of water molecules, as schematically shown, and pK_a changes, which are indicated for the Schiff's base, asp 85 and asp 96. The retinal molecule probably passes through several different degrees of strain (twist) during the transfer from the K to the N state, and the transitions from one to another, as well as the isomerizaton back to the all-*trans* state that accompanies the N to O transition, will all contribute to driving the sequence of conformational changes in the protein that occur during the photocycle. The change in shape of the retinal pocket is shown to reflect that the loss of thermal energy (strain) from the retinal drives all but the first steps of the photocycle; the diagram should not be taken to mean that a single retinal conformation is retained throughout the cycle. Some of the approximate pK_a values are shown for the key functional groups involved in proton translocation. The proton translocation mechanism is not simple and, indeed, other side chains apart from those shown play some role in the process (see text for further details).

the effect of lowering the pK_a of residue 96 and promoting proton transfer to the Schiff's base. Unfortunately, the resolution of the electron crystallography method does not allow one to see bound water molecules or amino acid side chain positions. Thus, it will be of great interest to see a high-resolution structure of this state. In principle, the X-ray structure of the illuminated D96N mutant should also show the helix repositionings that are characteristic of the N state but only disorder was seen.

Another mutant that has been studied extensively by the electron diffraction method is the triple mutant D96G/F219L/F217C. In this case the wider channel is open on the cytoplasmic side even in the dark state and it is argued to reflect fairly closely the structure of the N state.

A complementary approach to use of illuminated mutant proteins for assessing the structures of the intermediates of the photocycle is to trap these intermediates for the wild-type protein at very low temperatures. Success with this approach has been reported to date for K, L and M intermediates. Whilst the K state shows the beginnings of conformational changes that could lead to the M state discussed here, the L state structure has some features that are not readily compatible with the present discussion; there may be technical problems at play. On the other hand, the M state structure obtained in this way is broadly consistent with the description given here.

The changes in water accessibility mean that by the time the N state is reached D96 has transferred its proton to the Schiff's base. Reprotonation of D96 occurs next. This probably requires more water molecules to enter the channel between D96 and the cytoplasmic surface of the protein, although this must be coupled to the subsequent restoration of the original high pK value of D96 so as to ensure its reprotonation. The structure of a D85S mutant is believed to model the O state. This shows that the retinal has reisomerized to the all-*trans* state; large-scale structural changes relative to the bR state are confined to an opening of the channel leading to the P side, although increased hydration towards the N surface suggests how D96 might have become reprotonated during the formation of the O state; retinal reisomerization is presumed to be driven by the protonation of D96. R82 is still in its downward pointing configuration (Figs 6.15 and 6.16), consistent with its return to its bR position (Figs 6.15 and 6.16) requiring reformation of the carboxylate anion side chain on D85.

The final step in the transport process (Plate H) is the transfer of the proton from D85 to the surface cluster of E194/E204 plus water. The return of the initial electrostatic environment around the Schiff's base nitrogen and D85 will also mean that the side chain of R82 will now return to its original position and the P-side channel become narrower. The conformational changes seen in the 'O' structure in this region are presumably critical to these events.

It is important to appreciate that the complexity of the mechanism of proton pumping by bacteriorhodopsin means that several, not necessarily exclusive, hypotheses are still under consideration for the basis of the key switch whereby the accessibility of the Schiff's base nitrogen changes. For example, in one scheme the switch is essentially dictated by the retinal for which a change in curvature of the polyene chain would in turn move the retinylidene nitrogen towards the cytoplasm. Such movement is envisaged as hindering the return of the proton from D85 via the latter's H-bond with T89. One reason for supporting this mechanism is that the triple mutant (D96G/F219L/F217C) will catalyse an attenuated rate of proton translocation without the major conformational changes that are envisaged for the wild-type protein, leaving reorientation of the retinal as the only plausible switch mechanism. According to this view, the protein conformational changes would serve to enhance the rate of vectorial proton movement. Eventually, the acquisition of very high resolution crystal structures will establish whether or not sufficient changes in curvature occur at the appropriate stages of the photocycle to substantiate this hypothesis, which has a different basis than the view that underpins Figs 6.14 and 6.15.

6.5.3 Halorhodopsin

Review   Kolbe *et al.* 2000

Bacteriorhodopsin has a close relative called halorhodopsin, which acts as an inwardly directed chloride pump. The high-resolution structure of this protein has shown that it is remarkably similar to bacteriorhodopsin. There are important differences, especially the absence of residues equivalent to the carboxylate side chains of D85 and D96 and that the Schiff's base nitrogen remains protonated throughout the photocycle. The crystal structure of halorhodopsin in the dark state shows that a chloride ion is bound where the carboxylate side chain of asp 85 is located in bacteriorhodopsin. Thus it is proposed that transport is initiated by a light-driven conformational change in the retinal, which results in movement of the Schiff's base nitrogen towards the cytoplasmic side of the membrane and drags the chloride with it, so that the chloride is moved to the cytoplasmic side of the protein. As with bacteriorhodopsin, no obvious route exists for ion movement between the Schiff's base nitrogen and the cytoplasm in the structure of the unilluminated protein. In due course, therefore, it is expected that structures of photocycle intermediates, obtained by studying mutant proteins or low-temperature species, will show an opening up of the channel to the cytoplasmic side. It may also be possible to track the pathway of the chloride down this route, something that cannot be done for the proton. Thus insight into the mechanism of bacteriorhodopsin may also ensue from the studies of halorhodopsin, especially when it is recalled that it is possible that some steps in the bacteriorhodopsin reaction may be hydroxyl rather than proton migration.

In particular, comparison of halorhodopsin with bacteriorhodopsin suggests that the deprotonation of the Schiff's base in the latter may reflect movement of hydroxide towards the nitrogen atom rather than of the proton away from this atom of the Schiff's base. Indeed, as it is very difficult to distinguish movement of protons in one direction from that of hydroxides in the opposite, it cannot be rigorously excluded that bacteriorhodopsin is a hydroxide pump. It is usual to postulate proton movement rather than of hydroxide because the former has much faster mechanisms for moving through proteins. However, although we can be fairly sure that protons flow through the ATP synthase (Chapter 7, Section 7.5.2) and are taken up and released by the cytochrome bc_1 complex (Chapter 5), there are many cases where we cannot be so certain. The mitochondrial phosphate transporter (Chapter 8) is almost certainly a hydroxide antiporter, rather than a proton symporter, as it belongs to a whole set of proteins that are anion antiporters.

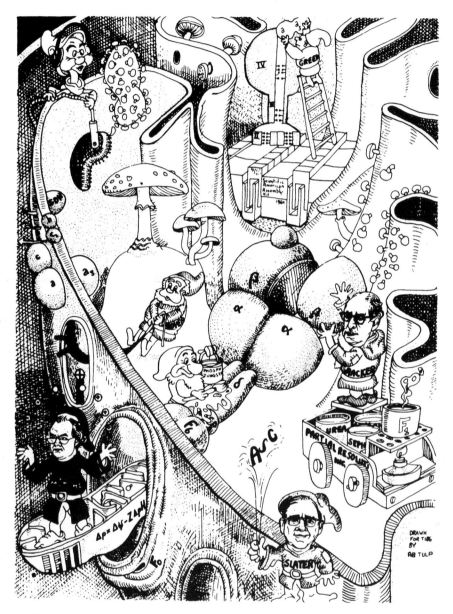

Some of the senior figures in the history of oxidative phosphorylation in action. Racker assembles the cold-labile F_1-ATP synthase (negatively stained with phosphotungstate), while Mitchell juggles protons and charges, Slater attempts to grasp the elusive squiggle (the non-existent chemical intermediate that was anticipated before the advent of the chemiosmotic theory), and Boyer induces a conformational strain which has proved to be more important in the ATP synthase itself rather than in coupling electron transport to ATP synthesis

7 THE ATP SYNTHASE

7.1 INTRODUCTION

Reviews and further reading   Abrahams *et al.* 1994, Boyer 1997, Noji *et al.* 1997, Nakamoto *et al.* 1999, Stock *et al.* 2000, Walker 2000a, b; Arechaga and Jones 2001, Menz *et al.* 2001, Tsunoda *et al.* 2001a, Yoshida *et al.* 2001

In contrast to the great variety of mechanisms found in different organisms for the respiratory or photosynthetic generation of Δp, the major consumer of Δp, the ATP synthase, is highly conserved. It is present in mitochondria, chloroplasts, both aerobic and photosynthetic bacteria, and even those bacteria that lack a functional respiratory chain and where the enzyme generates Δp at the expense of hydrolysing ATP produced in glycolysis. For reasons that will soon become clear, the H^+-translocating ATP synthase is known as an $F_1.F_o$-ATPase, distinguishing it both from $E_1.E_2$-ATPases such as the (Na^+/K^+)-translocating ATPase in eukaryotic plasma membranes, whose catalytic cycle involves a covalent attachment of the phosphate from ATP, and also from the more recently described class of V (for vacuolar) ATPases, which can pump protons across internal membranes (e.g. tonoplasts in plants and synaptic vesicles in neurons).

The function of the ATP synthase is to utilize Δp to maintain the mass-action ratio for the ATPase reaction 7–10 orders of magnitude away from equilibrium, or in the case of fermentative bacteria, such as *S. facaelis*, to utilize ATP to maintain Δp for the purpose of transport. Although the function of the complex in all except this last case is to synthesize rather than hydrolyse ATP, it is sometimes referred to as the proton-translocating ATPase.

7.2 F₁ AND Fₒ

The general features of the ATP synthase were deduced from early studies with submitochondrial particles (SMPs) (Section 1.3.1). These inside-out membrane vesicles catalyse either ATP synthesis in the presence of a respiratory substrate or ATP hydrolysis in the absence of such a substrate. Both respiration and ATP hydrolysis generate a Δp (with the positive, P-phase, in the lumen of the vesicle) and H^+/ATP ratios can be determined, albeit imprecisely.

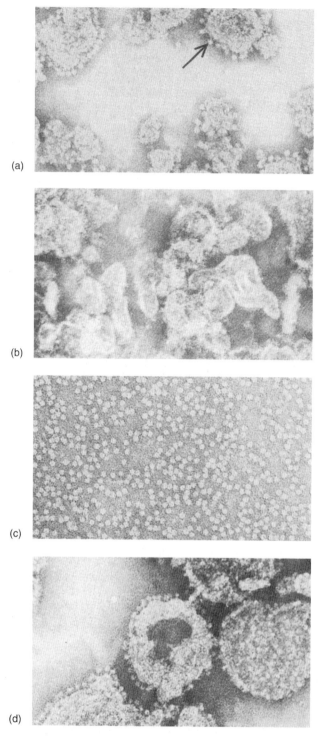

(a)

(b)

(c)

(d)

Figure 7.1   The F_1 and $F_1.F_0$ mitochondrial and *E. coli* ATP synthase visualized by electron microscopy of sub-mitochondrial particle membranes or purified enzyme preparations.

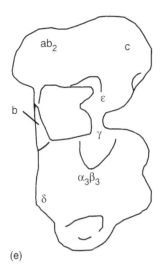

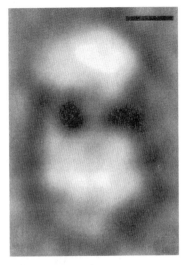

(e)

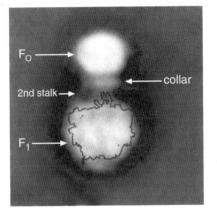

(f)

(a) Electron micrograph of submitochondrial particles (magnification $175\,000\times$) shows F_1 molecules (arrow) on the surface. (b) Electron micrograph showing particles from which F_1 has been removed by urea treatment; the particles have greatly diminished ATPase activity. (c) Molecules of F_1 visualized by electron microscopy after removal from submitochondrial particles. (d) The appearance of urea-treated particles following reconstitution with F_1; the surface particles are again evident and ATPase activity is restored to the membranes. Note that (a)–(d) were all visualized by negative staining. (Reproduced with permission from G. Weissman and R. Claiborne (eds) *Cell Membranes: Biochemistry, Cell Biology and Pathology* HP Press, (1975).) (e) Cryo-electron microscopy (at much higher magnification than (a)–(d)) of a negatively stained sample of *E. coli* $F_1.F_0$-ATP synthase shown alongside an interpretation of the arrangement of the different subunits (cf. Fig. 7.2). Note the evidence for a second stalk to the left of figure, connecting F_0 (top) to F_1 (bottom). (Adapted from Wilkens and Capaldi (1998) *Nature* **393**, 29.) (f) A projection by single particle imaging by electron microscopy of the bovine heart mitochondrial ATP synthase. Note both the second stalk and the collar, which is seemingly absent from the *E. coli* enzyme. The three-dimensional structure of F_1 is superimposed. (Adapted with permission from Karrasch and Walker (1999) *J. Mol. Biol.* **290**, 379–384.)

In good preparations of SMPs the rate of ATP hydrolysis can be stimulated by addition of a protonophore to collapse Δp. Both the synthesis and hydrolysis of ATP by mitochondrial membranes can be inhibited by several reagents, including oligomycin.

The ATP synthase can be visualized under the electron microscope in preparations of SMPs that have been negatively stained with phosphotungstate. The complexes appear as roughly spherical knobs projecting from the original matrix side of the membrane (Fig. 7.1). When SMPs were washed with urea, chelating agents or low ionic strength media the knobs were lost from the membrane. At the same time ATPase activity was solubilized from the SMPs. This activity was not inhibited by oligomycin and was termed the F_1-ATPase ('fraction 1'). Naturally, the soluble F_1 was incapable of ATP synthesis. The stripped SMP membranes had lost ATP synthase and ATPase activity, but interestingly also behaved in an 'uncoupled' manner, with no respiratory control with NADH as substrate and with evidence of a high proton permeability. When these depleted SMPs were pretreated with oligomycin, some respiratory control was reintroduced, and the proton permeability, estimated, for example, from the rate of decay of a pH gradient after cessation of respiration, was reduced to the considerably lower value seen with untreated SMPs.

These seminal observations suggested that oligomycin binds to and inhibits a component in the membrane of depleted SMPs, which could conduct protons, and that in intact SMPs the free passage of protons through this component was in some way controlled by F_1. This proton channel was termed F_o (fraction oligomycin) and can be inhibited by oligomycin in both depleted and untreated SMPs. Purification of the intact $F_1.F_o$-ATPase complex requires detergents to maintain the solubility of the highly hydrophobic F_o.

These findings can be generalized to bacterial and thylakoid membranes, although oligomycin, as well as another inhibitor of F_o, venturicidin, only inhibit the $F_1.F_o$-ATPases from mitochondria and a limited number of bacterial genera. In all energy-conserving membranes, ATP is always hydrolysed or synthesized on the side of the membrane from which the knobs project (the N-phase), while during ATP synthesis protons cross from the side which lacks knobs (the P-phase; Fig. 1.1). Thus F_1 faces the mitochondrial matrix, the bacterial cytoplasm and the chloroplast stroma.

7.3 THE SUBUNITS OF THE $F_1.F_o$-ATPase

The most definitive information available is for the *E. coli* $F_1.F_o$-ATP synthase. F_1 preparations have five types of polypeptide, usually known as α, β, γ, δ and ε, whilst three further polypeptide species, a, b and c, have been identified in F_o. The genes for the *E. coli* $F_1.F_o$-ATP synthase are in a single operon containing eight cistrons that correspond to the eight polypeptides. The stoichiometry of the subunits of the *E. coli* $F_1.F_o$-ATPase is $a.b_2.c_{9-12}.\alpha_3.\beta_3.\gamma.\delta.\varepsilon$. Uncertainty surrounds the stoichiometry of the c subunits for several reasons, including the difficulty of accurately assessing its stoichiometry from radiolabelling of the enzyme *in vivo*; small experimental errors are sufficient to generate an uncertainty in the range 9 to 12.

The estimated molecular weights of F_o and F_1 are estimated to be 160 kDa and 370 kDa, respectively. Amino acid sequences of the five F_1 polypeptides are consistent with a hydrophilic globular structure, whilst those of each of the F_o polypeptides have hydrophobic regions consistent with the presence of transmembrane α-helices.

ATP synthases from other sources have very similar subunit structures, although that from mitochondria is more complex, possessing extra subunits, the functions of which are generally unclear. There is not an exact correspondence between the roles of the smaller subunits in the enzymes from different sources. Thus the *E. coli* δ-subunit is equivalent to the mitochondrial oligomycin-sensitivity conferring factor (OSCP), a subunit of F_1 that is required for the mitochondrial $F_1.F_0$ to be sensitive to oligomycin. The mitochondrial δ-subunit is related to the ε-subunit of the *E. coli* enzyme, while rather oddly the ε-subunit in the mitochondrial enzyme has no counterpart in the bacterial ATP synthase. Figure 7.2 provides a comparison of the overall organization of subunits in the enzymes from the two sources, omitting, with one exception, the extra subunits of the mitochondrial enzyme for which little information about location or function is known. The evidence for this organization will be provided in the next sections. The exception is the inhibitor polypeptide for the mitochondrial enzyme, which has no direct counterpart in the bacterial enzyme.

We shall return to the mode of action of this inhibitor later in this chapter.

7.4 THE STRUCTURE OF $F_1.F_O$

Further reading   Abrahams *et al.* 1994, Stock *et al.* 1999, Gibbons *et al.* 2000

In 1994, a 2.8 Å structure of the α- and β-chains and part of the γ-chain for F_1 from bovine heart mitochondria was obtained. This landmark achievement showed that the α- and β-chains were arranged alternately around the γ-chain, rather like an orange with six segments surrounding a central core. The part of the γ-chain that could be seen in the structure was mainly folded as two α-helices coiled around one another such that both the N- and the C-terminals were at the same end of the central core of the structure (Fig. 7.3 and Plate I). This can safely be deduced to be the side of the F_1 molecule, which, when part of the complete $F_0.F_1$, is the furthest distance away from the membrane bilayer, because the remainder of the γ-chain protrudes from the $\alpha_3\beta_3$ assembly in the form of a stalk-like structure. The general features of the protein folds in the each of the six α- and β-chains are broadly similar, but there are significant differences in the detail that, as we shall discuss shortly (Section 7.6), are very important for understanding the mechanism of the enzyme.

Much of the γ-chain that protrudes from the $\alpha_3\beta_3$ core, along with the δ- and ε-subunits, was disordered in the original crystals, but a second crystal form later allowed the entire structure to be obtained. Most of the remainder of the γ-polypeptide has unexpectedly proved to adopt a globular fold, comprising α-helices and β-sheets, which is often associated with nucleotide binding in other proteins. No such binding is seen in the ATPase, but rather this feature of the γ-chain, together with the δ- and ε-subunits that interact with γ in this region, contributes to a foot-shaped structure that can interact with the F_0 part of the molecule (Fig. 7.2 and Plates I and J). The overall length of the γ-subunit from its C-terminus to its foot is 114 Å, while the length of the protruding part of γ is 47 Å, consistent with the dimensions of the central stalk-like structure seen in the original electron micrographs (Fig. 7.1).

The foot-like structure at the base of the γ-subunit has also been seen in a structure of a form of the yeast mitochondrial ATP synthase in which the resolution (3.9 Å) was too low to discern much about side-chain locations. However, this structure had a remarkable feature, the presence of a ring of 10 c subunits from the F_0 sector, attached to this foot of the

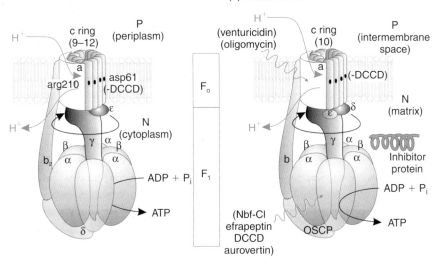

(a) *E. coli*

c ring (9–12)

P (periplasm)

H^+

a

arg210 ••••• asp61 (-DCCD)

ε

N (cytoplasm)

H^+

b_2 β γ α β α α

ADP + P_i

ATP

δ

F_o

F_1

(b) Mitochondria

H^+

(venturicidin) (oligomycin)

c ring (10)

P (intermembrane space)

a

••••(-DCCD)

ε δ

N (matrix)

H^+

b β γ α β α α

Inhibitor protein

ADP + P_i

(Nbf-Cl efrapeptin DCCD aurovertin)

OSCP

ATP

Figure 7.2   Schematic diagrams of current information about the overall structure of $F_1.F_o$. (a) *E. coli* enzyme and (b) mitochondrial enzyme showing the similarities, equivalencies and differences.
The diagrams are based on the X-ray diffraction data available for the $\alpha_3\beta_3\gamma\delta\varepsilon$ and c_{10} subunit assemblies of the yeast and bovine heart mitochondrial enzymes. Locations of other subunits have been deduced by biochemical methods as discussed in the text. Note that (i) OSCP in the mitochondrial enzyme is equivalent to the δ-subunit in *E. coli*; (ii) that the mitochondrial ε-subunit has no equivalent in bacteria; and (iii) for the mitochondrial enzyme, the a and c subunits are sometimes called subunits 6 and 9, respectively. The locations of binding of principal inhibitors of the ATP synthase are also shown. Nbf and DCCD bind to one of the three β-subunits, aurovertin to two of the three β-chains and efrapeptin binds in the central cavity between the three α- and three β-subunits into which both the C- and N-terminal ends of the coiled coil γ-subunit penetrate. The exact sites of venturuicidin and oligomycin binding within F_o are not known but very low concentrations of DCCD specifically react (i.e. no reaction with β-subunit under such conditions) with an aspartate or glutarate residue (D61 in the *E. coli* enzyme) on the c subunit. Only one β-subunit is represented as catalysing ATP synthesis but as discussed in the text all three play this role. The diagram shows a postulated route of proton flow through the F_o sector. It is envisaged (see text) that the protons flow part way across the membrane through the a subunit before binding to the carboxylate side chain of the key c subunit aspartate residue that reacts with DCCD; this residue is also implicated by inactivation of the enzyme following its mutation to asparagine. This protonation releases an attraction between the carboxylate form of the aspartate and the positively charged side chain of arginine 210, causing rotation of the set of c subunits in the sense shown (going behind the plane of the paper). Nine to 12 copies of the c subunit are shown (there is uncertainty over the exact stoichiometry – see text). The last step in a 9–12 step process results in discharge of the proton through a different part of the a subunit. Thus 9–12 protons passing through the entire c subunit assembly would cause a rotation of 360°. Note that, in the case of the mitochondrial enzyme, there are 10 c polypeptides and thus movement of 10 protons would cause the 360° rotation. Whereas the b subunit is thought to be dimeric in *E. coli*, it is curiously argued to be monomeric in mitochondria. The inhibitor protein is a dimer, which has a coiled coil structure that is believed to be able to bind simultaneously to two F_1 molecules on the same membrane. The binding site on F_1 for the inhibitor molecule is shown as close to the DELSEED region (cf. Fig. 7.3). The rotation of the γ-chain is shown as clockwise for ATP synthesis when viewed from the membrane side.

F_1 molecule (Plate J). Each c subunit is folded as two transmembrane α-helices connected by a loop, which is believed to interact with the foot region. This requires that the N and C termini of each c subunit are located at the P-side of the membrane. There is biochemical evidence for this orientation. As explained earlier, the stoichiometry of the c subunit has proved difficult to obtain by biochemical methods, but it had been assumed that the number of copies per ATP synthase molecule would be invariant irrespective of the source of the enzyme. Remarkably, evidence is emerging from structural studies that contradicts this expectation. The c subunit of the thylakoid enzyme has been shown to form rings containing 14 polypeptides, while a very stable c subunit preparation from a sodium translocating ATP synthase (as found in *P. modestum*, see Section 5.15.8) has a ring of 11 c subunits. A possible important basis for this variation in c subunit stoichiometry is discussed later (Section 7.6).

Until this structural information became available, there had been a growing belief, based on biochemical and genetical studies, that each ring contained 12 c subunits. This conclusion was reached mainly on the basis of work with the *E. coli* enzyme but re-evaluation and new experiments in light of the structural data now indicate that the ring in this bacterium may contain 10 c subunits. It is wise to sound a note of caution at this point. It is possible that the c polypeptide can rearrange itself. Thus, unlikely as it may seem, it is just possible that the loss of the a and b subunits from the yeast enzyme has been accompanied by the loss of some (two?) c subunits and a shrinking of the ring size to 10. Variation, or otherwise, of c subunit ring size will only be established for sure when crystal structures of complete $F_1.F_o$ molecules from different sources are obtained.

To date, X-ray diffraction studies have not provided any direct information about either the structure or the locations of one of the F_1 subunits, OSCP in mitochondria, δ in bacteria (although there is an NMR structure for the isolated *E. coli* δ-subunit) or the a and b subunits of F_o. However, δ/OSCP can be confidently positioned in the dimple (Fig. 7.2) at the 'top' of the $\alpha_3\beta_3$ assembly on the basis of immuno-electron microscopy experiments. There is evidence from cross-linking that the b subunit, part of which is indicated by NMR studies to be a coiled coil dimer in *E. coli*, but possibly organized as a monomer for the mitochondrial enzyme (Fig. 7.2), extends from the membrane and contacts OSCP (δ) (Fig. 7.2). This means that there is a second stalk connecting F_o to F_1, and indeed this has been seen in higher resolution electron micrographs of $F_1.F_o$ as shown in Fig. 7.1; the higher resolution data also indicate a central stalk length of 45 Å, consistent with the length of the central stalk estimated from crystallographic and earlier electron microscopy observations. There is biochemical evidence that the remainder of the each b subunit is folded as one transmembrane α-helix in the membrane.

Finally, it has proved difficult to deduce the organization of the a subunit by biochemical or genetic means. Currently, it is thought to be folded into five transmembrane α-helices with short connecting loops and thus entirely confined to the bilayer. Biochemical and genetic data place the a and b subunits on the outside of the ring of c subunits, the interior of which is presumably filled with phospholipids in the membrane. The a and b subunits must make asymmetric interactions with the ring of c subunits, owing to the different stoichiometries and the lack of symmetrical sequence repeats in a and b. Having established the main overall structural features of the $F_1.F_o$ enzyme in Fig. 7.2, we shall consider next what has been discerned from biochemical experiments about the enzymology of ATP synthases before returning to see how these can be correlated in a satisfying way with a more detailed analysis of the structure.

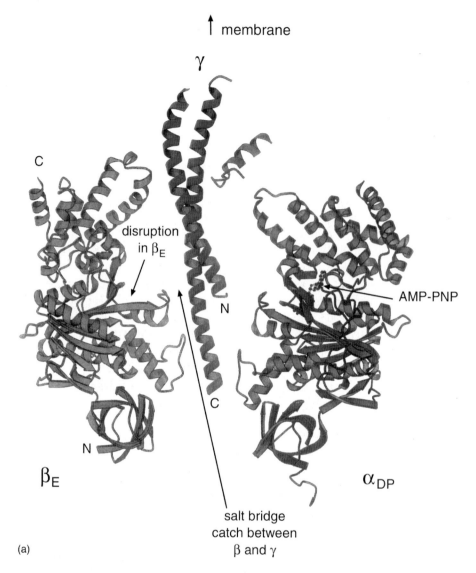

(a)

Figure 7.3   Selected important features of the organization of the α-, β- and γ-chains in the F₁ part of the ATP synthase.

The first crystal structure of the enzyme was obtained from a solution containing ADP and the non-hydrolysable analogue of ATP known as AMP-PNP. The three β-chains were called β_E, β_{DP} and β_{TP}, depending upon whether they respectively had no ligand, bound ADP or bound AMP-PNP. The α-chains, which each had a bound AMP-PNP were named α_E, α_{DP} or α_{TP}, depending on which type of β-chain was their neighbour, according to an arbitrary convention concerning the direction in which neighbour was defined. The alternate organization of the α- and β-chains (Fig. 7.2) means that a cross-section through the molecule will show the β_E in juxtaposition to α_{DP} (a) and β_{DP} to α_E (b). Comparison of (a) and (b) shows that, whereas the conformation of the α-chains is essentially identical, the β_E has considerable differences from β_{TP}, including the disruption of some β-sheet structure adjacent to the P-loop region, which interacts with the terminal

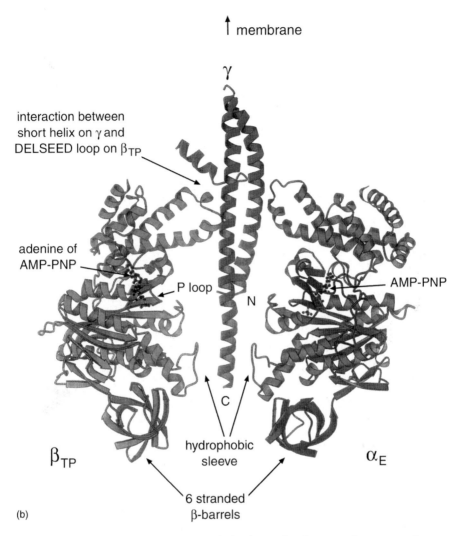

↑ membrane

γ

interaction between
short helix on γ and
DELSEED loop on β$_{TP}$

adenine of
AMP-PNP

P loop

N

AMP-PNP

β$_{TP}$

C

hydrophobic
sleeve

α$_E$

6 stranded
β-barrels

(b)

phosphates on ATP (see text), and the interaction between the conserved
DELSEED region and a short (residues 73–90) α-helix of the central γ-subunit
with its N and C termini between one (of three) pair of α- and β-subunits. (The
three β-chains are not structurally identical.) This would break as γ moves to
another β-chain. Four out of six β-barrels that hold the three α and three β
chains together are shown in (a) and (b). Another notable feature is the hydrophobic sleeve that may allow rotation of γ within the central cavity in the core of the
α$_3$β$_3$ assembly. Adapted with permission from Abrahams *et al.* (1994).

There is a final component of the mitochondrial ATP synthase, the inhibitor protein. It is
a dimer in its active form. Its structure has been solved by NMR and X-ray diffraction, and
found to be an unusual coiled coil, sufficiently elongated to be able to bind to two adjacent
F$_1$ molecules. A stoichiometry of one inhibitor polypeptide per F$_1$ is sufficient for inhibition
and it is thought to bind to the C-terminal region of a β-subunit of F$_1$, although this has not
been demonstrated by crystallography.

7.5 ENZYMOLOGY OF ATP SYNTHASE

7.5.1 Studies with inhibitors show that enzyme is a co-operative unit and implicate β-chains as the site for formation or hydrolysis of ATP

The complex subunit structure of the ATP synthase initially raised questions as to which subunit carried the active site and the extent to which the enzyme exhibited co-operativity. Apart from oligomycin, one of the earliest inhibitors discovered was DCCD (dicyclohexyl-carbodiimide), which at an extremely low concentration will specifically label the c subunit of ATP synthase in mitochondrial, thylakoid or bacterial membranes. In common with oligomycin it also blocks the proton conductance of F_0. Unlike oligomycin, or venturicidin, DCCD inhibits all ATP synthases. DCCD reacts covalently with the carboxylate side chain of a conserved aspartate (or often a glutamate) that is part of one of the transmembrane helices in the c subunit.

It is generally thought that this carboxylate side chain plays a key role in the translocation of protons (or Na^+ for the *P. modestum* enzyme, Section 5.15.8) by F_0. A key observation in this respect is that the reaction with DCCD is blocked by the presence of Na^+ in the sodium-translocating *P. modestum* enzyme (Section 5.15.8), suggesting that the translocated ion comes into close proximity to the carboxylate side chain. Incorporation of approximately one DCCD molecule per assembly of ten or more c subunits is sufficient to block the activity of the ATP synthase completely, strongly suggesting that the assembly of c subunits works co-operatively and not independently. The exact binding site for oligomycin is not known for certain but is clearly amongst the F_0 subunits. At first sight, therefore, it is odd that the OSCP protein is at a considerable distance from the F_0 part of the enzyme (Fig. 7.2); it is clearly *not* the oligomycin-binding protein as is often thought. We shall return later to a possible explanation for the separation of OSCP from the site of oligomycin (Section 7.6).

Several inhibitors of F_1 are known (Fig. 7.2). Perhaps most instructive is the reagent Nbf-Cl, which reacts specifically with the hydroxyl group of one tyrosine residue of one β-chain to inactivate the enzyme completely. This observation implicates the β-chain as the catalytic site and also establishes that three separate active sites on the three β-subunits cannot function independently. The reason for the inhibitory action of Nbf-Cl has proved to be steric hindrance caused by the bulky modifying group. The Nbf group from Nbf-Cl can undergo an intramolecular shift within a β-chain from its initial attachment at Tyr_{311} to Lys_{162}, indicating that the residues must be adjacent. The enzyme remains inactive. This was notable, since Lys_{162} is found at a sequence motif -Gly-X-X-X-X-Gly-Lys- (this is commonly called the P loop or Walker A motif), which is found in many ATP-utilizing enzymes and participates in binding of phosphate groups. In accord with this, P_i will protect F_1 against the initial labeling by Nbf-Cl.

At higher concentrations than are needed for reaction with the c subunit, DCCD will react with just one glutamate side chain in one β-subunit of the enzyme. As with modification with Nbf, activity of the enzyme is completely lost, pointing again to co-operativity between the β-chains. Two other inhibitors, both naturally occurring, aurovertin and efrapeptin, also bound to F_1 at a stoichiometry of less than 3 mol per mol enzyme, again suggestive of co-operativity. In at least some cases, the asymmetric interaction of the γ-chain with the $\alpha_3\beta_3$ assembly explains these stoichiometries. However, this cannot be the whole story, as a form of the enzyme that possesses only the $\alpha_3\beta_3$ assembly also binds only 1 mol of Nbf to give complete inhibition. This suggests that there is an element of intrinsic or easily induced asymmetry within the three α- and β-chains.

The role of β-chains as the locus of catalytic sites was also implicated by an approach known as affinity labelling in which derivatives of ATP or ADP carrying a chemically reactive group were found to incorporate specifically into β-chains. In other experiments, the β-chains were also shown to bind adenine nucleotides, which could be exchanged with unbound radio-labelled nucleotides in solution. This property is expected for ligands at an active site. In contrast, the α-chains were each found to contain an ATP or ADP that could not be exchanged; the function of these tightly bound nucleotides is obscure but they can safely be concluded to be bound to non-catalytic sites.

7.5.2 Mechanistic enzymological experiments establish co-operative features and that affinity of ATP synthase for ATP is attenuated by protonmotive force; the binding change model of ATP synthase

ATP hydrolysis is more easily studied than ATP synthesis, although, of course, no direct information on the mechanism of coupling to proton translocation can be obtained by using the F_1 form of the enzyme, which is easier to handle than the complete $F_o.F_1$ enzyme.

The hydrolysis of ATP involves breakage of the bond between the bridging oxygen and the γ-phosphorus atom and thus an oxygen on the liberated P_i comes from water (Fig. 7.4). This can be shown by carrying out the hydrolysis in $H_2^{18}O$ and quantifying the label appearing in P_i. Under some conditions, the hydrolysis of ATP by SMPs results in more ^{18}O incorporation into P_i than can be accounted for by ATP hydrolysis alone. One explanation for this is that P_i, produced at the active site with one ^{18}O by hydrolysis of ATP, can rotate within the active site before being released. If the ATP hydrolysis reaction is reversible, and if the rate of release of products from the catalytic site is slow relative to the rate of resynthesis of ATP, then the P_i reincorporated into ATP will lose ^{16}O rather than ^{18}O. Subsequent ATP hydrolysis will lead to further incorporation of ^{18}O into P_i, which will ultimately be released to the medium. Most surprisingly, this extra incorporation of ^{18}O was not abolished when Δp was dissipated by a protonophore.

The occurrence of this exchange in the absence of a Δp was one of the first pieces of evidence that the protonmotive force was not directly used to condense ADP and P_i. In subsequent experiments, the analogous exchange reaction was studied with the F_1 fragment, but with the difference that the substrate ATP was labelled on its γ oxygens with ^{18}O and the deficit of ^{18}O in the product phosphate was analysed. This procedure was more sensitive. When the hydrolysis of extremely low concentrations of ATP (about 10^{-8} M) was studied, the loss of ^{18}O label from product P_i was more than expected from the hydrolysis reaction, again indicating that exchange had occurred within a catalytic site. This provided further evidence that the hydrolysis of ATP to ADP and P_i by the soluble F_1 was to some extent reversible even with no input of energy from Δp (Fig. 7.4). This conclusion could be reinforced by similar experiments in which loss of ^{18}O originally in P_i was observed in ATP synthesized in steady-state oxidative phosphorylation. More complex experiments, in which these exchange reactions were studied with variable substrate concentrations, indicate that interactions occur between more than one catalytic site in the ATP synthase.

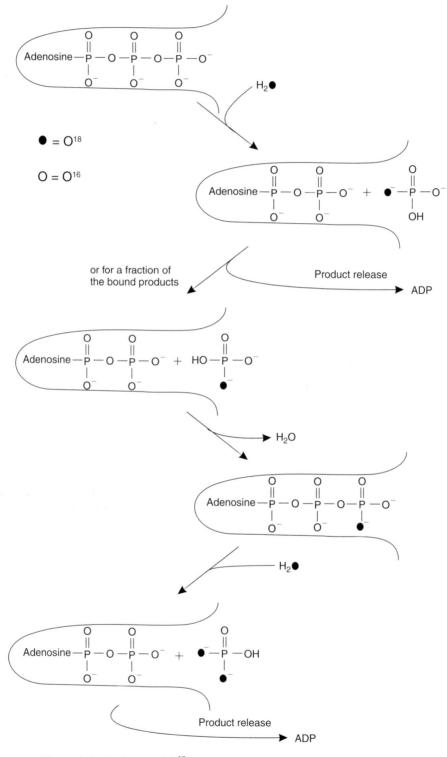

Figure 7.4   Exchange of O^{18} into or from H_2O during ATP hydrolysis by soluble F_1-ATPase indicates that the reaction is reversible.

An equilibrium constant is equal to the ratio of the forward and reverse rate constants. Since the K' for ATP hydrolysis in free solution is approximately 10^5 M (Section 3.2.1), the reverse reaction should be undetectably slow. F_1 therefore appears to alter the equilibrium constant for ATP hydrolysis to make the rate of the reverse reaction significant. How can this occur without offending against the first law of thermodynamics, since we appear to be making ATP with no energy input? The answer is that we are not making *free* ATP, but rather ATP bound to F_1. If that ATP is very firmly bound, then considerable energy will have to be used to release the ATP into solution. Consider the ΔG values associated with the following steps:

1. $$ADP_{free} \rightleftharpoons ADP_{bound} \qquad\qquad \Delta G_1$$
2. $$P_{i\,free} \rightleftharpoons P_{i\,bound} \qquad\qquad \Delta G_2$$
3. $$ADP_{bound} + P_{i\,bound} \rightleftharpoons ATP_{bound} \qquad\qquad \Delta G_3$$
4. $$ATP_{bound} \rightleftharpoons ATP_{free} \qquad\qquad \Delta G_4$$

The overall reaction is the sum of these steps:

$$ADP_{free} + P_{i\,free} \rightleftharpoons ATP_{free} \qquad \Sigma\Delta G = +40\,kJ\,mol^{-1}*$$

* Typical value for the mitochondrial matrix.

What is being observed in the ^{18}O exchange reaction is not the overall reaction, but reaction 3, which occurs with a ΔG close to zero. Nearly all the input of $+40\,kJ\,mol^{-1}$ is required for the final step, the removal of very tightly bound ATP from the catalytic site. As we shall see below, it is the conformational change driven by the protonmotive force (see above), which releases the bound ATP. Confirmation of this model has come from measurements of the dissociation constant for the F_1.enzyme.ATP complex, where a value of about 10^{-12} M was obtained.

With very low concentrations of ATP ($<10^{-10}$ M) labelled with ^{32}P on the γ-phosphate, hydrolysis proceeds very slowly. If, however, a higher concentration of cold ATP is subsequently added in a chase experiment, the rate of hydrolysis of the already bound ATP-$\gamma^{32}P$ considerably increases. It appears that the higher concentration of ATP occupies one or more lower affinity ATP binding sites on different β-subunits of the enzyme and that this causes a conformational change, allowing release of the products at an accelerated rate. Thus there are site-to-site interactions, perhaps mediated through subunit interfaces. This negative co-operativity of binding, but positive co-operativity of catalysis (k_{cat}), explains why the ^{18}O exchange experiment discussed above must be performed with F_1 at very low ATP concentrations: at higher concentrations of ATP, the ADP formed would be released rather than remaining to allow the reverse reaction (and thus ^{18}O exchanges) to occur. It is notable that mutations in the α-chain of the *E. coli* enzyme attenuate such co-operativity and impair $\alpha\beta$ interface interactions.

The study of ATP hydrolysis by soluble F_1 is of value to the extent that it provides information about ATP synthesis by the membrane-bound $F_1.F_o$-ATP synthase. The ^{18}O exchange experiment discussed above can also be observed with the intact $F_1.F_o$ complex *in situ* in SMPs in the absence of Δp. However, for reasons that are not clear, it is no longer necessary to use a very low ATP concentration in order to see the exchange.

If the molecular basis of ATP synthesis is the reverse of the F_1 hydrolysis discussed above, it follows that the major ΔG changes are associated with the binding of ADP and

P_i and/or the release of ATP, and that these are the steps which must in some way be coupled to Δp through a conformational change. A crucial experiment using SMPs involved loading the very high affinity ATP binding site on F_1 in the absence of a Δp and then initiating respiration. The generated Δp caused a release of this tightly bound ATP, clearly demonstrating that Δp decreases the binding affinity of ATP. The change in binding affinity has to be dramatic: from a value of about 10^{-12} M in the absence of Δp to a sufficiently loose binding such that ATP can dissociate in the presence of the normal N-phase concentration of the nucleotide. Experiments with a bacterial vesicle system and SMPs have suggested that, during ATP synthesis, the binding of ATP to a catalytic site has a dissociation constant in the range 10^{-6} to 10^{-5} M, a change of 10^6 to 10^7 induced by the protonmotive force.

The combination of the co-operative properties and the evidence for bound ATP formation from ADP and P_i at an active site led to the formulation of the binding change mechanism in which each β-subunit would be in a different conformation at any given instant, the three conformations reflecting different affinities for ATP, ADP and P_i. Thus the three catalytic sites for nucleotide exist in O (open), L (loose) and T (tight) conformations and protonmotive force-induced conformational change causes a T-site with bound ATP to become an O-site, and thus release its bound ATP, while at the same time causes a second site to change from a L-site, with loosely bound ADP and P_i, to a T-site, where the substrates are tightly bound, allowing bound ATP to be formed (Fig. 7.5). Thus each of these catalytic sites has at any instant a distinct conformation, but all the sites pass sequentially through the same set of at least three conformations.

It should be noted that unlike bacteriorhodopsin (Section 6.5) whose reaction mechanism involves proton flux through the catalytic site (i.e. the Schiff's base of the retinal), evidence is against a direct involvement of the translocated protons in the ATP synthesis reaction itself. Thus the proton flow through the F_o is envisaged as driving conformational changes at a distance in the F_1 part of the enzyme. Evidence for this view includes: (a) the reversible nature of the ATP $\rightleftharpoons$ ADP + P_i reaction even at the active site of isolated F_1, which cannot be linked to proton translocation; (b) the thinness of the central stalk connecting F_o to F_1 argues against it providing a proton conducting pathway; (c) the enzyme from *P. modestum* uses Na^+ instead of H^+ but the specificity for Na lies in the F_o part of the enzyme, making it improbable that any part of F_1 conducts H^+ or Na^+. The properties of this cation translocating *P. modestum* enzyme also strongly imply that protons rather than hydroxides normally pass through the ATP synthase.

The stereochemistry of ATP hydrolysis can be investigated by isotopic labelling. We know from the ^{18}O exchange experiment that it is the bond between the γ-P and oxygen that is broken because ^{18}O from $H_2^{18}O$ is incorporated into P_i rather than ADP. The attack of water on the γ-phosphate could be similar to a classic SN_2 reaction at a saturated carbon atom with the phosphate inverting its configuration (Fig. 7.6). This cannot be tested with ATP itself, since the γ-phosphate is not a chiral centre; however, an ATP analogue, Fig. 7.6, does have such a centre and is a substrate for F_1. Using this analogue together with $H_2^{17}O$ confirms this inversion. In contrast to F_1, which does not form a phosphoenzyme intermediate, an $E_1.E_2$-ATPase such as the Na^+/K^+-ATPase, which does form such an intermediate, retains the stereochemistry of the terminal phosphate. In the latter enzyme, two inversions corresponding to phosphorylation and dephosphorylation steps occur leading to net retention of configuration.

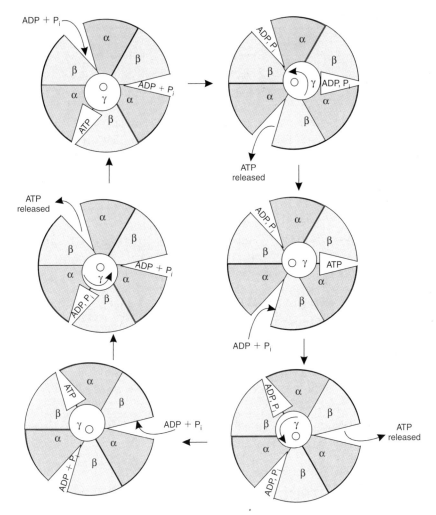

Figure 7.5 The three-site alternating binding site mechanism for ATP synthase.

This diagram shows an anticlockwise rotation of the γ-chain, looking from below F_1 and towards the bilayer, that is associated with ATP synthesis. It is envisaged that the α- and β-chains are held stationary by a stator. Starting from the top left, the scheme shows binding of ADP and P_i to an empty site (β_E), whilst ADP and P_i are already bound at a second site, tentatively but not definitely correlated with the β_{DP} subunit seen in the crystal structure; tightly bound ATP is at the third site, tentatively β_{TP}. (Note the sites are shown at interface regions where they have been located by crystallography but features of those sites in the structure, see Fig. 7.3, do not allow definite correlation of AMP-PNP bound with the tight ATP site during ATP synthesis.) Rotation of γ would cause a conformational change in β_{TP} such that ATP is released and the conformation would change towards β_E. Concomitantly, the original β_E is transforming towards β_{DP} and the initial β_{DP} to β_{TP}. Thus in the middle diagram (right-hand side) the structure has returned to the original (top left) conformation except that three conformations have migrated round the ring of α- and β-subunits. Repeat of these steps releases a further two ATP molecules and returns the structural arrangement to the original (top left). This scheme is an oversimplification as an additional β-chain conformation has recently been observed (Menz *et al.* 2001), which corresponds to a half-closed state of the β_E conformation.

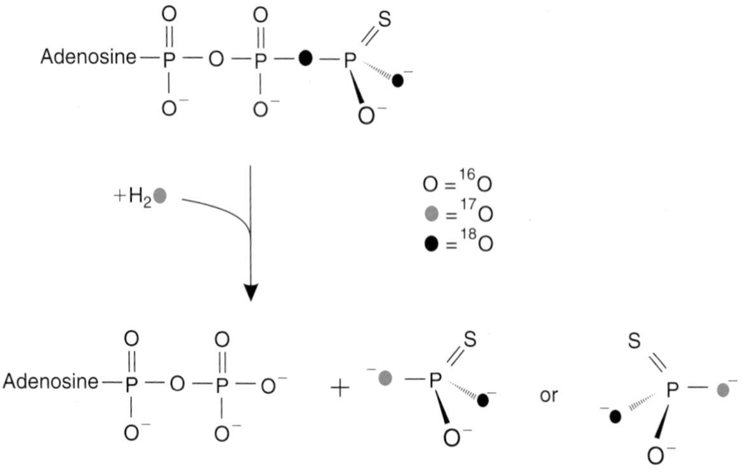

Figure 7.6 Stereochemistry of ATP hydrolysis.
As shown, adenosine 5'-(3-thiotriphosphate), stereospecifically labelled with ^{18}O in the γ-position was hydrolysed in $H_2^{17}O$ to give a chiral $^{16,17,18}O$ P_i product. The configuration of the product showed that hydrolysis had proceeded with inversion at the γ phosphorus atom. The crystal structure of the mitochondrial enzyme shows that the attacking water molecule is polarized by the side chain of β-subunit glutamate 188, and the developing negative charge on a putative pentaco-ordinate transition state stabilized by β-chain lysine 162, together with arginines 189 (β-chain) and 373 (α-chain). A critical difference between the catalytic β- and non-catalytic α-chains is that the latter lack a residue corresponding to the glutamate.

7.6 RELATING THE STRUCTURE TO FUNCTION FOR ATP SYNTHASE

Reviews and further reading Krulwich 1995, Ferguson 2000, Hutcheon *et al.* 2001, Kaim 2001, Yasuda *et al.* 2001, Capaldi and Aggeler 2002

Close inspection of the crystal structure of F_1 allows one to correlate structure with function. The crystals of the enzyme were obtained in the presence of a non-hydrolysable analogue of ATP (AMP-PNP, which has an NH instead of an oxygen atom between the β and γ phosphates). One of the three β-chains has this ligand, and thus is termed β_{TP}; one has ADP bound (β_{DP}) and the other has no bound ligand (β_E), thus correlating with the binding change model (Fig. 7.5). The β-chain folds into three domains. At the N terminus, located furthest from the membrane, is a six-stranded β-barrel, which links to the same structure in adjacent α-subunits so as to contribute to the stability of the ring of alternating α- and β-subunits.

The central domain, comprising a nine-stranded β-sheet and nine associated α-helices, contains the nucleotide binding site. The Walker A motif is found in a P loop that joins one helix to one β-sheet (Fig. 7.3), while the glutamate whose carboxylate reacts with DCCD is suitably positioned to polarize a water molecule for attack on the γ phosphate of ATP in the ATP hydrolysis direction. The third domain, closest to the membrane, comprises a

C-terminal bundle of six α-helices. The adenine nucleotide sites on the β-chains are each close to an interface with an α-subunit. The α-subunits have very similar structural features to the β, but a key difference is that the adenine nucleotide sites, each close to an interface with a β-subunit, are all fully enclosed, consistent with the non-exchangeability of the ligands; the absence of a glutamate or other suitable residue in a crucial catalytic position also suggests that these sites do not participate in catalysis.

The differences in conformations of β-subunit occur in the central and C-terminal domains (Fig. 7.3). Conformational differences in the latter region result in significantly different interactions of the three β-chains with the central γ-subunit. This subunit is organized within the core of the $\alpha_3\beta_3$ structure such that both the N- and C-termini of the polypeptide are oriented towards the top of the $\alpha_3\beta_3$ unit. The C-terminus extends into the β-barrel region made up from all six α- and β-subunits. The N- and C-terminal parts of the chain are coiled around one another. In addition, there is in this region another short stretch of α-helix (18 amino acid residues). There is no three-fold repeating symmetry in the γ-chain and so it is inevitable that different contacts are made between γ and the three β-chains. A probably crucial interaction is that between the 18 residue helix on γ and a short stretch of the β-chain on the β_{TP} subunit, known as the DELSEED sequence (the letters represent the one letter amino acid code), which is conserved between ATP synthases from different sources. This interaction between γ and β_{TP} can be regarded as a 'catch'.

As explained in Section 7.5, the alternating site mechanism envisages that each of the three β-chains adopts at least three conformations sequentially, such that a site with ATP bound is conformationally driven to release ATP and become a loose site (Fig. 7.5). The engagement of the catch with any one β-subunit at a time could be a crucial event in triggering these conformational changes. One way in which this could be achieved is for the γ-chain to move in a single direction sequentially between the three β-chains; this would correspond to a rotation of the γ-chain, but to be effective, something would have to stop the whole $\alpha_3\beta_3\gamma$ assembly moving together with no relative movement of γ from one β to another. This 'something' could be the second stalk, which could act as a stator (Figs 7.1 and 7.2). Another feature of the structure strengthened the possibility of rotation. At the top of the $\alpha_3\beta_3\gamma$ assembly (i.e. furthest from the membrane bilayer), the several β-sheets provide what is in effect a hydrophobic sleeve, which could allow the γ-subunit to rotate within the core of the $\alpha_3\beta_3$ assembly. (It is worth noting at this point that there is experimental evidence from exchanges of radioisotopically labelled subunits for movement of γ relative to $\alpha_3\beta_3$ subunits in a preparation of $E.\ coli$ F_1. Quite how this can occur without a stator to fix $\alpha_3\beta_3$ relative to γ is not clear unless δ alone can contribute to stopping $\alpha_3\beta_3$ moving with γ.)

The combination of the biochemical and structural data led to the design of a sophisticated experiment that demonstrated directly the rotation of the γ-subunit relative to the $\alpha_3\beta_3$ assembly (Fig. 7.7). An enzyme from a bacterial source that could be prepared as an $\alpha_3\beta_3\gamma$ complex was used, having first removed all the naturally occurring cysteine residues and inserted one in the γ-subunit at a position distant from the $\alpha_3\beta_3$ core. In addition, the β-chains had at their N terminus a polyhistidine tag that is usually added to proteins by genetic engineering for purification purposes. The tag enabled the protein to be anchored to a microscope slide that was coated with nickel ions that bind strongly to polyhistidine. The single introduced cysteine was used for attachment of a long actin filament (much bigger than the enzyme!) to which a fluorescent label had been added.

(a)

(b)

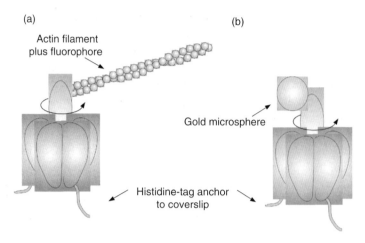

Actin filament
plus fluorophore

Gold microsphere

Histidine-tag anchor
to coverslip

Figure 7.7   Experimental demonstration of the rotation of the ATPase.
The anticlockwise rotation of γ driven by ATP hydrolysis refers to looking down
from the membrane towards the F_1 molecule. In each case (a) and (b), the enzyme
was immobilized on a microscope slide, coated with Ni, via polyhistidine tags
attached to the N terminii of the three β-chains by recombinant DNA methodology.
In (a) a fluorescently labelled actin molecule actin filament (1–4 μm in length
and therefore out of scale in the diagram) carrrying a fluorophore was attached
via a streptavidin linker, which was bound via a covalent bond to the thiol of a
cysteine introduced at the end of the γ-chain most distant from the $\alpha_3\beta_3$ assembly. In (b) a gold bead, diameter 40 nm and thus longer than the α and β chains,
was linked via bovine serum albumin and streptavidin via the same engineered
cysteine of the γ-chain as in (a). The gold bead, being smaller than the actin filament, permitted higher speed rotation, owing to less viscous drag. A maximum
rotation of 130 revolutions per second would correlate with an ATP hydrolysis
rate of 390 per second, similar to k_{cat} for the enzyme. The use of the gold bead
allowed the steps in the rotation to be split into 90° and 30° substeps. Videos of
the rotation can be seen at www.res.titech.ac.jp/seibutu/main_html

Having assembled this modified ATPase, it was viewed under a fluorescent microscope
and ATP plus magnesium added. The actin filament was observed to rotate in one direction
only with a 120° stepping action (Fig. 7.7). The rotation could be correlated with the exchange
in position of γ relative to the three different β-subunits as each in turn passed through the
different conformations and hydrolysed a molecule of ATP. At saturating concentrations of
ATP, the full speed of rotation is estimated as 130 revolutions per second. In a development
of this approach, the ε-subunit was similarly labelled and also found to rotate. Later studies
using very high speed imaging have shown that a 120° step comprises 90° and 30° substeps, each taking only a fraction of a millisecond. ATP binding may drive the 90° step,
while the 30° substep is probably connected to release of a hydrolysis product.

Whilst strongly supporting a rotary mode of catalysis for the $F_1.F_o$ enzyme, this type of
experiment prompted further experiments, in particular those designed to show whether the
ring of c subunits also rotate. One approach to this question has been to label the complete
$F_1.F_o$ enzyme with actin on a c subunit, whilst anchoring the β-chains as before to a microscope slide. Although such an approach has given evidence for rotation, it is technically
very difficult and thus an independent line of confirmation has been sought. This has been

done by cross-linking the ε-subunit to both γ and c subunits within the *E. coli* enzyme and demonstrating that the enzyme is still capable of ATP synthesis when it is in the cytoplasmic membrane. If it is accepted that γ and ε rotate during ATP synthesis (but note that strictly speaking this has not been experimentally demonstrated), then the failure of the cross-link to the c subunit to stop ATP synthesis means that the c subunit must also rotate in synchrony with γ and ε. The notion that the c ring assembly functions as a co-operative unit had been expected from the stoichiometry (one) of DCCD binding to each c subunit assembly (see Section 7.5.1).

The next issues to address are what causes the c ring-γ assembly to move and what stops the $\alpha_3\beta_3$ assembly moving with it? The answer to the latter question is thought to be that the a, b and OSCP (δ in bacteria) subunits form a stator that holds the $\alpha_3\beta_3$ assembly effectively in place as γ rotates between the three β-subunits. What exactly stops the ab assembly rotating in the bilayer along with the c subunit assembly is not clear. Experimental evidence that c does move relative to a and b has come from experiments in which the cross-links between a or b and c subunits were studied. The formation of such cross-links themselves indicate proximity between the a subunit and at least one b or c subunit.

Exchange of c subunits at the interface of the c ring with the dimer of b subunits in the *E. coli* enzyme has supported rotation of the c ring relative to the b subunit. Proximity of subunit a to one c subunit renders the latter resistant to a certain radiochemical labelling reaction relative to the other c subunits in the same ring in the *E. coli* enzyme. Thus chemical cross-linking of the a and c subunits gives a product with little radioactivity. However, if after the labelling the enzyme is allowed to hydrolyse ATP, subsequent cross-linking gives radiolabelled product, indicating that the c subunit that was initially adjacent to a and thus protected against labelling has exchanged with another c subunit. Although this does not directly demonstrate rotation of the c ring relative to a, the observation is supportive of such a mechanism.

The proposal of a, b plus OSCP (δ) as the stator would also explain why the OSCP protein is not the site of oligomycin binding; the latter is in F_o, some way distant from OSCP (Fig. 7.2). Breakage of the link between b and the $\alpha_3\beta_3$ part of F_1, as a result of loss of OSCP, should allow the $\alpha_3\beta_3$ unit to function and undergo conformational changes independently of oligomycin residing in the F_o part of the enzyme. Similarly, one can explain why blockade of the F_o proton channel by either oligomycin or DCCD drastically attenuates the high affinity binding of ATP to $F_o.F_1$; a necessary conformational change would be blocked if OSCP was interacting, via subunit b, with an inhibited F_o, which therefore could not itself undergo an obligatory reciprocal conformational change that would normally occur upon high-affinity ATP binding tens of angstroms distant in F_1.

Having demonstrated that the cγ assembly rotates, the next point to be addressed is how the rotation can couple the movement of protons through F_o to the synthesis of ATP? Or to put it another way, what is the mechanism whereby movement of protons causes the rotation? Here there is considerable uncertainty. A model that has gained popularity, but which has only limited supporting evidence, envisages that the DCCD-reactive carboxyl group on c subunits is located approximately in the middle of the bilayer and the carboxyl group on one of these subunits can make an interaction with an essential arginine (residue 210 in the *E. coli* enzyme) of the a subunit (Fig. 7.2). A proton arriving from the periplasm via a channel that provides a path only to this depth in the membrane, could bind to the carboxylate group and thus relieve a previous electrostatic interaction between the negative carboxylate side chain and the positively charged arginine. This would cause a rotation of the c subunit assembly such that the fixed arginine side chain would now gain a carboxylate anion partner from another c subunit. The previously protonated carboxyl group of the c subunit would subsequently at some stage in the rotation of the ring come into alignment with a channel leading to the cytoplasm and then discharge its proton. A complete rotation would bring this particular carboxylate back into interaction with the arginine of subunit a; the number of protons translocated per 360° rotation would clearly correspond to the number of c subunits in the ring.

For obvious reasons the above mechanism for linking proton movement to rotation is often termed the two-channel model. A variation on this theme is the one-channel model, which places the carboxyl group of the c subunit much nearer the N-side of the membrane. Thus, once the protonated (or sodium-liganded) carboxylate had moved away from its interaction with the arginine of subunit a, it would be able to discharge its bound cation more or less directly into the N phase.

This one-channel model is supported by experiments on the sodium translocating ATP synthase from *P. modestum* in which radiolabelled Na ions bound in the F_o can be measured. Far fewer are found than predicted by the two-channel model discussed above in which one might reasonably expect several protonated (or sodium-bound) c subunits between receiving a proton from the periplasmic channel and discharging it to the cytoplasm (i.e. at any one time several protons or sodium ions might be expected to be bound in F_o). The ability to measure bound sodium (there is no corresponding approach for protons because tritium [^{3}H] can exchange into many sites, e.g. peptide bonds, in proteins) may prove to be a very advantageous feature of the *P. modestum* enzyme for mechanistic studies on F_o. For example, such studies indicate that, in the absence of a $\Delta\psi$, the F_o component can enter an idling mode and allow sodium ions to exchange across the membrane. This could be achieved by brownian motion of the c ring, allowing key sites alternate access to the two sides of the membrane without complete rotations. The presence of $\Delta\psi$ stopped the exchange reaction, presumably by imposing a single direction of rotation on the c ring and thus unidirectional flow of Na^+ ions through it. Understanding the basis of this change is at the heart of the fundamental problem as to how the Δp (and perhaps, see below, especially $\Delta\psi$) generates torque in the rotor components of F_o.

A rather different model for the rotation of the c subunit rotation derives from NMR studies of the conformation of the isolated c subunit in solution. It has been found that protonation of the key carboxylate group causes the C-terminal α-helix to rotate by 140° around its helix axis. The structure of the c ring in the yeast enzyme structure shows that the C-terminal helices form an outer ring surrounding the N-terminal helices. It is possible that the 140° rotation of this outer helix drives the rotation of the c ring. This mechanism still requires proton access pathways to and from the carboxylate.

One complete rotation of a ring of c subunits is envisaged to cause a complete rotation of γ within the $\alpha_3\beta_3$ unit and thus the formation of three molecules of ATP. An important parameter for understanding the ATP synthase is the H^+/ATP ratio. This is a matter of experimental uncertainty (Chapter 4). Most current models for the action of the c ring assembly suppose that a complete 360° turn will require the passage of a number of protons equal to the number of c subunits. Thus, for the mitochondrial enzyme with 10 c subunits, translocation of 10 protons would generate three ATP molecules, giving an H^+/ATP ratio of 3.3. Thus such a rotational mechanism of ATP synthesis leads to the generally unanticipated consequence that the H^+/ATP ratio for the $F_1.F_o$ enzyme need not be an integer. A value of 3.3 would not be inconsistent with the difficult and indirect approaches that have otherwise been used to assess this important ratio (Chapter 4). Clearly, if we had rigorous independent experimental evidence that the ratio was exactly 3, then the model we are discussing here for c subunit assembly rotation could not be correct, at least in all its details.

A further development is that the size of the c subunit ring appears to vary according to the source of the enzyme. Thus, in thylakoids, it appears to have 14 polypetides from electron and atomic force microscopy studies; the corollary is the H^+/ATP ratio for this enzyme is predicted to be higher, 4.6 approx. There is, indeed, some experimental basis for supposing that this ratio is higher for the thylakoid enzyme (Chapters 4 and 6). Variation of the c subunit ring size is not easy to accept given the high sequence similarity between c rings from different sources but at present the experimental evidence does support the variation. Indeed the c ring size for the enzyme from *P. modestum* is estimated to be 11 by the experimental approach used with the thylakoid enzyme. An interesting aside here is the problem of alkaliphilic bacteria in which the Δp is very small because of the larger pH gradient, alkaline on the P (external) side. An unusually high H^+/ATP, and large number of c subunits, might explain how these organisms function.

The presence of 12 c subunits in $F_1.F_o$ has been intuitively attractive as this would permit a symmetrical relationship relative to the $\alpha_3\beta_3$ assembly. Furthermore, for the reasons outlined above, movement of 12 protons through this assembly might be consistent with an H^+/ATP ratio for the enzyme of 4, a value for which, however, there is experimental support for thylakoids but not for mitochondria. Contrary to intuition, symmetry matching between the c subunit assembly of F_o and the $\alpha_3\beta_3$ assembly might be a disadvantageous feature. The symmetry mismatching that ensues from the c10, c11 or c14 structures may help facilitate rotation by avoiding relative conformations with symmetry that may represent energy minima.

A long-standing tenet of the chemiosmotic mechanism is that the ΔpH and $\Delta\psi$ components of the protonmotive force are equivalent. Recently, it has been argued that this is not true and that a certain minimum $\Delta\psi$ is needed in order to generate the torque for rotation of the $c\varepsilon\gamma$ part of the $F_1.F_o$ molecule. This is one of the reasons why it is now argued that the classic acid bath experiment (Chapter 4) did not demonstrate ATP synthesis driven solely by ΔpH, contrary to the interpretation that has lasted over 30 years! If true, it follows that, in the steady-state, thylakoids cannot function in ATP synthesis with ΔpH as the sole component of Δp. It will be very important for the understanding of the AT synthase to ascertain whether indeed $\Delta\psi$ and ΔpH values of the same magnitude are not mechanistically equivalent for the ATP synthase, or indeed any other consumer of Δp (Fig. 1.8). Irrespective of this point, we can be rather certain that the final understanding of the ATP synthase will need a structure of the entire $F_o.F_1$ complex, probably in several conformations, as well as more biophysical experiments on its properties as a remarkable rotary motor.

Finally, however the rotation in the γ-subunit is driven, the problem remains as to how this rotation drives conformation transitions in the three β-subunits in order to drive the binding change mechanism. The DELSEED region appears from the structure to be important but significant mutation in this region does not have major deleterious effect. However, three residues, a his-gly-gly sequence, close to the P-loop region have distinct dihedral angles in the different β-subunits and may form an important hinge for permitting conformational transitions between the different β-subunits. Consistent with this interpretation, mutation of all three of these residues is very inhibitory.

7.7 NON-THERMODYNAMIC REGULATION OF THE ATP SYNTHASE

Further reading   Perez and Ferguson 1990, Tsunoda *et al.* 2001b

The ATP synthases from various sources are regulated in several ways. The mammalian mitochondrial enzyme has an additional subunit known as the inhibitor protein. The binding of one molecule to a β-subunit is sufficient to block ATP hydrolysis in the absence of a Δp. The inhibition is removed in the presence of a Δp, and thus the inhibitor protein may act to prevent unwanted ATP hydrolysis by mitochondria under conditions such as anoxia when the ATP synthase would otherwise reverse. It is possible that the latter conditions cause the mitochondrial matrix to become more acidic, a condition that is known to promote binding of the inhibitor protein to F_1.

The regulation in bacteria is less well understood. In the case of *P. denitrificans* and many other organisms in which respiration and hence oxidative phosphorylation is obligatory, ATP hydrolysis and an ATP-P_i exchange reaction are both very feeble in the absence (and even presence) of a Δp. There is no evidence that these bacteria possess separate inhibitor proteins. However, for the *E. coli* enzyme there is evidence that the ε-subunit might confer this unidirectional behaviour by acting like a ratchet. It appears to take up two very different conformations. In one of these, the C-terminal domain extends towards the F_1 part and allows the enzyme only to synthesize ATP. An interaction of this domain with one $\alpha\beta$ pair might be broken only by rotation of γ relative to the $\alpha_3\beta_3$ assembly in the direction associated with ATP synthesis; this would be a ratchet-like action.

In thylakoids an essential requirement is to avoid ATP hydrolysis in the dark. In the light, activation of the ATP synthase follows exposure of a disulphide bridge in the γ-subunit and its reduction by a thioredoxin. This bridge is near the interface with F_o and can be modelled into the mitochondrial enzyme structure. The thioredoxin is reduced in turn by ferredoxin and the activity of photosystem I. In the dark, ATP synthase relaxes back into an inactive state because reduced ferredoxin and thioredoxin are no longer formed and the disulfide bridge, formed between cysteines six residues apart in the sequence, regenerates under the more oxidizing conditions. Indeed, it has been possible to induce this thiol sensitivity into an $F_1.F_o$ enzyme from a non-photosynthetic source by incorporating the two cysteine residues by recombinant DNA methodology.

7.8 PROTON TRANSLOCATION BY OTHER ATPases AND PYROPHOSPHATASES

Reviews   Arechaga and Jones 2001 (Drozdowicz and Rae 2001, Gruber *et al.* 2001)

The $F_1.F_o$-ATPase is related to the $V_o.V_1$-ATPase that is a proton pump involved in acidification of the interior of vacuoles and structures such as secretory granules in chromaffin cells. Although the organization of the polypeptides is a little different, and in particular individual c subunits appear to be fused into dimeric units, each forming four transmembrane α-helices, there is little doubt that this enzyme also works by a rotary mechanism. An important point about this class of enzyme is that the H^+/ATP ratio is often predicted to be <2 so that the available ΔG of the ATP hydrolysis can maintain very large pH gradients (up to 5 units) across its membrane. If we follow the argument about linking the rotation of a set of n c subunits to the translocation of n protons by $F_o.F_1$-ATPase, then it is predicted that some V-type ATPases will have a small number of c subunits (e.g. two protons per ATP would require six conventional F_o c subunits or three copies of the V-type dimeric c subunit). It remains to be seen if this prediction will be verified.

Archaebacteria have an $A_o.A_1$-ATP synthase, which from sequence comparisons is effectively a hybrid of the F- and V-type enzymes. Finally, there are proton translocating pyrophosphatases. These are also often found as inwardly directed proton pumps on vacuolar membranes. However, they are also found in the cytoplasmic membranes of some bacteria, e.g. *Rhodospirillum rubrum*, where they may function as pyrophosphate synthases. These enzymes are much simpler than the ATP synthase. A single polypeptide with between 14 and 16 transmembrane helices is all that is required. Thus, in this case, phosphoanhydride bond formation/cleavage is seemingly not associated with a rotary machine. This focuses attention on the intriguing question as to how the rotary type of ATPase evolved.

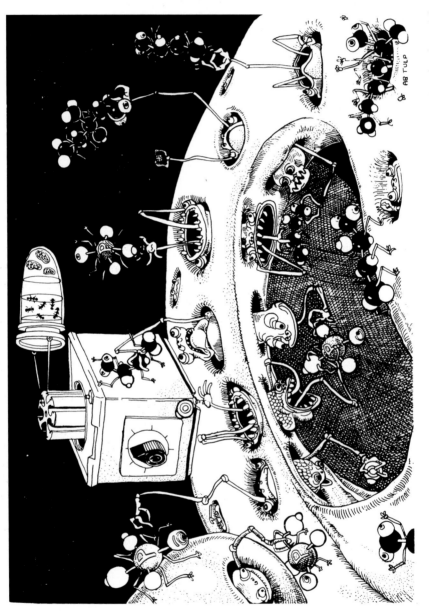

Distinct carriers in the mitochondrial inner membrane exchange phosphate for hydroxyl, phosphate for malate, and malate for citrate (plus a proton)

METABOLITE AND ION TRANSPORT

8.1 INTRODUCTION

Mitochondria and bacteria continuously exchange metabolites and end products with the cell cytoplasm or external environment. At the same time the membranes maintain a high Δp for ATP synthesis. Since most metabolites are charged and/or weak acids, it follows that their distribution will be affected by $\Delta\psi$ or ΔpH (Chapter 3). In practice, transport mechanisms are not only designed to operate under the constraints of a high $\Delta\psi$ and/or ΔpH gradient, but may also exploit these gradients to drive the accumulation of substrates or the expulsion of products across the membrane. Some of the more common strategies (Fig. 8.1) are:

(a) Proton symport with a neutral species leading to an accumulation driven by the full Δp, e.g. the *lac* permease of *E. coli*.
(b) Electroneutral proton symport or hydroxyl antiport leading to an accumulation driven by ΔpH alone, e.g. the mitochondrial P_i^-/OH^- exchanger.
(c) Electrical uniport of a cation, driven by $\Delta\psi$, e.g. mitochondrial Ca^{2+} accumulation.
(d) Electroneutral or electrogenic exchange of two metabolites, e.g. the mitochondrial ATP^{4-}/ADP^{3-} antiporter. This is a common device to allow the entry of a polyanionic species such as citrate^{3-} or malate^{2-}, which would be effectively excluded from the mitochondrial matrix owing to the high negative membrane potential if it attempted to enter the mitochondrion by a uniport mechanism.
(e) Electroneutral antiport of an ion or metabolite with protons, e.g. the mitochondrial Na^+/H^+ antiporter, which expels any excess Na^+ which leaks into the matrix in response to the membrane potential.

Mitochondria and bacteria have very distinctive transport properties and these will be considered separately.

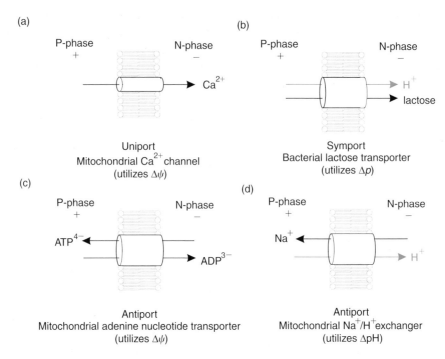

(a)

P-phase
+

N-phase
−

Ca^{2+}

Uniport
Mitochondrial Ca^{2+} channel
(utilizes $\Delta\psi$)

(b)

P-phase
+

N-phase
−

H^+
lactose

Symport
Bacterial lactose transporter
(utilizes Δp)

(c)

P-phase
+

N-phase
−

ATP^{4-}

ADP^{3-}

Antiport
Mitochondrial adenine nucleotide transporter
(utilizes $\Delta\psi$)

(d)

P-phase
+

N-phase
−

Na^+

H^+

Antiport
Mitochondrial Na^+/H^+ exchanger
(utilizes ΔpH)

Figure 8.1   Strategies for ion and metabolite transport across energy-transducing membranes.
The mitochondrial Ca^{2+} uniporter (a) accumulates Ca^{2+} in the matrix in response to $\Delta\psi$; the bacterial lactose transporter (b) utilizes the full Δp to accumulate the sugar. Two mitochondrial antiporters (c,d) differ in that the adenine nucleotide transporter displays charge imbalance and utilizes $\Delta\psi$ to drive ADP uptake and ATP expulsion, while the electroneutral Na^+/H^+ antiporter is driven by ΔpH.

8.2 MITOCHONDRIAL CATION TRANSPORTERS

8.2.1 Mitochondrial monovalent cation transporters

Reviews and further reading   Mitchell and Moyle 1969b, Nicholls *et al.* 1972, Li *et al.* 1992, Kowaltowski *et al.* 2001

The 150 mV membrane potential implies that the slightest leakage of Na^+ or K^+ across the inner mitochondrial membrane would lead to an osmotic explosion of the mitochondrion as the cation accumulates in the matrix down its electrochemical gradient. As originally demonstrated by Mitchell, there has to be a mechanism to counter this. Mitochondria possess a transporter capable of exchanging either K^+ or Na^+ for H^+. Since mitochondria operate with a ΔpH of about -0.5, equivalent to a three-fold gradient of H^+ out/in, this exchange is capable of lowering the matrix K^+ or Na^+ to about 30% of the concentration in the cytoplasm and can control matrix volume. Passive swelling experiments of the type detailed in Chapter 2 show that Na^+/H^+ activity is much higher than K^+/H^+ exchange. A transporter has been partially characterized as an 82 kDa protein.

An electrical monovalent cation uniport showing similar pharmacology to the plasma membrane ATP-sensitive K^+ channel (i.e. inhibition by sulfonyl ureas such as gliben-clamide) has been identified in the inner mitochondrial membrane. Since a K^+ uniport leads to matrix swelling and an electroneutral exchanger to contraction, the two together would provide a means of osmotic regulation of matrix volume. Specifically, lowered cytoplasmic ATP levels would activate the uniport and increase the steady-state volume of the matrix compartment. This in turn could control the activity of matrix enzymes such as NAD^+-linked dehydrogenases, whose ability to transfer NADH to complex I can be viscosity-limited when the matrix is condensed. As in the case of Ca^{2+} transport discussed below, ion cycling comes at an energetic cost and this K^+-cycle is driven by the inward flux of protons through the K^+/H^+ exchanger.

8.2.2 Mitochondrial Ca^{2+} transport

Reviews and further reading   Nicholls and Åkerman 1982, Brand 1985, Rottenberg and Marbach 1990, Gunter *et al.* 1994, 2000, Jung *et al.* 1995

Mitochondria from vertebrate sources accumulate Ca^{2+} by a uniport mechanism when exposed to Ca^{2+} concentrations in excess of 0.5–1 μM. As little as 1 pmol mg^{-1} of the glyco-protein stain ruthenium red can inhibit the uniporter, indicating that its abundance in the mitochondrial inner membrane is minute, consistent with the failure to date to identify or clone the protein. At equilibrium the Ca^{2+} uniporter would develop a concentration gradient across the inner membrane of no less than 10^5 (ten-fold for each 30 mV of $\Delta\psi$, equation 3.40). That this does not occur in practice is due to the presence of an independent efflux pathway, which in most mitochondria operates as a Ca^{2+}/nNa^+ antiport. Its stoichiometry is controversial, being either electroneutral, $Ca^{2+}/2Na^+$ or electrogenic, $Ca^{2+}/3Na^+$.

Na$^+$/Ca^{2+} exchange is found in mitochondria from most tissues, including heart, brain and brown adipose tissue. The Na^+/Ca^{2+} exchanger can be inhibited by the drug CGP-37157 and by tetraphenylphosphonium, TPP^+. This last clearly complicates experiments where TPP^+ is used simultaneously to monitor $\Delta\psi$. Liver mitochondria have a Na^+-independent $Ca^{2+}/2H^+$ antiporter. Either mechanism drives the efflux of Ca^{2+} from the matrix, the Na^+-coupled pathway requiring the additional intervention of the Na^+/H^+ exchanger introduced above (Fig. 8.2a). The existence of the independent efflux pathway can be demonstrated most simply by the selective inhibition of the uptake pathway by ruthenium red once steady-state conditions have been obtained. As the inhibitor does not affect the efflux pathway, a net efflux of Ca^{2+} from the matrix occurs. The uniporter may also be inhibited by other multivalent cations, such as Mg^{2+} and lanthanides.

The apparently symmetrical cycling of Ca^{2+} across the inner membrane is driven by the proton circuit (Fig. 8.2a) and might therefore be thought of as an energy-dissipating 'uncoupling' process. In practice, the rate of Ca^{2+} cycling is restricted by the relatively low activity of the efflux pathway, such that no more than a few percent of the resting state 4 respiration is used to maintain Ca^{2+} cycling. As will now be discussed, the distinctive dependencies of the uptake and efflux pathways on cytoplasmic and matrix Ca^{2+} concentration, respectively, allow the mitochondria to act as sinks of extramitochondrial Ca^{2+} and also to regulate the activity of matrix Ca^{2+}-activated enzymes.

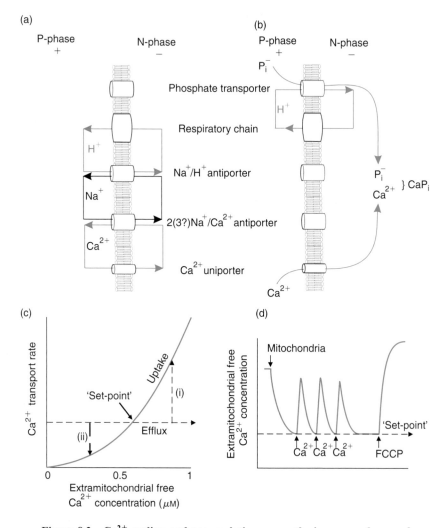

Figure 8.2 Ca²⁺ cycling and accumulation across the inner membrane of heart mitochondria.

(a) Steady-state cycling of Ca^{2+}: a circuit of protons (via respiratory chain and Na^+/H^+ antiporter) drives a sodium circuit (via the Na^+/H^+ antiporter and the $2Na^+/Ca^{2+}$ antiporter), which in turn drives a calcium circuit (via the Na^+/Ca^{2+} antiporter and Ca^{2+} uniporter). (b) Net accumulation of Ca^{2+}: massive amounts of Ca^{2+} can be accumulated in parallel with P_i; the two form an osmotically inactive but rapidly dissociable calcium phosphate 'gel'. (c) The concept of the 'set-point' at which uptake and efflux balance. If the calcium phosphate 'gel' is formed in the matrix, matrix free Ca^{2+}, and hence the activity of the efflux pathway, is independent of total matrix Ca^{2+}. If extramitochondrial Ca^{2+} is above the set-point, the mitochondria will load with Ca^{2+} (i); conversely, mitochondria release Ca^{2+} when the extramitochondrial Ca^{2+} is below this value (ii). Hence mitochondria can act as perfect buffers of extramitochondrial free Ca^{2+} during multiple Ca^{2+} additions (d).

8.2.3 Ca²⁺ accumulation by isolated mitochondria

Reviews and further reading   Zoccarato and Nicholls 1982, Sparagna *et al.* 1995

The ability of mitochondria to accumulate Ca^{2+} from media containing $>0.5\,\mu M$ $[Ca^{2+}]$ is truly spectacular. Under appropriate conditions in excess of 1 μmol of Ca^{2+} per mg mitochondrial protein can be sequestered within the matrix with no deterioration of bioenergetic integrity, equivalent to a concentration of total Ca^{2+} approaching 1 M! Such accumulation requires the presence of P_i, which is taken up in parallel (Fig. 8.2b). Ca^{2+} entry lowers $\Delta\psi$, allowing more protons to be expelled by the respiratory chain. If this were the only process, Ca^{2+} accumulation would soon stop as $\Delta\psi$ is converted to ΔpH (Section 4.2.6). However, in the presence of external P_i, the increasing ΔpH causes P_i to enter the matrix via the phosphate transporter, which effectively exchanges $H_2PO_4^-$ for OH^- (see Section 8.5). Phosphate transport serves two functions in this context: first, it neutralizes the increase in internal pH; and, second, it complexes with the accumulated Ca^{2+} to form a calcium phosphate 'gel', which is not a conventional precipitate, since it instantly dissociates when $\Delta\psi$ is collapsed, allowing Ca^{2+} to be released via the uniporter and P_i via the phosphate transporter. Nevertheless, the 'gel' is osmotically inactive, preventing osmotic swelling as ion accumulation proceeds (Fig. 8.2b).

The activity of the Ca^{2+} uniporter increases as the second power of the free Ca^{2+} in the cytoplasm. When the extramitochondrial Ca^{2+}, $[Ca^{2+}]_e$, is high enough, the conductance of the uniporter is sufficient for the total respiratory capacity of most mammalian mitochondria to be entirely devoted to the accumulation of the cation. A rapid low-capacity Ca^{2+} sequestration pathway in liver mitochondria has been described and termed RAM (rapid accumulation mode); however, the nature of the charge-neutralizing process in this process is unclear.

Accumulation will continue until the mitochondrion succeeds in lowering $[Ca^{2+}]_e$ to a level at which the rate of uptake and efflux balance. When the mitochondrial matrix contains >10 nmol Ca^{2+} per mg protein in the presence of physiological concentrations of P_i, the efflux pathway becomes independent of the matrix Ca^{2+} content, apparently since the free Ca^{2+} in the matrix is essentially buffered by the formation of the calcium phosphate 'gel'. The value of $[Ca^{2+}]_e$ at which this kinetic balance occurs has been termed the 'set-point' and varies from 0.3 to 1 μM depending on incubation conditions (Fig. 8.2c). Isolated mitochondria in the presence of P_i seek to lower $[Ca^{2+}]_e$ to the set-point and thus appear to act as perfect 'buffers' of $[Ca^{2+}]_c$. The mitochondrion in the cell can thus serve as a temporary store of Ca^{2+} under conditions of elevated local Ca^{2+}.

There is considerable uncertainty as to the true range of values for the free matrix Ca^{2+} concentration. Under conditions of minimal Ca^{2+} loading when $[Ca^{2+}]_e$ is below the set-point, the activities of the Ca^{2+}-activated enzymes in the matrix are consistent with a matrix free Ca^{2+} in the range 0.5–$2\,\mu M$, i.e. a negligible concentration gradient exists across the inner membrane (Section 8.2.4). Similar low values are obtained using the fluorescent Ca^{2+}-indicator rhod-2, which by virtue of its positive charge appears to load within the mitochondrial matrices of intact cells when presented as the esterified precursor rhod-2AM. However, the range of this indicator is restricted by virtue of its high affinity for Ca^{2+}. Using matrix-targeted aequorins, which monitor free Ca^{2+} by chemiluminescence (Section 9.3), values from 10 to 100 μM are monitored transiently in cells under conditions of mitochondrial Ca^{2+} loading. Finally, the Ca^{2+}-phosphate complex only forms in the test-tube when the Ca^{2+} concentration is several millimolar. Since mitochondrial dysfunction has been linked to Ca^{2+} overload (Section 8.2.5), resolution of this uncertainty is important. Some consequences of mitochondrial Ca^{2+} transport for the cell will be discussed in Chapter 9.

8.2.4 Mitochondrial Ca^{2+} transport and the regulation of mitochondrial metabolism

Review   Hansford 1994

The activities of three matrix enzymes can be regulated by free matrix Ca^{2+} concentrations in the range 0.1–1 μM:

(a) Pyruvate dehydrogenase phosphatase removes P_i from the inactive, phosphorylated form of the pyruvate dehydrogenase complex, thus allowing the V_{max} of the complex to increase.
(b) The K_m for isocitrate of NAD^+-linked isocitrate dehydrogenase is decreased by Ca^{2+}, allowing a given flux through the citric acid cycle to be achieved at a decreased substrate concentration.
(c) The affinity for substrate of 2-oxoglutarate dehydrogenase (OGDH) is also increased by Ca^{2+} over this concentration range.

There are a number of conditions where a hormonal response requiring an increased ATP demand results in an increase cytoplasmic free Ca^{2+}, which is relayed, via steady state cycling (Fig. 8.2a), into an increased matrix free Ca^{2+} (Fig. 8.2). This in turn will increase citric acid cycle activity and thus NAD^+ reduction to minimize the drop in Δp, which would otherwise accompany the increased ATP turnover. Broadly speaking, this aspect of mitochondrial Ca^{2+} transport is more significant in non-excitable cells whose cytoplasmic Ca^{2+} does not undergo large excursions, whereas Ca^{2+} buffering may be more relevant in excitable cells, particularly neurons, where ion channels can induce dramatic acute elevations in cytoplasmic Ca^{2+} (Section 9.3).

8.2.5 The mitochondrial permeability transition

Reviews   Crompton *et al.* 1988, Bernardi *et al.* 1998, Halestrap *et al.* 1998, Crompton 1999

The ability of mitochondria to accumulate Ca^{2+} (Section 8.2.3) is large but not infinite. When the limit is exceeded, there is a dramatic collapse of Δp, release of the accumulated Ca^{2+}, swelling of the matrix (revealed by a decrease in light-scattering, Section 2.5) and rupture of the outer membrane. This is the mitochondrial permeability transition (MPT). By determining the minimum molecular weight of solutes that continue to provide osmotic support, it was possible to determine that a pore had appeared in the inner membrane, which was non-selectively permeable to solutes up to about 1.5 kDa. Naturally the presence of such a pore is not compatible with the retention of Δp. The first suggestion that the MPT might be more than a non-specific membrane rupture came with the discovery that the onset of the MPT could be prevented by the immunosuppressive drug cyclosporin A (CsA), which is, incidentally, an inhibitor of the Ca^{2+}-dependent phosphatase calcineurin. In the present context, CsA binds to a protein termed cyclophilin D (CyP-D) localized in the mitochondrial matrix. CyP-D has the ability to catalyse the isomerization of proline residues in proteins and can thus have a profound effect on protein structure.

A number of observations implicate the adenine nucleotide translocator (ANT; Section 8.5) in the MPT. First, the Ca^{2+} loading required to induce the transition is decreased by atractylate, which stabilizes the 'c' (cytoplasmic) conformation of the translocator but increased by the 'm' (matrix) conformation-stabilizing bongkrekate. Secondly, the MPT is most readily observed with isolated mitochondria incubated in the absence of adenine nucleotides – indeed, in the presence of physiological concentrations of ADP or ATP, high total Ca^{2+} loads are required to induce the transition. It has been proposed that ANT and CyP-D can form a complex that is prevented by CsA. An affinity column with bound CyP-D fusion protein was found to retain ANT; interestingly, the outer membrane porin VDAC (voltage-dependent anion channel; Section 1.3.1) was also present, and may thus be an additional component of the permeability transition pore at contact sites between the inner and outer membranes.

Oxidative stress is a major factor in MPT induction; thus acetoacetate, which oxidizes NADH in liver mitochondria, or t-butyl hydroperoxide, which oxidizes GSH, each facilitate the transition. However, while the MPT is observed at high matrix Ca^{2+} loads, its induction does not appear to correlate directly with matrix free Ca^{2+}, $[Ca^{2+}]_m$, since it is potentiated by elevated phosphate concentrations, which should lower $[Ca^{2+}]_m$ by formation of the calcium phosphate complex.

The physiological (or pathological) role of the MPT will be discussed in the cellular context in Chapter 9.

8.3 MITOCHONDRIAL METABOLITE TRANSPORTERS

Reviews   Palmieri *et al.* 1996, 2000

The genomic revolution has led to the characterization to date of at least 25 distinct inner mitochondrial membrane transporters. Those that have been cloned show a number of characteristic features: a recurrent amino acid motif repeating about every 100 amino acids, six putative transmembrane α-helices and a molecular weight in the range of 30–40 kDa (300–400 amino acids). Key members of the family are listed in Table 8.1.

The metabolites transported across the inner membrane are predominantly anionic and their transporters are often referred to as *anion transporters*. The range of transporters expressed in the inner mitochondrial membrane varies from tissue to tissue. All mitochondria possess the adenine nucleotide and phosphate transporters, which are responsible for the uptake of ADP + P_i and the release of ATP to the cytoplasm. Virtually all mitochondria oxidize pyruvate and so possess the pyruvate carrier. However, there is a tissue-specific expression of the other carriers listed in Table 8.1, which correlates with the range of metabolic pathways present in the cell. Thus the liver with its plethora of metabolic pathways has mitochondrial transport pathways for most of the citric acid cycle intermediates, for a number of amino acids, and for carnitine and its fatty acyl ester. The more specialized metabolic role played by mitochondria in the heart is reflected in the more restricted variety

Table 8.1   Representative mitochondrial inner membrane transporters. A/B = antiport

Transporter	Mechanism
Phosphate	P_i^-/OH^-
Pyruvate	pyr^-/OH^-
Dicarboxylate	$malate^{2-}/Pi^{2-}$
Tricarboxylate	$citrate^{3-} + H^+/malate^{2-}$
Carnitine/acylcarnitine	$carn^+/acylcarn^+$
Adenine nucleotide	ADP^{3-}/ATP^{4-}
2-Oxoglutarate	$2OG^{2-}/mal^{2-}$
Glutamate	glu^-/OH^-
Uncoupling protein 1 (UCP1)	H^+ or Cl^- uniport

of carriers, while the mitochondria of a highly specialized tissue such as brown fat can only transport the metabolites acyl-carnitine, succinate and pyruvate.

8.3.1 The ammonium swelling technique for the detection of mitochondrial anion carriers

This technique is of historic interest, since it provided the first demonstration of the major metabolite carriers and their functioning as linked systems of antiporters. When non-respiring mitochondria are suspended in an isotonic solution of the ammonium salt of a bilayer-permeant weak acid such as acetic, the mitochondria undergo very rapid osmotic swelling, followed as a decrease in the light scattered as the matrix refractive index decreases to that of the medium (Section 2.5). Swelling can occur because NH_4^+ crosses as the neutral NH_3 and acetate$^-$ crosses as the protonated acetic acid, both passive diffusions leading to the NH_4^+ and CH_3COO^- concentrations in the matrix becoming equal to those in the suspending medium. There is therefore no charge or pH imbalance during transport (Fig. 8.3) and massive swelling can occur as the matrix expands to regain osmotic equilibrium with the suspending medium.

In the case of ammonium phosphate, swelling is again observed, indicating that the phosphate anion crosses as the neutral species: which can be visualized as a $H^+ : H_2PO_4^-$ symport or the indistinguishable P_i^-/OH^- antiport (Fig. 8.3). However, when mitochondria are suspended in isotonic ammonium malate, no swelling occurs until a low concentration of P_i is added, since malate permeates the inner membrane on a carrier protein in exchange for P_i. Phosphate can thus cycle across the membrane between the malate and phosphate carriers. The situation with a tricarboxylic acid such as citrate is still more complex: low concentrations of both phosphate and malate are required, since the citrate–isocitrate carrier exchanges with malate, which in turn exchanges with P_i (Fig. 8.3).

Careful examination of the stoichiometries in Fig. 8.3 reveals that the net accumulation of $H_2PO_4^-$ is accompanied by the entry of one proton; that malate uptake is accompanied by two protons and that three protons enter with each citrate. In thermodynamic terms this means that the phosphate gradient in/out can in theory equal the proton concentration gradient out/in (membrane potential has no effect), that the malate gradient can equal the square of the proton concentration gradient and that citrate or isocitrate can be accumulated

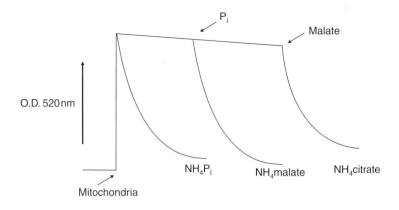

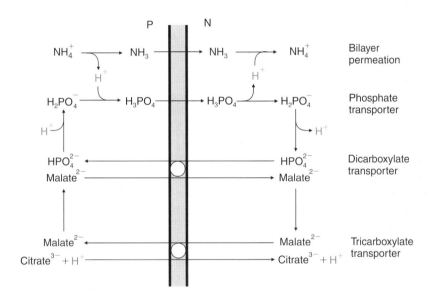

Figure 8.3   Ammonium swelling and mitochondrial metabolite carriers.
The swelling technique is described in Section 2.5.

up to the cube of the proton concentration gradient. Conversely, these transporters prevent ΔpH from increasing much above 0.5 units in an intact cell by redistributing the anions to dissipate any increased gradient.

8.3.2 The structure and function of mitochondrial anion transporters

Review   Palmieri *et al.* 2000

With the notable exception of the adenine nucleotide transporter and the brown fat uncoupling protein UCP1 (Sections 4.6.1 and 8.6), the low abundance of the metabolite transporters in mitochondria has hindered their structural characterization. Thus by 2000 the only transporters that had been characterized by classic protein identification and purification, in addition to UCP1 and the three isoforms of the ANT, were the citrate and phosphate transporters. In contrast, the genomic approach has revealed no less than 35 sequences in the yeast *S. cerevisiae* genome that encode putative mitochondrial carriers. Overexpression of these gene products in *E. coli*, followed by purification and reconstitution has helped to establish their function, while examination of sequence homology with the mammalian genome has indicated the occurrence of homologues for many of these putative carriers in mammalian mitochondria.

By 2000, seven yeast mitochondrial transporters and their corresponding *S. cerevisiae* genes had been identified. These included the *succinate–fumarate* carrier, which catalyses the antiport of succinate for fumarate in yeast mitochondria, allowing succinate generated in the cytoplasm from isocitrate by the glyoxylate cycle under some growth conditions to enter the mitochondrion in exchange for fumarate. The yeast *ornithine carrier* catalyses neutral ornithine transport in exchange for a proton, and allows ornithine synthesized in the matrix to be exported to the cytoplasm, driven by Δp, for conversion to arginine and polyamines. The ornithine carrier from rat liver mitochondria also transports lysine, arginine and citrulline by an electroneutral antiport mechanism. The exchange of cytoplasmic ornithine for matrix citrulline is essential for the urea cycle.

Fatty acids are activated to acylCoA on the outer mitochondrial membrane and converted to acylcarnitines via carnitine *N*-acyltransferase I. Acylcarnitines of various chain lengths enter via the *carnitine carrier* in exchange for carnitine liberated by carnitine *N*-acyltransferase II, which regenerates acylCoA for β-oxidation and liberates the carnitine required for the exchanger. The carrier is not specific for the acyl chain length and will transport acetylcarnitine and short- and long-chain acylcarnitines. Yeast expresses a short-chain *acylcarnitine/carnitine antiporter* required for the import of acetylcarnitine generated in peroxisomes. The mammalian transporter has presumably evolved to transport longer chain acylcarnitines with high efficiency.

The *dicarboxylate carrier* allows the net export of citric acid cycle intermediates from the matrix for gluconeogenesis. The carrier catalyses the electroneutral exchange of malate^{2-} or succinate^{2-} for HPO_4^{2-}; however, the low abundance of the dicarboxylate carrier precluded sequence determination from purified protein. Since the yeast and mammalian carriers are too divergent to allow direct sequence comparison, an intermediate comparison with *C. elegans* (a worm) was carried out. The sequence of this carrier was then compared with candidate partial sequences identified in the mouse. Synthetic oligonucleotides based on these sequences were used to amplify rat and human full-length cDNAs. The carrier is most abundant in liver and kidney, consistent with its role in gluconeogenesis, but is also present in heart and brain mitochondria. There is no yeast homologue for the mammalian *2-oxoglutarate transporter*, which exchanges the oxodicarboxylic acid for malate as an electroneutral exchange of dianionic species. Together with the electrogenic *glutamate–aspartate transporter*, the 2-oxoglutarate transporter is a component of the malate–aspartate shuttle (Section 8.4).

The *tricarboxylate carrier* exchanges citrate^{3-} or isocitrate^{3-} for malate^{2-} and is important for the export of the tricarboxylic acids for fatty acid synthesis. The carrier catalyses an electroneutral exchange, since it cotransports a proton together with citrate (Table 8.1). The carrier, which can be inhibited by 1,2,3-benzyltricarboxylate, has been cloned and sequenced.

The *glutamate⁻/OH⁻* exchanger is present in liver and kidney mitochondria and provides an alternative to the glutamate–aspartate carrier for the transport of glutamate. Because pyruvate is a monocarboxylic acid, it could be argued that it would cross bilayer regions without the need for a carrier, following the precedent of acetate. However, cyanohydroxy-cinnamate inhibits pyruvate transport specifically indicating that a carrier protein is involved, and this has been confirmed by the subsequent purification and functional reconstitution of the *pyruvate carrier*.

8.4 THE TRANSFER OF ELECTRONS FROM CYTOPLASMIC NADH TO THE RESPIRATORY CHAIN

NADH, which is produced in the cytoplasm of mammalian cells, for example, by glycolysis, does not have direct access to complex I, whose NADH binding site is located on the inner face of the inner membrane. The inner membrane is impermeable to NADH and two strat-egies are employed to transfer the electrons to the respiratory chain (Fig. 8.4).

The 2-oxoglutarate and glutamate–aspartate carriers coexist in many mitochondria. One function of these carriers is as components of the *malate–aspartate shuttle* (Fig. 8.4), a device allowing the oxidation of cytoplasmic NADH by the respiratory chain. A thermodynamic

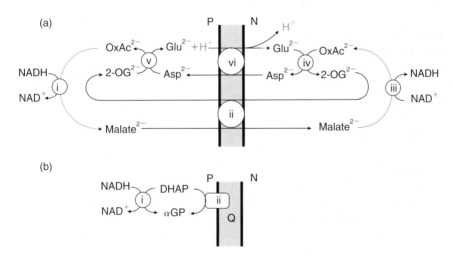

Figure 8.4   The oxidation of cytoplasmic NADH by mitochondria.
(a) The glutamate–aspartate shuttle: (i) cytoplasmic NADH is oxidized by cytoplasmic malate dehydrogenase; (ii) malate enters matrix in exchange for 2-oxoglutarate; (iii) malate is reoxidized in the matrix by malate dehydrogenase, generating matrix NADH; (iv) matrix oxaloacetate transaminates with glutamate to form aspartate and 2-oxoglutarate (which exchanges out of the matrix); (v) 2-oxoglutarate transaminates in the cytoplasm with transported aspartate to regenerate cytoplasmic oxaloacetate and to give cytoplasmic glutamate, which (vi) re-enters the matrix by proton symport in exchange with aspartate. (b) The *s,n*-glycerophosphate shuttle for the oxidation of cytoplasmic NADH: (i) cytoplasmic *s,n*-glycerophosphate dehydrogenase (NAD⁺-linked); and (ii) inner membrane *s,n*-glycerophosphate dehydrogenase (flavoprotein-linked).

problem posed by this process is that the cytoplasmic $NAD^+/NADH$ couple is considerably more oxidized (i.e. less reducing) than the equivalent matrix couple. The thermodynamic impasse for the import of electrons is overcome by the electrical imbalance of the glutamate–aspartate carrier, which exchanges glutamate$^-$ plus a proton for aspartate$^-$ and is, therefore, driven in the direction of glutamate uptake and aspartate expulsion when a $\Delta\psi$ exists across the membrane.

The *s,n-glycerophosphate shuttle* provides an alternative means for the oxidation of cytoplasmic NADH (Fig. 8.4). This makes use of the two *s,n*-glycerophosphate dehydrogenases present in most cells, a cytoplasmic NADH-coupled enzyme reducing dihydroxyacetone phosphate to *s,n*-glycerophosphate and an enzyme on the outer face of the inner mitochondrial membrane reoxidizing this and feeding electrons directly to UQ. In this case the directionality is induced by feeding electrons to the quinone pool at a potential close to zero millivolts (Section 5.7). Yeast mitochondria, which lack the glutamate/aspartate shuttle (Section 8.3.2), oxidize cytoplasmic NADH by an NADH-dehydrogenase on the outer face of the inner membrane as in plant mitochondria (Fig. 5.20).

8.5 THE PHOSPHATE AND ADENINE NUCLEOTIDE TRANSPORTERS

Further reading Heimpel *et al.* 2001

The *phosphate carrier* catalyses the electroneutral transport of $H_2PO_4^-$, either in exchange for OH^- or by symport with a proton, the two being functionally indistinguishable. The carrier is inhibited by mercurial reagents such as *p*-mercuribenzoate and mersalyl, and also by *N*-ethylmaleimide, although none of the inhibitors is completely specific. The phosphate carrier is extremely active. Because of the proton symport, the distribution of P_i across the membrane is influenced by ΔpH. Two isoforms of the phosphate transporter (A and B) exist in mammals, encoded by the same gene but generated by alternative splicing of two exons. Isoform A, found in heart and skeletal muscle, has a higher catalytic activity and affinity for phosphate than the more generally expressed B isoform.

In bacteria and chloroplasts, the ATP synthase produces the ATP in the same compartment in which it is utilized, but mitochondria synthesize ATP in the matrix and then export the nucleotide to the cytoplasm. Two carriers are involved: the phosphate carrier for the uptake of P_i, and the *adenine nucleotide transporter* for the uptake of ADP and export of ATP (Fig. 8.5). The adenine nucleotide carrier catalyses the 1:1 exchange of ADP^{3-} for ATP^{4-} across the inner membrane, but does not transport AMP or adenosine. The total pool size of adenine nucleotides in the matrix (i.e. ATP + ADP + AMP) does not change, as the uptake of a cytoplasmic nucleotide is automatically compensated by the efflux of a nucleotide from the matrix. Thus even if mitochondria are suspended in a nucleotide-free medium (as they are during preparation), no loss of nucleotide normally occurs.

A number of inhibitors are specific for the translocator. Atractyloside, a glucoside isolated from the Mediterranean thistle, *Atractylis gummifera*, is a competitive inhibitor of adenine nucleotide binding and transport. The closely related carboxyatractylate binds more firmly (K_d 10^{-8} M) and cannot be displaced by adenine nucleotides. Bongkrekic acid is produced by *Pseudomonas cocvenenans* and derives its name from its discovery as a

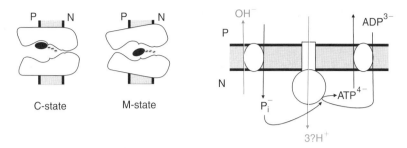

Figure 8.5   The adenine nucleotide translocator (ANT).
The translocator undergoes a conformational change between the C-state, in which the nucleotide binding site faces the cytoplasm and the translocator binds the inhibitor carboxyatractylate, and the M-state, where the binding site is exposed to the matrix and the translocator binds bongkrekic acid. This type of conformational changes is known as a 'rocking bananas' mechanism. ANT is electrogenic, allowing the entry of ADP^{3-} and export of ATP^{4-}. Together with the phosphate transporter, a cycle of ADP + P_i entry and ATP exit utilizes an additional proton. This energy is not lost but allows ΔG_p in the cytoplasm to exceed that in the matrix (Section 3.6.2). Note that P_i^-/OH^- antiport is energetically equivalent to, and experimentally indistinguishable from, $H^+:P_i^-$ symport.

toxin in contaminated samples of the coconut food product bongkrek. It is an uncompetitive inhibitor of the translocator.

The polypeptide of the adenine nucleotide carrier was initially identified by its ability to bind [^{35}S]carboxyatractylate with sufficient affinity to survive solubilization of the membrane with the non-ionic detergent Triton X100 and subsequent purification. A second approach involved the use of photoaffinity analogues of adenine nucleotides, which are competitive inhibitors of transport and bind to the catalytic site of the carrier in the dark. In ultraviolet (UV) light they lose N_2 to form highly reactive nitrene-free radicals, which attach covalently to the nearest amino acids of the polypeptide to which they are bound, i.e. the translocator. The primary sequence of the bovine T1 isoform of the 30 kDa protein was initially determined by peptide sequencing, and revealed a tripartite structure formed from three closely related 100 amino acid repeats, each of which may fold into two transmembrane α-helices. A similar structure has been subsequently deduced for each of the mitochondrial transporters, indicating their membership of a single family. The even number of transmembrane segments means that the N- and C-terminals are on the same side of the inner membrane, and these are found to be on the P-face of the membrane. Three isoforms of the adenine nucleotide translocator have been described in mammals (ANT1, ANT2 and ANT3). They are expressed differently in various tissues, but the significance of this distribution is unclear.

In the model of the translocator proposed by Klingenberg, the carrier would function as a 60 kDa dimer with just one nucleotide binding site. The transport path would follow the central two-fold axis. The dimer would exist in two states differing in the orientation of this single nucleotide binding site (Fig. 8.5), which can either be accessible from the cytosol (C-state) when it can bind carboxyatractylate (one molecule per dimer), or from the matrix (M-state) when bongkrekic acid can bind. As these two inhibitors have completely different structures, it indicates that the properties of the nucleotide binding site must differ in the two orientations. The two inhibitors bind sufficiently tightly for the carrier to be fixed in a

C- or M-state. Binding of adenine nucleotide is less tight because, it is argued, intrinsic binding energy is realized to drive conformational changes in the relatively fluid dimeric protein in order to facilitate transport. These changes do not occur following the much stronger binding of the inhibitors that are not transported.

Note that the model proposes that slight conformational changes make the substrate binding site alternatively accessible from either side of the membrane (Fig. 8.5). Direct experimental evidence for such a conformational change has come from monitoring the accessibility of lysine residues to the hydrophilic reagent pyridoxyl-5-phosphate when added to either the cytoplasmic or matrix face of the membrane: distinct sets of residues are exposed in the M- and C-conformations. This model for the translocator and related transporters has been graphically termed the 'rocking bananas' model. Formation and breakage of salt bridges, particularly involving conserved arginines, between helices are believed to be important in this mechanism. The ability now to express the translocator in *E. coli* will permit analysis of the roles of such positively charged residues by mutagenesis.

8.5.1 Interrelations between the ATP synthase, phosphate carrier and adenine nucleotide carrier

The complete system for mitochondrial ATP synthesis and export requires the ATP synthase, ANT and phosphate carrier. While ANT transports ADP and ATP symmetrically when there is no membrane potential, under normal respiring conditions uptake of ADP and efflux of ATP are preferred, corresponding to the physiological direction of the exchange. The reason for this asymmetry lies with the relative charges on the two nucleotides. ATP is transported as ATP^{4-}; ADP is transported as ADP^{3-}. The resulting charge imbalance means that the equilibrium of the exchange is displaced ten-fold for each 60 mV of membrane potential. The combined effect of the phosphate carrier and adenine nucleotide carrier is to cause the influx of one additional proton per ATP synthesized (Fig. 8.5). Note that, although the additional proton apparently enters with P_i, this is an electroneutral process and the charge of the additional proton is used to drive the exchange of ADP^{3-} for ATP^{4-}.

The thermodynamic consequences of this are considerable. First, at least four protons appear to be used to synthesize a cytoplasmic ATP but one less for a matrix ATP. Thus, up to 25% of the Gibbs energy of the eukaryotic cytosolic ATP/ADP + P_i pool comes, not from the ATP synthase itself, but from the subsequent transport. Thus isolated mitochondria can maintain a ΔG of up to 64 kJ mol^{-1} in contrast to a value of less than 50 kJ mol^{-1} for submitochondrial particles.

8.6 THE UNCOUPLING PROTEIN FAMILY

Reviews and further readings   Enerback *et al.* 1997, Arsenijevic *et al.* 2000, Echtay *et al.* 2000, Bouillaud *et al.* 2001, Pecqueur *et al.* 2001

The prototypic uncoupling protein, UCP1, expressed in brown adipose tissue has the clear thermogenic role of allowing respiration to proceed in the absence of stoichiometric ATP synthesis (Section 4.6.1). Thus mice with a knockout of the *ucp1* gene have difficulty coping with low ambient temperatures. The protein is a 32 kDa polypeptide with six transmembrane segments with both N- and C-termini located in the intermembrane space and a triplicated structure similar to that found in mitochondrial anion transporters. The C-terminal domain shares a nine amino acid sequence with ANT and the DNA binding domain of several transcription factors, and this has been proposed to represent the purine nucleotide binding site of the uncoupling protein. UCP1 has rather low transport activity, which accounts for its presence as a major inner membrane protein in thermogenic brown adipose tissue mitochondria; however, site-directed mutagenesis at any of the three matrix loops converts the protein into a high-conductance non-selective pore, reminiscent of the permeability transition pore (Section 8.2.5).

The mechanism of UCP1 has been the subject of some debate. Models must incorporate the observation that both protons (or hydroxyls) and chloride are transported by the protein in a purine nucleotide-sensitive manner and that proton, but not chloride, conductance is activated by fatty acids. The simplest model is one in which the fatty acids facilitate the delivery of protons to the transport site. Interestingly, the purified, reconstituted UCP1 displays the nucleotide-sensitive but fatty acid-insensitive Cl⁻ conductance characteristic of the *in situ* protein, yet no significant fatty acid-dependent proton conductance. This is restored by the addition of ubiquinone (but hardly by ubiquinol), suggesting that the oxidized coenzyme may facilitate proton delivery to the transporter.

An alternative mechanism proposes that UCP1 acts as a non-selective anion channel allowing the fatty acid anion to be transported out of the mitochondrion, completing a putative dissipative cycle including the entry of the protonated fatty acid, i.e. a classical protonophore mechanism (Section 2.3.5). Such a mechanism can be demonstrated with a range of anion transporters in the presence of high, probably non-physiological, levels of fatty acids. However, UCP1 is extremely sensitive to nanomolar fatty acid concentrations and also retains activity (in the absence of GDP) in the strict absence of fatty acids.

In 1997 two genes with about 70% sequence identity to UCP1 were found, proposed as candidate uncoupling proteins and named UCP2 and UCP3. Later two further genes were identified in brain, but showing substantially less sequence similarity to UCP1. These two latter were termed UCP4 and BMCP1 (brain mitochondrial carrier protein 1, sometimes referred to as UCP5). It is, of course, not sufficient simply to propose a function for a protein on the basis of a sequence homology, and it was necessary to investigate whether these proteins did actually possess proton translocating activity. It must be emphasized that the normal levels of expression of these candidate uncoupling proteins are far lower than UCP1.

UCP2 has a broad tissue distribution with highest concentration in spleen mitochondria, although even here its expression level is only 1% of that for UCP1 in brown fat. UCP2 is induced in a variety of stress conditions. The UCP2 knockout mouse does not display enhanced cold-sensitivity or obesity (which might have been predicted if the protein was a significant contributor to basal metabolic rate). However, the mice show an interestingly increased resistance to infection, which was correlated with an increased generation of reactive oxygen species by macrophages, consistent with the hypothesis that candidate UCPs may be involved in the control of membrane potential, and hence reactive oxygen species generation, in view of the high $\Delta\psi$-dependency of superoxide generation (Chapter 5). Retinoids activate proton conductance of yeast mitochondria expressing UCP2.

UCP3 is primarily expressed in skeletal muscle and exists in long- and short-form transcripts, the latter lacking the sixth transmembrane domain. UCP3 (and UCP2) increase in muscle during starvation, although no change in mitochondrial proton conductance can be detected.

Expression of candidate UCPs in yeast has yielded confusing results, owing to the difficulty of distinguishing a genuine proton conductance from a non-selective increase in membrane proton permeability due to overexpression of a foreign protein disturbing the lipid bilayer. Additionally, a negative result could be a consequence of incorrect insertion or post-translational modification of the protein. An additional complication is the acute control of the protein; thus, even UCP1 is latent until nucleotides are removed or fatty acids added, and it cannot automatically be assumed that the novel UCPs have the same regulatory mechanism as UCP1, namely inhibition of conductance by GDP and activation by fatty acid. However, recently it has been reported that high levels of exogenously generated superoxide reveals a 'classical' GDP-sensitive, fatty acid-activated proton conductance in mitochondria naturally expressing UCP2 (kidney or spleen) or UCP3 (skeletal muscle) but not in mitochondria lacking these isoforms.

UCP4 is expressed in brain. A marginal depolarization was detected in mammalian cells transfected with UCP constructs. Even low levels of BCMP1 (UCP5) expression in yeast lead to a large increase in proton conductance; however, both UCP5 and UCP4 have limited sequence homology to UCP1; indeed, the sequence is as homologous to the oxoglutarate transporter as it is to UCP1.

Finally, it must be borne in mind that liver mitochondria lack detectable levels of UCPs and yet have a typical non-ohmic proton leak (Section 4.6.2). Thus this inherent property of the mitochondrial inner membrane cannot be solely due to the presence of one or more of the candidate UCPs.

8.7 BACTERIAL TRANSPORT

Review   White 2001

Bacteria survive in environments, which are far more variable, and usually more hostile, than anything experienced by a mitochondrion. As a consequence they have developed a variety of mechanisms for the transport of metabolites, such as amino acids and sugars, from the external medium where such molecules occur at very low concentrations to the cytoplasm where the concentrations must be considerably higher to sustain metabolism. The origins of the chemiosmotic theory lay in Mitchell's desire to explain such 'active transport' in bacteria, and there is now extensive evidence that many such transport processes are directly driven by Δp. However, there are also active transport processes that are powered by a Na^+ electrochemical gradient or by the direct hydrolysis of ATP, while a fourth class relies upon phosphoenol–pyruvate as immediate energy source. Finally, there are examples of anion exchange reactions, e.g. hexosephosphate–phosphate antiports. We shall consider each of these in turn.

8.7.1 Proton symport and antiport systems

Reviews and further reading   Marger and Saier 1993, Fu *et al.* 2000, Green *et al.* 2000, Sahin *et al.* 2000, Williams 2000, Kaback *et al.* 2001, Kelly and Thomas 2001, Sahin and Kaback 2001

The first evidence that active transport across a bacterial cytoplasmic membrane could be linked to Δp was obtained for lactose uptake by *E. coli*. This was known to involve the product of the *lac Y* gene, which is traditionally known as the lactose *(lac)* permease and whose expression is inducible. The key experimental observation was that addition of lactose to an anaerobic and lightly buffered suspension of cells, in which *lac Y* expression was induced, resulted in the pH of the external medium showing a small alkaline shift. The pH change is not seen if *lac Y* expression was not induced. These experiments, which were analogous to the oxygen pulse experiment for detecting respiration-driven proton translocation (Chapter 4), thus indicated that lactose entered the cells in symport with one or more protons. Quantification is difficult but all available evidence is that the stoichiometry is $1H^+$/lactose. As lactose is uncharged this means that the $lactose_{in}$/$lactose_{out}$ gradient at equilibrium is maintained by Δp and can attain a value of about 3000 at the usual values of Δp.

The purified *lac* permease protein has been inserted in phospholipid bilayers and shown to catalyse the expected symport. Such experiments demonstrated that the single polypeptide chain was sufficient for the movement of both the H^+ and the lactose. To drive accumulation an artificial Δp (interior alkaline and negative) was generated across the vesicle membrane by transferring a suspension of vesicles held in the presence of valinomycin and high K^+ into a medium at slightly more acidic pH containing radiolabelled lactose and low K^+ (Fig. 8.6). A Δp of the correct polarity would be instaneously established (K^+ efflux generates the $\Delta\psi$) and the uptake of lactose could be assessed by rapid filtration of the vesicles. In principle this experiment would work best if all the lactose/proton symporter molecules were to insert into the vesicles with the same orientation as they have in intact cells. In practice uniform orientation can be very difficult to achieve, but in this case carriers with the opposite orientation will be inert because they are not exposed to the appropriate Δp.

The lactose : proton symporter has probably been more extensively examined, by a whole armoury of biochemical techniques, than any other transporter. This has given some insight into how symport might operate. There is little doubt that the single polypeptide LacY, which has 417 amino acids, functions as a monomer and folds into 12 transmembrane α-helices with only relatively short loops between them. The exact boundaries of these helices and their relative spatial positions cannot be determined with certainty in the absence of a crystal structure but a variety of biochemical approaches suggest that a rough plan of the protein is as shown in Fig. 8.7. Thus helices 4, 5, 7, 8, 9 and 10 are thought to provide a central channel through which transport is argued to occur.

Extensive site-directed mutagenesis studies, and other biochemical approaches, have identified only six amino-acid side chains that are essential for active transport. These include glu 126 on helix 4 and arg 144 on helix 5, which can be deduced to form a salt bridge and therefore to be spatially adjacent; they are believed to play a key role in binding lactose. Four other residues, glu 269 (helix 8), arg 302 (helix 9), together with his 322 and glu 325 (helix 10) have been deduced to be in close proximity, facing inward into a pore between several of the helices as shown in Fig. 8.7, and implicated in proton translocation steps.

Particular emphasis in understanding the mechanism of the lactose transporter has been placed on the properties of the protein when glu 325 has been replaced by certain neutral residues. These are defective in all translocation modes that involve *net* proton movement but catalyse exchange reactions at least as well as the wild-type protein. The scheme shown in Fig. 8.8 allows us to interpret the properties of the protein carrying amino acid alterations at postion 325. Exchange of lactose between the inside and the outside of the cell could occur using steps 2, 3 and 4; these steps involve no uptake or release of protons at either side of the membrane. Thus the continuation of exchange when glu 325 is absent, but

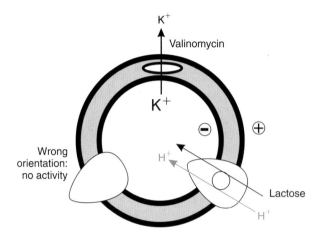

Figure 8.6   Reconstitution of the lactose symporter from *E. coli*.
The lac transporter is purified from cells in which it is overexpressed by selective
solubilization of cytoplasmic membranes with octylglucoside followed by ion-
exchange chromatography. Identification of the permease throughout the original
purification was aided by covalent attachment of a radiolabelled and highly
specific photoaffinity label. Unilamellar vesicles (sometimes called proteolipo-
somes) containing permease polypeptide are prepared by dilution of octylglucoside in the presence of phospholipids, freezing and subsequent sonication. Uptake
of lactose can be studied by first incubating the vesicles in 50 mM potassium
phosphate (pH 7.5), addition of valinomycin and then dilution into sodium phosphate, pH 7.5, containing [^{14}C]lactose. A $\Delta\psi$, negative inside is thus generated
and the resulting uptake of lactose can be determined at various times by rapid
dilution of the vesicles followed by immediate filtration. The radioactivity retained
within the vesicles trapped on the filter can then be determined. Such experiments
showed that the turnover number and K_m for lactose were similar to the values
observed with cytoplasmic membranes, establishing that the purified protein was
in an unperturbed state. In a variation of the experiment, ^{86}Rb$^+$ can be included
in the lumen of the vesicles and its efflux, which is stimulated by the presence of
lactose, monitored in parallel with the uptake of [^{14}C]lactose. Such studies indicate
the movement of 1Rb$^+$ per lactose, consistent, for reasons of charge balance, with
a stoichiometry of 1H$^+$ moving per lactose.

not of any transport mode involving the net movements of protons, can be taken as a clue
that this residue plays a central role in the proton translocation process.

A complete translocation cycle requires that the protein can undergo a conformational
transition (step 6, Fig. 8.8) in the absence of either the bound proton or lactose; the absence
of glu 325 is argued to prevent the crucial deprotonation step 5 (Fig. 8.8). Further information about the current view on the roles of other essential amino acid residues is given in
Fig. 8.7 and its legend. A full understanding of the lactose : proton symporter, or any other
similar transporter will eventually need the combination of information of the type outlined
here with a high resolution structure. Studies on bacteriorhodospin (Chapter 6) suggest that
the scheme outlined here (Fig. 8.8) will prove to be a considerable oversimplification.

In *E. coli* a range of other sugars, including arabinose, xylose and galactose, are transported in symport with protons, while other bacterial species will have comparable systems. Surprisingly, although the proteins from these three transport systems of *E. coli*

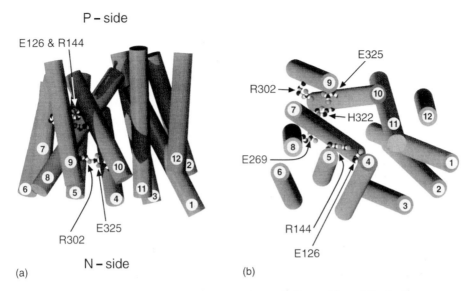

P – side

E126 & R144

E325

R302

H322

E269

E325

R302

N – side

R144

E126

(a)   (b)

Figure 8.7   A model for the transmembrane helix packing of the lactose: proton symporter. (a) Side view. (b) View from cytoplasm.
The 12 α-helix model for the secondary structure (reproduced with permission from Kaback *et al.* (2001) Nature. *Mol. Cell. Biol. Reviews* **2**, 610–620) of the lactose : proton symporter (also known as lac permease) based on the hydropathy profile of the amino acid sequence and supported by circular dichroism studies (indicating high α-helical content), antibody binding, fusions of the *lac Y* gene with the gene (phoA) for a periplasmic alkaline phosphatase, cross-linking and spectroscopic techniques. The latter two approaches in particular provided 100 distance constraints that were used for the model. The positions of six residues that are generally found to be irreplaceable for proper function (i.e. accumulation of lactose against its concentration gradient involving protonation or deprotonation events) are shown: glu 126 (helix 4); arg 144 (helix 5); glu 269 (helix 8); arg 302 (helix 9); his 322 (helix 10); and glu 325 (helix 10). The disaccharide lactose molecule, approximately 9 Å in length, is envisaged as moving, one sugar ring before the second, through a channel provided by many of the helices with glu 126 and arg 144 providing ligands at the initial (exterior-facing or P-side) binding site. Although a considerable body of evidence supports the model shown in the figure, there are other studies on the lactose : proton symporter that suggest that not all six residues shown here are absolutely indispensible for some activity of the protein (e.g. replacement of his 322 by different residues does not always completely block activity). It may be that, under certain conditions, the absence of a centrally important residue can be partially compensated by another residue. It has also been argued that transmembrane helix 2 plays a more important role than would be envisaged on the basis of this model (see e.g. Green *et al.* (2000) for further details of alternative models of the protein).

resemble each other in sequence, they differ from the lactose symporter very significantly, with the exception they too are proposed to form 12 transmembrane helices. The sequences of the arabinose, xylose and galactose symporters have a remarkable similarity to several sugar transport proteins from eukaryotic cells, for example, the glucose transporter from the cytoplasmic membranes of liver cells. The latter is a passive transporter and thus an intriguing challenge now is to identify the features of the bacterial proteins that confer on them

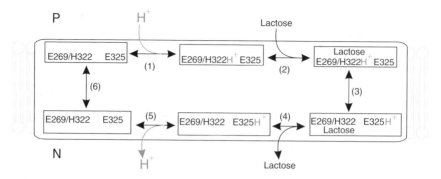

Figure 8.8   A six-step model for lactose : proton symport activity.
Transport can be envisaged as commencing at the outer surface (P-side) of the membrane with the protein having neither substrate, proton or lactose, bound. Proton binding contributes to a charge-pairing between glu 269 and his 322, which is followed by lactose binding to a site oriented towards the P-side of the membrane (step 2). This must be followed by reorientation of this single site (cf. discussion of the adenine nucleotide translocator) such that it now faces the N-side of the membrane. This, step 3, is envisaged to involve the concomitant transfer of the proton from the glu 269/his 322 pair to glu 325. In step 4 the lactose is released to the cytoplasm followed by transfer of the proton to the cytoplasm in step 5. Step 6 is the reorientation of the sugar binding site in the unloaded transport protein. The properties of mutant proteins, in particular variants at residue 325 (see text), suggest that the relative order of lactose and proton binding and release must be as shown; exchange of internal for external lactose is possible by following steps 2, 3 and 4, and then the reverse. Even if further work should show that there is not a compulsory order of proton and lactose binding/release at the two surfaces, a mutant protein that can catalyse exchange but not net transport of lactose must be unable to access step 6. Conversion of this active transporter into a facilated diffusion protein would require that steps 1–6 could be completed with no uptake or release of a proton.

the capacity to transport the sugars in symport with a proton. Very many bacterial transporters possess 12 transmembrane helices and belong to the 'major superfamily of transmembrane facilitators' (MSF).

The components of Δp can, in principle, drive the uptake of any type of nutrient molecule (Fig. 8.1). In the case of a positively charged molecule, for example, lysine at neutral pH, a possible mechanism would be a uniport driven by $\Delta\psi$ and the logarithm of the equilibrium accumulation ratio would be a function of $\Delta\psi$. If a proton were to be cotransported with lysine, then both ΔpH and $\Delta\psi$ would contribute to the driving force and the equilibrium distribution would be a function of $(\Delta pH + 2\Delta\psi)$. A monoanionic species in symport with one proton would be driven only by ΔpH. However, as Δp in most species of bacteria is usually predominantly present as $\Delta\psi$, such a symport is improbable. On the other hand, symport of an anion with $2H^+$ would allow an equilibrium to be attained which would be proportional to $(2\Delta pH + \Delta\psi)$. These should be seen as illustrations because there is no *a priori* reason why other proton stoichiometries should be excluded.

Great efforts have been made to obtain crystals of a proton symporter or antiporter in order to elucidate the coupling mechansim. At the time of writing the most success had been obtained with the sodium/proton antiporter of the *E. coli* membrane, known as NhaA.

Two-dimensional crystals of NhaA have allowed a structure with $7\,\text{Å}$ resolution in the plane and $14\,\text{Å}$ perpendicular to the plane to be obtained. As expected for proton symporters and antiporters of the MSF superfamily in general, there are 12 transmembrane α-helices. However, contrary to expectation from work on the lactose : proton symporter, these are not organized into two groups of six with an approximate two-fold symmetry relationship between them. Rather there is one compact bundle of six helices but the other six helices are arranged in approximately linear fashion. The implication of this organization for ion-coupled transport is not clear at present.

The only high-resolution structure available for a bacterial cytoplasmic membrane transporter is that for the protein that facilitates the passage of glycerol across the *E. coli* membrane. This is *not* an active transporter and does not comprise 12 transmembrane α-helices being a member of the aquaporin family, which has ten such helices. Nevertheless, it gives a first insight into how a hydrophilic molecule can pass through the bilayer.

A variant on the 12 transmembrane helical transporter, as exemplified by the *E. coli* lactose : proton symporter, has been discovered recently because a group of proton symporters have two additional subunits. One is a second subunit that comprises four transmembrane α-helices, while in Gram-negative bacteria the third is a water-soluble periplasmic protein. The latter is often called an extracytoplasmic solute-binding receptor and belongs to a class of proteins more widely found in a class of transporters in which ATP hydrolysis directly powers transport across the bacterial cytoplasmic membrane (see Section 8.7.4). In fact it had been assumed that such receptors were not found in association with proton-driven symporters. The new class of such symporters is known as TRAP (tripartite ATP-independent periplasmic) transporters; one of the best characterized is that for C4-dicarboxylates (e.g. malate) in *R. capsulatus.*

The advantage conferred on a transport system by having the periplasmic receptor protein has been difficult to identify even for the long-established ATP-dependent transporters (Section 8.7.4). In Gram-positive bacteria this type of receptor protein is anchored to the external face (P-side) of the cytoplasmic membrane.

8.7.2 Δp driven transport across the bacterial outer membrane

Reviews and further reading Ferguson *et al.* 1998, Buchanan *et al.* 1999, Koronakis *et al.* 2000, Putman *et al.* 2000, Sharff *et al.* 2001

The outer membranes of Gram-negative bacteria contain large receptor molecules that participate in the active transport of iron–siderophore complexes and vitamin B_{12}. Remarkably, the energy for this active transport is provided by the Δp across the adjacent cytoplasmic membrane. The precise details of how this is done are not yet clear, but it is believed that the outer membrane receptor interacts at its periplasmic side with a cytoplasmic membrane protein complex made up of three proteins known as TonB, ExbB and ExbD, at least one of which presumably undergoes a conformational change in response to protons flowing through it from the periplasm and thus back into the cytoplasm. The crystal structure of one outer membrane receptor, FepA, shows that, at its periplasmic surface, there is a globular domain that folds into the major β-barrel structural feature of the protein and thus blocks access from the interior of the barrel into the periplasm. The globular domain also provides two loops for potential ligand binding and interaction with the TonB–ExbB–ExbD complex. Thus Δp-dependent conformational change in the latter complex could in turn drive a conformational change in the globular domain of FepA and thus permit ligand, e.g. vitamin B_{12}, entry into the periplasm. This would thus achieve the coupling of Δp across the cytoplasmic membrane to a transport process across the outer membrane. Similar principles apply to two other outer membrane transporters of known structure, FhuA and TolC. The latter has a spectacularly long α-helical structure, which spans the periplasm from the outer membrane to the cytoplasmic membrane.

8.7.3 Sodium symport systems

Review   Häse *et al.* 2000

Protons were once thought to be the only cations to move in symport with sugars and amino acids into *E. coli*. It is now known that melibiose and proline transport both occur in symport with Na^+, driven by the Na^+ electrochemical gradient $\Delta\tilde{\mu}_{Na^+}$ (note that the term 'sodium motive force' is not in general use). Unlike *P. modestum* (Section 5.15.7), there is no primary Na^+ pump. Instead $\Delta\tilde{\mu}_{Na^+}$ is maintained by an electroneutral Na^+/H^+ exchanger in the cytoplasmic membrane, which equilibrates the Na^+ and H^+ concentration gradients. $\Delta\psi$ is, of course, a delocalized parameter and so the $\Delta\psi$ generated by proton pumping will be a component of $\Delta\tilde{\mu}_{Na^+}$.

There are now many instances where the role of a Na^+ circuit has been established for various species of bacteria including some pathogens. In some cases, e.g. active transport into alkaliphilic bacteria, this can be rationalized on the basis that Δp is too small to drive active transport. It is not clear why *E. coli* should use both proton and sodium symports, but nor is it appreciated why methane synthesis (Section 5.15.6) seemingly involves both Na^+ and H^+ translocation.

8.7.4 Transport driven directly by ATP hydrolysis

Reviews   Hung *et al.* 1998, Holland and Blight 1999, Putman *et al.* 2000, Chang and Roth 2001, Higgins and Linton 2001, Sharff *et al.* 2001

This mode of transport will be illustrated by reference to two types of system.

The concentration of K^+ in the bacterial cytoplasm is generally much higher than in the surrounding medium; this gradient of K^+ has to be actively maintained, and its mechanism has been studied in depth for *E. coli*. Doubtless related systems operate in other genera. One of the K^+ uptake systems in *E. coli* is known as Kdp and is induced when the external K^+ is very low. Three gene products, KdpA, KdpB and KdpC, have been identified as having characteristics of integral membrane proteins and together they constitute a K^+-dependent ATPase. The KdpA protein is thought to be responsible for the initial binding of K^+ at the periplasmic surface. The site of ATP hydrolysis resides on the KdpB protein; an acyl-phosphate intermediate forms, in contrast to the $F_1.F_0$-ATP synthase, but similar to mammalian $E_1.E_2$ ion pumps such as the Na^+/K^+-ATPase. Indeed, there are significant regions of sequence similarity between the latter and the KdpB protein. It is not known whether the Kdp system catalyses electrogenic import of K^+ or whether K^+ entry might be in exchange for another ion. A second major K^+ transport system in *E.coli is* the constitutive

TrkA system. This has a lower affinity than Kdp and its energetics have not been fully elucidated. It appears to require both ATP hydrolysis and the presence of a protonmotive force for activity but, as the latter drives the synthesis of ATP, it is conceivable that Δp is not directly involved. In principle K^+ uptake could be via a uniport in response to $\Delta\psi$ but this seems not to be a mechanism that has been adopted.

A major group of transport systems in Gram-negative bacteria involve water-soluble proteins in the periplasm that first bind the transport substrate after it enters from the external medium through the porin of the outer membrane. The substrates handled in this way, at least for enteric bacteria such as *E. coli* and *Salmonella typhimurium*, include P_i, SO_4^{2-}, ribose, maltose and histidine. The periplasmic binding proteins have a high affinity for their substrates; their specificity is known from a series of high-resolution X-ray diffraction structures to be conferred by a set of hydrogen bonds, which interact with the substrate. If the contents of the periplasm are released, for example, by exposing the cells to an osmotic shock, transport is greatly inhibited.

Genetic analysis of many of these transport systems has shown that three or four other polypeptides are involved in addition to the periplasmic binding protein. The histidine and maltose uptake systems serve as examples (Fig. 8.9). Two of these additional polypeptides, known as **M** and **Q** for histidine and **F** and **G** for maltose, are membrane-spanning on the basis of their high content of hydrophobic amino acids and appear to recognize a substrate even in the absence of the periplasmic binding protein. Thus a mutant lacking the periplasmic binding protein still transports maltose, albeit with very low affinity.

A third protein **P** (histidine) or **K** (maltose) has a sequence consistent with its attachment via **Q**(**F**) and/or **P**(**G**) to the inner surface of the cytoplasmic membrane. **P** and related polypeptides of other systems have a regions of sequence, including the Gly-X-X-X-X-Gly-Lys- motif seen in $F_1.F_o$-ATP synthases (Section 7.5.1) but in addition a signature motif Leu-Ser-Gly-Gly-Gln, that indicate the presence of an A̲TP-b̲inding motif or c̲assette (hence the proteins are considered members of the ABC family of transporters). The ABC cassette is now known to be part of a nucleotide binding domain (NBD) in which ATP hydrolysis occurs. Several crystal structures have now been obtained for the NBD parts of these ABC transporters, whilst for the MsbA protein from *E. coli* (see legend to Fig. 8.9), a structure at moderate resolution has been obtained for the entire molecule.

The involvement of ATP hydrolysis in the functioning of ABC transporters has been confirmed with a reconstituted histidine transport system and also with a strain of *E. coli* in which ATP hydrolysis with a stoichiometry of about 2ATP/molecule transported could be demonstrated to be associated with transport of maltose. This high stoichiometry (remember that 2ATP are equivalent to about $6H^+$ translocated) could account for the very high accumulation ratios, up to 10^5, that are achieved by these transport systems. Indeed, if a dianion, e.g. sulfate or phosphate at pH values above 7, were to be taken up by a chemiosmotic mechanism, then up to four protons would need to move in symport at a cytoplasmic membrane potential of 180 mV to achieve accumulation ratios of this order.

Thermodynamic considerations neither account for the occurrence of periplasmic binding proteins in the ATP-dependent transport systems nor their absence from proton symport systems. Any advantage to be gained from the avid binding of the substrate would be balanced by the need of the membrane-bound components to abstract the substrate from its binding site. This is an unresolved issue, but a factor to be taken into consideration is that analogous ATP-dependent transport systems have recently been identified in Gram-positive

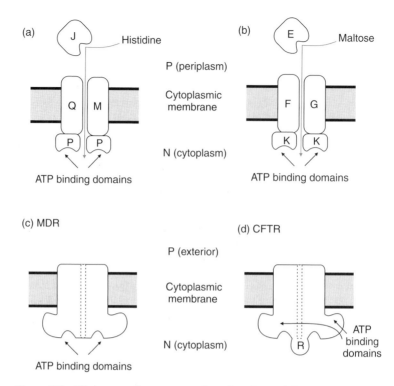

Figure 8.9   Diagrammatic representation of two bacterial transport systems, for histidine in *S. typhimurium* and maltose in *E. coli*, that involve periplasmic binding proteins and comparison with two related proteins from mammalian systems.

Both the histidine (a) and maltose (b) systems involve four polypeptides and it is speculated that the polypeptide responsible for hydrolysis of ATP (P in histidine system and K in maltose transporter) may be present in two copies for each copy of the other subunits. Analysis of other systems, e.g. that for ribose in *E. coli* shows that the subunit equivalent to P or K is larger and contains two ATP binding domains. The multidrug resistance protein (c) and the product of the gene in which a defect gives rise to cystic fibrosis, known as the CFTR protein (d), have comparable ATP-binding domains as the bacterial transport systems but are made up of a single polypeptide, approximately equivalent to fusing two copies of P with one each of Q and M of the histidine system, and lack an equivalent of the periplasmic binding proteins. CFTR also has a fifth domain known as R, which has a regulatory role. A 4.5 Å structure of the homodimeric multidrug resistance transporter homologue MsbA from *E. coli* shows the protein has 12 transmembrane α-helices, which form a chamber that opens out and is surrounded on the N-side of the membrane by the two nucleotide binding domains where the sequence motifs characteristic of an ABC protein are found (Chang and Roth 2001). Higgins and Linton (2001) discuss how this structure might relate to ATP-dependent transport.

organisms that do not have a periplasm. In these instances the equivalent to the periplasmic binding protein appears to be anchored to the cytoplasmic membrane. By analogy it seems likely that in Gram-negative organisms the binding protein functions to deliver the bound substrate to the membrane-spanning components of the system.

Interest in the bacterial ATP- and periplasmic-binding protein-dependent transport systems has been significantly enhanced by the finding that there are considerable similarities with

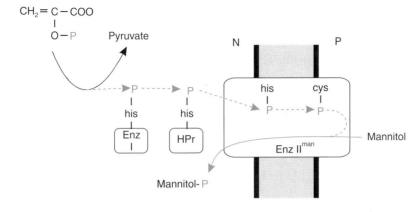

Figure 8.10 Schematic representation of the phosphotransferase system (PTS) for mannitol in *E. coli*.
In this case the mannitol specific enzyme II^{man} accepts phosphate on to a histidine residue from which it moves to a cysteine residue (a thus far rare example of a phosphocysteine catalytic intermediate in biology) before transfer to the incoming mannitol. As mentioned in the text, in other PTS transport systems, e.g. for glucose in *E. coli*, a separate enzyme III catalyses transfer of phosphate from HPr to the enzyme II^{glu}. Sequence analysis of the enzyme II^{man} shows that a cytoplasmic domain is in effect a fused type III enzyme. The N-terminal region contains the mannitol-binding domain. The enzyme II molecules (excluding where appropriate fused enzyme III domains) all have of the order of 650 amino acids, between 350 and 380 of which could fold to form transmembrane α-helices. There may be extensive regions of relatively hydrophilic polypeptide that extend from the bilayer. Such proteins must presumably fold to give a hydrophilic central channel in which phosphorylation occurs.

many mammalian proteins. Prominent amongst these is the multidrug resistance (MDR) protein, which is an ATP-dependent system of low specificity that exports drugs from cells and thus causes problems in certain drug treatments. A second gene product with resemblance to the bacterial ABC transporters is called the CFTR (cystic fibrosis transmembrane regulator) protein; a common mutation leads to a mistargeting of this protein with the consequence that cystic fibrosis develops. The comparison with the bacterial proteins initially suggested to some investigators that CFTR would be part of an ATP-dependent transport system. It is now accepted to be a chloride channel. The ATP binding site is present to permit a regulatory role, analogous to that exerted by GTP and GDP binding to G-proteins. The channel is opened maximally as a result of both phosphorylation of the protein and occupancy of two ABC sites by ATP. Slow hydrolysis of the latter (analogous to GTPase activity of G proteins) promotes closing of the channel. To complete the connection between the bacterial and mammalian systems, there is evidence for analogues of MDR in bacteria, for example, in a *Streptomyces* species that produces antibiotics to which it is resistant.

8.7.5 Transport driven by phosphoryl transfer from phosphoenolpyruvate

Reviews and further reading Postma *et al.* 1993, Rohwer *et al.* 2000

The fourth general class of transport mechanism is the phosphotransferase system (often abbreviated to PTS), which catalyses the transport of several hexose and hexitol sugars, e.g. glucose and mannitol, in many different bacterial genera, including *E. coli* and *Staphylococcus aureus*. A distinctive feature of this system (Fig. 8.10) is that an integral membrane polypeptide, specific for a particular sugar, and generally called enzyme II, binds the sugar from the external surface and phosphorylates it to give, e.g. glucose-6-phosphate, which is released into the cytoplasm. This has often been called a group translocation mechanism because substrates and products have to approach and leave the catalytic site of enzyme II along defined pathways (i.e. glucose should not have access to this site from the cytoplasm).

The ultimate source of the phosphate group for the phosphorylation is cytoplasmic phosphoenolpyruvate (PEP), which first transfers its phosphate to the N-1 nitrogen of a histidine residue in a soluble protein known as H-Pr. This transfer is catalysed by enzyme I, a water-soluble cytoplasmic protein that has no sugar specificity. Subsequently the phosphate group is transferred from H-Pr to a histidine residue on enzyme II, either by the catalytic activity of enzyme II itself or as a result of the activity of yet another enzyme known as enzyme III. It is known from crystallographic studies on soluble proteins that the incorporation of a phosphate group can cause conformational changes. Thus it is conceivable that phosphorylation of enzyme II could lead to a key conformational change that would prevent a sugar bound from the external media from returning there.

Under physiological conditions, the ΔG for PEP hydrolysis is more negative than for ATP hydroysis and is sufficient for phospho transfer to histidine. The driving force for transport is considerable; it has been calculated that at equal concentrations of pyruvate and PEP the equilibrium intracellular concentration of a phosphorylated sugar would reach approximately 100 M at 10^{-6} M external sugar. Such equilibration is apparently not achieved, indicating the requirement for tight control of these transport systems. Another energetic facet is that the bacterial cell expends one PEP molecule for the acquisition of an intracellular phosphorylated sugar. Other active transport processes (e.g. energy-consuming proton symports) are followed by intracellular phosphorylation at the expense of ATP. The conversion of PEP to pyruvate in the pyruvate kinase reaction yields only one molecule of ATP, in a reaction with a large negative Gibbs energy change under cellular conditions, and thus it is energetically advantageous to harness more fully the energy associated with PEP hydrolysis to drive both the transport and phosphorylation events.

8.7.6 Transport driven by anion exchange

There is evidence, especially for the Gram-positive organism *Streptococcus lactis*, that transport can occur by anion exchange systems analogous, at least in principle, to those found in the inner mitochondrial membrane (Section 8.4). The system identified is concerned with the uptake of glucose-6-phosphate (G6P) into the cell. In one mode 2 G6P$^-$ anions move into the cell in exchange for the export of 1 G6P^{2-}. This apparently curious exchange is therefore electroneutral and is equivalent to the net entry of one G6P^{2-} and 2H$^+$, and so is thermodynamically equivalent to a 2H$^+$:G6P^{2-} symport with ΔpH acting as the sole driving force. The exchange system can also operate by moving two H$_2$PO$_4^-$ anions out and one G6P^{2-} in. The exchange is electroneutral and the advantage to the cell is that growth on G6P provides too little carbon and too much phosphorus and therefore extrusion of excess phosphate in exchange for G6P is favourable. As presently understood, these exchange systems do not involve electrogenic movement of molecules across membranes.

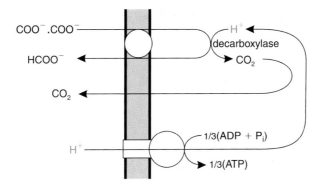

Figure 8.11 **Generation of Δp by anion exchange in *Oxalobacter formigenes*.** The cycle of influx, decarboxylation and efflux effectively constitutes a proton pump with a stoichiometry of $1H^+$ per turnover. If the H^+/ATP ratio for the ATP synthase is three, then the maximum stoichiometry of ATP synthesis would be one for each three oxalate molecules decarboxylated.

8.7.7 Generation of Δp by transport

Reviews and further reading Smith *et al.* 1993, Konings *et al.* 1995, Glass *et al.* 2000

Although transport processes are usually consumers of Δp, some species of bacteria have evolved strategies for coupling the export of end products of metabolism to the generation of Δp. For example, in fermenting bacteria, there is evidence that under some conditions the monovalent lactate anion leaves the cells together (i.e. in symport) with more than one proton, thus generating Δp. The driving force for this is the movement of lactate from a high concentration in the cell to a lower external concentration.

A different type of exchange system that generates Δp has recently been described. In the anaerobe *Oxalobacter formigenes*, oxalate is taken up as a dianion. Once in the cell, it is decarboxylated:

$$(COOH)_2^{2-} + H^+ \rightleftharpoons HCOO^- + CO_2 \qquad [8.1]$$

Formate as a monoanion exits from the cell in exchange for the oxalate, with the overall effect that the exchange is responsible for generating a membrane potential, positive outside (Fig. 8.11). There is also a tendency for a ΔpH to develop, alkaline inside, owing to the consumption of a proton during the decarboxylation reaction. This mechanism requires that the CO_2 leaves the cell in the gaseous form. If it were to be hydrated to HCO_3^- and exported as such from the cell, no $\Delta\psi$ would be generated. In line with this requirement it would be expected that carbonic anhydrase activity is very low or absent in these cells.

The mucosal pathogen *Ureaplasmsa urealyticum* provides a most unusual final example. This organism contains very high levels of urease and thus generates large amounts of ammonium cations in the cytoplasm as a result of the hydrolysis of urea. The cytoplasmic membrane contains transporters that catalyse efflux of ammonium from the cell. This generates $\Delta\psi$, positive outside, which provides a driving force for the $F_1.F_o$-ATP synthase that is known to be present from the complete genome sequence of the organism. This scheme supposes that the uncharged urea molecule enters the cell via facilitated diffusion and that the other product of urease action, carbon dioxide, leaves the cell as the uncharged gas.

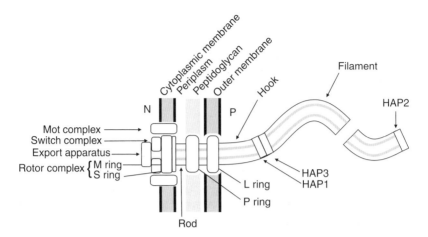

Figure 8.12   Schematic representation of a flagellum from an enteric bacterium.
The Mot complex, probably forming an 8–12 subunit circlet around the motor, is involved in transforming the protonmotive force into mechanical rotation. The MotA protein is a proton channel. Morphological features such as rings, rod, etc. are indicated. The S and M rings probably constitute the rotor; the switch complex controls the rotational direction. The filament is much longer (circa $10\,\mu$m) than shown here. A channel through which the extracellular protein subunits travel to reach their site of assembly is shown in dotted outline. HAP = hook associated protein. The stator and motor components have not been firmly identified.

8.8   TRANSPORT (MOVEMENT) OF BACTERIAL CELLS

Reviews and further reading   Berry and Armitage 1999, Kojima and Blair 2001

Many, but by no means all, species of bacteria are motile, usually as a consequence of the rotation of one or more helical flagella that extend from the surface of the cell (Fig. 8.12). The basal body of a flagellum is embedded in the cytoplasmic membrane, traverses the periplasm, the peptidoglycan and outer membrane, and connects to a filament that extends into the external phase. Movement of the filament, which propels the cell at several body lengths per second, is driven by a 'motor' embedded in the cytoplasmic membrane. The motor rotates at up to 3000 rpm about an axis perpendicular to the plane of the membrane and is usually driven by the protonmotive force, although a Na^+ electrochemical gradient is used in some organisms where sodium circuits are generally important, e.g. *Vibrio* species.

The bacterial flagellum was the first rotary electric motor identified in nature; it has been followed by the ATP synthase (Chapter 7). The molecular mechanism of torque generation by flagella is unknown. One hypothesis envisages that protons are channelled on to the rotor from outside the cell. Such proton movement through the MotA/MotB components of the stator may involve a conserved aspartate in MotB which binds protons (cf. F. a subunit, Chapter 7). (Interestingly, there may be a structural similarity between MotA/MotB and the ExbB/ExbD (Section 8.7.2) that transmits Δp to the outer membrane.) The rotor would then move, either under the effect of some force on the charged protons, or simply as consequence of thermal fluctuations. The protons would then be allowed to pass into the cytoplasm, completing the motor cycle, only after the rotor has moved a certain distance in the correct direction.

This could (compare hypotheses for ATP synthase rotation Chapter 7) involve protonation of (or binding of Na^+ ions to) a rotor component, causing repulsion from positive charges on a stator component. The latter, including the MotA and MotB proteins, is firmly anchored to the peptidoglycan layer in the cell wall. The availability of this rigid structure, which can anchor the stator, contrasts with the problem of identifying the stator in the membranes that contain ATP synthase (Chapter 7). At any given protonmotive force, the higher the number of protons translocated per rotation, the greater will be the torque. This proton stoichiometry is not known with certainty but calculated values are in the range 300–1000 with the efficiency varying from close to 100% at high load to 10% at low loads.

The flagella do not rotate continuously. The rotational movement of the external filament can be interrupted (or briefly stopped) and its direction reversed (i.e. clockwise or counter-clockwise). The direction and duration of rotation is controlled by a sensory system so that the direction of swimming is biased towards nutrients but away from toxic molecules and other disadvantageous environments. The mechanism of this control is beyond the scope of this book.

8.9 TRANSPORT OF MACROMOLECULES ACROSS BACTERIAL MEMBRANES

Reviews   Berks *et al.* 2000, Manting and Driessen 2000

Bacteria synthesize all of their proteins in the cytoplasm but the destination of some of the proteins may be the periplasm, the outer membrane or even the external medium. In these cases the newly synthesized polypeptide must be transported across the cytoplasmic membrane. Such transport appears in general to require firstly the participation of the *sec* gene products, secondly the retention by the polypeptides of a relatively unfolded state for which ATP may be required, and thirdly a leader or signal sequence of approximately 20–25 amino acids at the N-terminus of the polypeptide. The signal sequence is not conserved between proteins, but some features are always found, including the presence of several positively charged residues at the N-terminus and a hydrophobic sequence in the middle.

The connection with the subject matter of this book is that there is good evidence that Δp is involved in driving the translocation of nascent polypeptides across the cytoplasmic membrane. How this is done remains a matter of conjecture; a general electrophoretic effect is improbable because the charge on the leader sequence is of the wrong polarity and both $\Delta\psi$ and ΔpH are competent. It has, however, been suggested that the positive N-terminal end of the leader sequence may be anchored to the cytoplasmic side of the membrane by the $\Delta\psi$; however, it is evident that the membrane potential only acts across the membrane dielectric. Molecules that are confined to one side of the membrane are not affected by the membrane potential. Thus any such anchoring of the positively charged N-terminus of the leader sequence would probably have to be driven indirectly via an effect of membrane potential on a transmembrane protein.

Perhaps a more plausible possibility is that one of the components of the *sec* system, the Sec Y/EC_1 proteins, may act as a proton-polypeptide antiport system. In this context, protein movement within the secretory apparatus could be the basis of the dependence on Δp. In simple terms we could imagine an exaggerated 'breathing' movement of a channel-type protein in the membrane being responsible for squeezing the polypeptide across the membrane. The precedent for the mechanical movement of proteins driven by the proton-motive force is set by the flagella system and the ATP synthase.

It used to be considered axiomatic that proteins were translocated across membranes in an unfolded state. However, it has now been discovered that certain proteins, especially those containing complex redox cofactors, such as the MGD-Mo, are transported to the periplasm as folded proteins with such cofactors already bound. The transport system is known as Tat, twin arginine transporter, in recognition of the presence of two sequential arginine residues in the targeting sequences of proteins that are destined to be transported by this system. This Tat system turns out to be closely related to a longer known system, called the ΔpH transporter, which functions to import some proteins into the lumen of thylakoids. As the name suggests, the latter system is powered by ΔpH, which (recall Chapters 3 and 4), is the dominant component of Δp in thylakoid membranes. By analogy, therefore, and in view of the absence of any evidence for an ATPase activity being associated with the Tat system, it is very likely that the Tat system is driven by the Δp across the cytoplasm membrane. Quite how the movement of protons back into the cell through the Tat complex, or the thylakoid ΔpH transporter, drives movement of a folded polypeptide in the opposite direction is a mystery.

The import of proteins into mitochondria poses related problems and is addressed in Chapter 9.

MITOCHONDRIA
IN THE CELL

9.1 INTRODUCTION

Of the 34 000 publications since *Bioenergetics 2* that can be retrieved with the keyword 'mitochondria', the greatest growth has been in the application of bioenergetic principles to understanding the physiological or pathophysiological role of mitochondria in intact cells. This chapter will focus on four major topics of current interest: (i) the monitoring of mitochondrial membrane bioenergetics and ATP synthesis in intact cells; (ii) mitochondria and cellular Ca^{2+} homeostasis; (iii) mitochondria and oxidative stress; and (iv) mitochondria and the control of apoptotic and necrotic cell death. Each topic relies on the basic bioenergetic principles developed by studies of isolated mitochondria, but is complicated first by the much greater complexity of the cell compared with the isolated mitochondrion and second by the relative inaccessibility of the organelle, which is separated from the observer by a plasma membrane and a cytoplasm (Fig. 9.1).

The bioenergetic behaviour of the mitochondrion in the intact cell is governed by: (i) the supply of substrate from the cytoplasm; (ii) the turnover of ATP by cytoplasmic and plasma membrane energy-requiring processes; (iii) the ionic environment of the cytoplasm, particularly in relation to Ca^{2+}; and (iv) the redox state of the cell. Alterations in these parameters occur in response to changed energy demand, particularly in excitable cells such as those in muscle and brain. The bioenergetics of the *in situ* mitochondria can be influenced by a wide variety of cellular stimuli, including plasma ion channel activation, hormonal signalling and oxidative stress – the generation and/or inefficient detoxification of reactive oxygen and nitrogen species such as superoxide, hydrogen peroxide, hydroxyl ion, nitric oxide or peroxynitrite – with possible pathological consequences. Since all of these parameters are dependent on mitochondrial $\Delta\psi$ (which will be referred to as $\Delta\psi_m$ in this chapter to distinguish it from the plasma membrane potential $\Delta\psi_p$), we shall first consider how to monitor this component. There has been a general tendency to ignore the ΔpH component of Δp on the assumption that the pH gradient across the inner membrane is small in the presence of high concentrations of mitochondrially permeant anions in the cell.

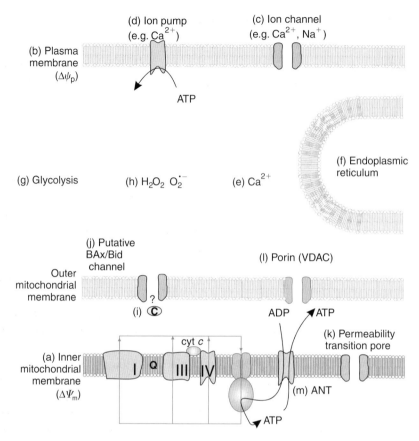

Figure 9.1   Schematic representation of some of the factors influencing mitochondrial function in the intact cell.

(a) The mitochondrial membrane potential ($\Delta\psi_m$) is not the only transmembrane potential in the cell; (b) the plasma membrane potential ($\Delta\psi_p$) is also critical for excitable cell function in controlling the activity of ion channels (c) and must also be taken into account for most estimations of mitochondrial membrane potential. Cytoplasmic Ca^{2+} (e) is a key regulator of cell and mitochondrial function, and is regulated not only by uptake into the mitochondrion, but also by the balance between entry into the cell and extrusion by ion pumps (d), and transport into and out of endoplasmic reticulum (f). Cytoplasmic metabolism, e.g. glycolysis (g), must be taken into account for substrate supply to the mitochondrion. (h) In cell pathology, the generation of reactive oxygen species both by mitochondrial and cytoplasmic pathways plays a major role, as does the release of pro-apoptotic cytochrome c in apoptosis (i) whether via a specific Bax/Bid-associated channel in the outer membrane (j) or through non-selective rupture of the outer membrane by activation of the permeability transition pore (k), which may be associated with the outer membrane porin, VDAC (l) and the adenine nucleotide transporter (m). Techniques are now available to cope with this increased complexity.

9.1.1 Some pitfalls in cellular bioenergetics

The isolated mitochondrion is a relatively simple preparation in that substrates, inhibitors and ionophores can be added to the preparation with confidence as to their site of action. With intact cells great care must be taken to ensure that an agent: (i) has access to the mito-chondrion; (ii) does not affect the mitochondrion in a way that differs from its action on the isolated organelle; and (iii) does not affect other cellular systems.

The plasma membrane of most cells is generally only sufficiently permeable to glucose and pyruvate–lactate to allow these substrates to be used as sole energy source for the cell. Access by substrates such as succinate or adenine nucleotides usually indicates that the plasma membrane has been permeabilized. Not all mitochondrial inhibitors can permeate across the plasma membrane; for example, atractyloside cannot access the ANT in intact cells whereas bongkrekic acid appears to permeate. Protonophores cannot be used in the same way as for isolated mitochondria: firstly, they insert in all membranes, affecting the transplasma membrane ΔpH and short-circuiting other membranes with H^+-translo-cating ATPases (lysosomes, synaptic vesicles, chromaffin granules etc.). Importantly, protonophore addition dramatically depletes cytoplasmic ATP, not only by preventing oxidative phosphorylation, but also by allowing the ATP synthase to reverse and hydrolyse glycolytically generated ATP. The latter can be prevented by using the combination of protonophore and oligomycin, as long as glycolysis is kinetically competent to supply the cell's ATP.

Potassium uniport ionophores such as valinomycin hyperpolarize the plasma membrane by clamping the plasma membrane potential, $\Delta\psi_p$, close to the K^+ diffusion potential (equation 3.41), but at the same time depolarize the mitochondria, since the cytoplasm and matrix each contain $\sim 100\,\text{mM}$ K^+. The K^+/H^+ exchanger nigericin (Chapter 2) on the other hand can be used in cells to decrease mitochondrial ΔpH and hyperpolarize the inner membrane.

Finally, the compartmentation within the cell means that whole-cell measurements of ATP/ADP ratios, glutathione reduction state, etc. do not accurately reflect values in the matrix. However, one approach that is underutilized in whole-cell bioenergetics, and that is accessible as long as cells can be obtained in suspension, is the oxygen electrode, which can obtain more precise information on the bioenergetic status of the *in situ* mitochondria than more 'high-tech' approaches such as fluorescent monitoring of $\Delta\psi_m$.

9.2 MONITORING $\Delta\psi_m$ AND ATP SYNTHESIS IN INTACT CELLS

Reviews   Nicholls and Budd 2000, Nicholls and Ward 2000, Buckman and Reynolds 2001

As discussed in Chapter 4, virtually all techniques for monitoring $\Delta\psi$ across the inner mito-chondrial membrane utilize membrane-permeant cations, which distribute across the membrane according to their Nernst equilibria (equation 3.40). The challenge at the cellular level is how to estimate the concentration gradient between matrix and cytoplasm. Since the probes permeate across the lipid bilayer, rather than through specific transport proteins, they show no membrane selectivity. If allowed to equilibrate with a concentration

$[R^+]_{out}$ in the medium, the concentrations of probe in the cytoplasm and matrix will be respectively:

$$\left[R^+\right]_{cyto} = \left[R^+\right]_{out} \times 10^{\left(\frac{\Delta\psi_p}{60}\right)} \tag{9.1}$$

$$\left[R^+\right]_{matrix} = \left[R^+\right]_{cyto} \times 10^{\left(\frac{\Delta\psi_m}{60}\right)} \tag{9.2}$$

Combining these equations:

$$\left[R^+\right]_{matrix} = \left[R^+\right]_{out} \times 10^{\left(\frac{\Delta\psi_p + \Delta\psi_m}{60}\right)} \tag{9.3}$$

Note that, when equilibrated with a fixed extracellular concentration of probe, the concentration in the matrix will be equally affected by changes in $\Delta\psi_m$ and $\Delta\psi_p$.

Studies with *isolated* mitochondria in suspension often make deliberate use of fluorescent dye aggregation and consequent quenching, which occurs above a critical concentration in the matrix; thus the total fluorescence of the mitochondrial suspension in a cuvette will change as probe is accumulated and quenched in the matrix, allowing changes in matrix accumulation to be monitored continuously without the need to separate mitochondria from the incubation medium (see Section 4.2.3).

Studies with intact cells may be performed under loading conditions, which are either above or below the threshold at which matrix quenching occurs. It is absolutely essential to realize the difference between these alternative experimental designs, since much confusion in the literature has resulted from inappropriate attention to loading conditions. The three techniques most commonly used for these experiments are flow cytometry, confocal imaging at single mitochondrial resolution and non-confocal imaging at single-cell resolution. The two first must be performed in non-quench mode, while with single-cell imaging there is an option of either mode.

9.2.1 Monitoring *in situ* mitochondrial membrane potential in non-quench mode

Reviews   Rottenberg and Wu 1998, Nicholls and Ward 2000

The threshold concentration for aggregation of most fluorescent membrane potential probes is in the range $1–100\ \mu M$ and must be determined empirically. With cells in suspension this is most easily done by decreasing the external probe concentration until addition of protonophore no longer produces an increase in total fluorescence in the cuvette. With cells attached to coverslips and imaged by digital fluorescence microscopy, probe loading can be decreased until protonophore addition no longer produces a transient increase in single-cell fluorescence owing to release of probe into the cytoplasm from the quenched matrix.

Non-quench mode is used for *flow cytometry*, where the fluorescence of single cells is determined as they pass through the exciting beam. Flow cytometry is used when one wishes to compare $\Delta\psi_m$ in two populations of cells, or to determine the heterogeneity of potentials in a population in which, for example, some cells are undergoing apoptosis.

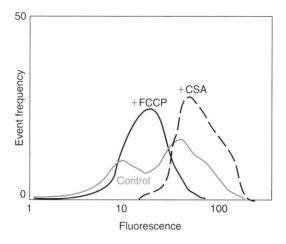

Figure 9.2   Flow cytometry to monitor the range of mitochondrial membrane potentials in lymphocyte populations from 20-month-old mice.
Lymphocytes were equilibrated with 0.6 nM $DiOC_6(3)$ followed by no further treatment, 50 μM of the protonophore CCCP or 3 μM cyclosporin A, an inhibitor or the permeability transition, as indicated. Note that the probe concentration avoids matrix quenching, and that the mitochondria within the lymphocytes show a suboptimal $\Delta\psi_m$, which is enhanced by the calcineurin and PTP inhibitor cyclosporin A (CSA). Data adapted from Mather and Rottenberg (2000).

The fluorescence will be proportional to the total probe accumulated within the cytoplasm and matrix of the cell, and will of course be responsive to changes in either plasma or mitochondrial potential. It is *extremely* important not to exceed the quench threshold at which the mitochondrial fluorescence ceases to be proportional to the accumulated dye. For the commonly used $DiOC_6(3)$, this means loading cells with no more than 1 nM probe. Unfortunately, many studies have been performed at excessive concentrations of this probe (up to 40 nM, when the whole-cell signal is almost totally insensitive to changes in $\Delta\psi_m$, while retaining sensitivity to $\Delta\psi_p$). Figure 9.2 shows an experiment where the probe is employed at the appropriate concentration to monitor the range of potentials in a lymphocyte population.

Non-quench mode is also used when single mitochondria, or clusters of mitochondria, are to be imaged by *confocal microscopy* or high-resolution non-confocal microscopy of cells with thin membrane processes, such as neurons. The ratio of fluorescence intensity between the mitochondria and adjacent mitochondria-free cytoplasm can in theory be put into the Nernst equation to derive $\Delta\psi_m$. In practice, the resolution of the confocal microscope is insufficient to image just the mitochondrial matrix and the actual intensity in the matrix is diluted by inclusion of cristae and blurring of the image. Nevertheless, qualitative real-time changes in $\Delta\psi_m$ in individual mitochondria can be detected by this technique.

9.2.2 Monitoring *in situ* mitochondrial membrane potential in quench mode

Reviews   Nicholls and Ward 2000, Ward *et al.* 2000

As discussed above, the redistribution of a membrane potential probe across the inner membrane of isolated mitochondria in response to changes in $\Delta\psi$ can be monitored in a cuvette without resolving or separating the individual organelles, if the probe is loaded at a concentration sufficient to cause matrix quenching. An analogous approach can be taken with the mitochondria inside a single cell, with a low resolution, *non-confocal digital imager* capable of integrating the total fluorescence from the cell without requiring the ability to resolve the individual mitochondria. The additional complication, however, is that the 'cuvette' is now the cell plasma membrane, which is also permeant to the probe. Two biophysical principles govern the signal obtained under these circumstances. Firstly, the probe is non-selectively permeant across both plasma and mitochondrial inner membranes (since it crosses via lipid bilayer regions) and seeks to achieve a Nernst equilibrium across both membranes (equation 9.3). However, because of the vastly greater surface to volume ratio of the inner membrane/matrix compared with the plasma membrane/cytoplasm, the probe re-equilibrates much more rapidly between matrix and cytoplasm than between cytoplasm and external medium. Secondly, the probe appears to retain its full fluorescence efficiency in the matrix until the quench threshold is attained, but the excess is non-fluorescent. A simple mathematical model incorporating these principles aids interpretation of single-cell signals obtained from cells in quench mode (see Fig. 9.3) and allows changes in plasma and mitochondrial membrane potentials to be resolved, and their approximate magnitudes to be estimated. For the program, see http://www.academicpress.com/bioenergetics/.

The uncertainties in the parameters required to precisely quantify $\Delta\psi_m$ in cells are such that it is virtually impossible to arrive at a value for $\Delta\psi_m$ which is sufficiently reliable to be of use in thermodynamic calculations. However, what can be done with considerable sensitivity, owing to the logarithmic nature of the Nernst equation, is to detect small changes in $\Delta\psi_m$, particularly in quench mode, when a mitochondrial depolarization leads to an increase in whole-cell fluorescence, following release of probe from the quenched matrix to the unquenched cytoplasm. It is easily possible to detect a rapid change of 3–5 mV in $\Delta\psi_m$, for example, as a consequence of altered utilization of the proton circuit. One powerful application of this approach is to assess whether mitochondria, under conditions of stress, retain the ability to generate ATP. An 'oligomycin null-point assay' detects whether mitochondria hyperpolarize on addition of this ATP synthase inhibitor (indicating that they were synthesizing ATP prior to addition of the inhibitor) or whether the mitochondria were maintaining a membrane potential by hydrolysing cytoplasmic (glycolytic) ATP, perhaps as a consequence of respiratory inhibition or a leaky inner membrane, in which case oligomycin would cause enhanced mitochondrial depolarization. This assay has been used to establish the function of the mitochondrial populations within single neurons during the initial stages of cell death induced by activation of glutamate receptors (see Fig. 9.3).

9.2.3 Some complications

Reviews   Bunting 1992, Scaduto and Grotyohann 1999, Duchen 2000

Oxidative stress (Chapter 5) is a major contributor to mitochondrial dysfunction and care must be taken to minimize the artifactual generation of free radicals as a result of photon absorption by the fluorescent probes themselves. Generally, this problem is greatest with confocal microscopy, where an intense laser spot is scanned across the image, but overillumination with non-confocal imaging can also generate sufficient free radicals to compromise mitochondrial function. It is important to employ minimal illumination and maximal detection sensitivity.

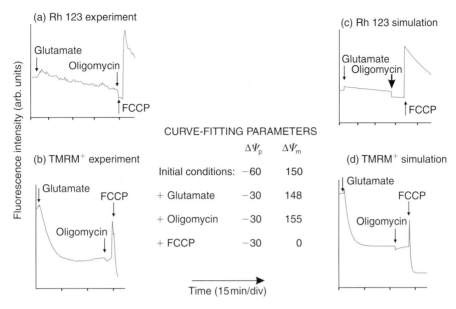

Figure 9.3 Changes in total fluorescence in the cell body of a rat cerebellar granule neuron loaded with the membrane potential probes TMRM⁺ or rhodamine-123 under quench conditions.

Cells were exposed sequentially to glutamate to activate NMDA-selective receptors, oligomycin to inhibit the ATP synthase and FCCP to collapse $\Delta\psi_m$. The actual fluorescence traces recorded from representative cell bodies loaded with TMRM⁺ or rhodamine-123 are shown, respectively, in (a) and (b). The virtual cell simulation (see text) produces acceptable curve fits with the experimental traces (c, d), if the assumption is made that in both experiments glutamate depolarized the plasma membrane by 30 mV and the mitochondrial membranes by 5 mV (due to Ca^{2+} loading), that oligomycin hyperpolarized the mitochondria by 10 mV (state 3–4 transition) and that FCCP collapsed $\Delta\psi_m$. Note that TMRM⁺ equilibrates across the plasma membrane 20 times more rapidly than rhodamine-123 and that this accounts for the slower re-equilibration across the plasma membrane. Data adapted from Ward *et al.* (2000).

A second problem stems from the massive accumulation of the probes themselves within the matrix (equation 9.3). Even a weak inhibition of a matrix enzyme can be greatly amplified by the matrix accumulation. All potentiometric probes are to some extent capable of affecting mitochondrial function, most commonly by restricting respiration. DiOC₆(3) is one of the worst offenders in this context, being a potent inhibitor of complex I unless loaded at subnanomolar concentrations.

9.3 MITOCHONDRIA AND CELLULAR Ca²⁺ HOMEOSTASIS

Reviews   Crompton 1999, Duchen 2000, Rizzuto *et al.* 2000

The cytoplasmic pool of Ca^{2+} is very limited and its free concentration is normally maintained in the region 0.05–0.5 μM in most cells. Three membrane systems bound the cytoplasm (Fig. 9.1), the mitochondrial inner membrane, the endoplasmic reticulum and the plasma membrane. Each is equipped with Ca^{2+} transport pathways and can regulate cytoplasmic free Ca^{2+}, $[Ca^{2+}]_c$.

The plasma membrane possesses both a Ca^{2+}/H^+-ATPase and a $3Na^+/Ca^{2+}$ exchanger. The rapidity and precision with which $[Ca^{2+}]_c$ stabilizes at its resting level in cells suggests that plasma membrane Ca^{2+} transport comprises an outward 'pump' and an inward 'leak' analogous to that at the inner mitochondrial membrane (Section 8.2.3), allowing a dynamic balance between uptake and efflux to be attained, rather than a slow asymptotic approach to thermodynamic equilibrium. The set-point for net mitochondrial Ca^{2+} accumulation (Section 8.2.3) is approximately $0.5\ \mu M$ in most cells. It is important that this is set higher than the ambient resting cytoplasmic free Ca^{2+} concentration, since otherwise the mitochondria would inexorably load with the cation, with the likely activation of the permeability transition (Section 8.2.5).

In contrast, the peak cytoplasmic free Ca^{2+} concentrations, $[Ca^{2+}]_c$, attained, for example, in neurons loading with Ca^{2+} as a result of the activation of voltage- or ligand-activated Ca^{2+} channels, can greatly exceed the set-point, leading to net accumulation of the cation within the matrix. This can be detected: (i) by determining the ^{45}Ca accumulated in the cells after digitonin permeabilization of the plasma membrane; (ii) by detecting a transient 'spike' in $[Ca^{2+}]_c$ on depolarizing the mitochondria with a protonophore; and (iii) by determining an increase in free matrix Ca^{2+} with a matrix-located fluorescent Ca^{2+} indicator or by determining the rate of photon emission by matrix located aequorin (Section 9.3.2). It is important to distinguish total from free matrix Ca^{2+} ($[Ca^{2+}]_m$). There is currently great uncertainty as to the range of $[Ca^{2+}]_m$ experienced by mitochondria $in\ situ$ (Section 9.3.2).

A second physiological condition that results in matrix Ca^{2+} loading is the release of Ca^{2+} from endoplasmic reticulum, which in many cells is in close proximity to mitochondria. There is strong evidence for an amplified uptake of such Ca^{2+} by mitochondria, in excess of that predicted from increases in bulk $[Ca^{2+}]_c$ detected by cytoplasmic Ca^{2+} indicators (Fig. 9.4). This would allow the mitochondrion rapidly to respond to such release by increasing the activity of the citric acid cycle in concert with the cytoplasmic metabolic changes caused by the increase in cytoplasmic free Ca^{2+}.

9.3.1 The measurement of cytoplasmic free Ca^{2+} concentrations

Review   Takahashi $et\ al.$ 1999

Because of the profound effects that ambient free Ca^{2+} concentrations have on mitochondrial bioenergetics, it is important to be able to monitor the cytoplasmic free Ca^{2+}, $[Ca^{2+}]_c$, in studies of mitochondrial function in intact cells. The dominant technique, introduced by Roger Tsien, relies on the altered fluorescent properties of a series of Ca^{2+} chelators when they bind the cation. The most commonly used indicator is fura-2, a fluorescent derivative of the Ca^{2+} chelator BAPTA. Free fura-2 and fura-2 complexed with Ca^{2+} have different excitation spectra, and determining the ratio of emission intensity following excitation at $340\ nm$ and $380\ nm$ allows the free Ca^{2+} to be estimated. Since fura-2 is a pentacarboxylic acid, it is membrane impermeant. Tsien therefore esterified the carboxyl groups to form acetoxymethyl esters, which, being uncharged, can cross the plasma membrane. Most cells possess non-selective esterases in their cytoplasm, and these can hydrolyse the ester to regenerate the free fura-2. The charged species is thus entrapped in the cytoplasm and,

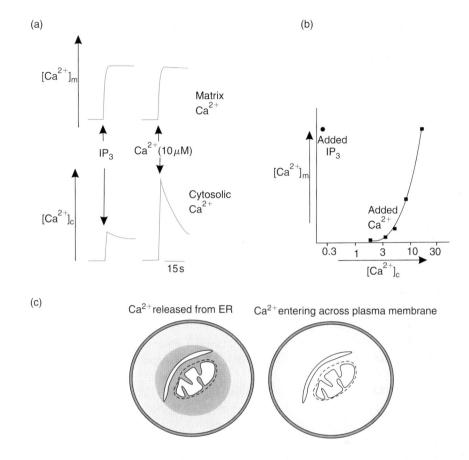

Figure 9.4   Mitochondria preferentially accumulate Ca²⁺ released by endo-plasmic reticulum.
Fura2FF (a low-affinity fluorescent Ca²⁺ probe) was loaded into the matrices of mitochondria in a mast cell line. The cells' plasma membranes were selectively permeabilized with digitonin and a second Ca²⁺ probe rhod-2 used to monitor 'cytosolic' (extramitochondrial) Ca²⁺. (a) Release of ER stores with inositol-triphosphate (IP₃) produces a large [Ca²⁺]ₘ signal but a small [Ca²⁺]꜀ signal. To generate the same matrix signal from the simple addition of Ca²⁺ to the incubation, [Ca²⁺]꜀ must be raised to a much higher value. (b) The different relationship between [Ca²⁺]ₘ and [Ca²⁺]꜀ depending on whether Ca²⁺ is released from ER or added to the incubation. (c) Interpretation: the local cloud of Ca²⁺ released from ER is accumulated by adjacent mitochondria, while more distant plasma membrane Ca²⁺ entry sites cannot generate a high concentration at the inner mitochondrial membrane. Data adapted from Csordás *et al.* (1999).

by simple diffusion of the ester followed by trapping, much of the fura-2AM can be accumulated from the incubation medium.

Fura-2 has a K_d for Ca²⁺ of about 0.3 μM in the cytoplasm and is thus suitable for the measurement of [Ca²⁺]꜀ in the range 30 nM to 3 μM. This encompasses much of the normal range of [Ca²⁺]꜀ encountered in the cell, although for some purposes where very large increases in [Ca²⁺]꜀ occur, lower affinity analogues are required.

9.3.2 The measurement of matrix free Ca^{2+} concentrations

Reviews   Rizzuto *et al.* 1992, Rutter *et al.* 1996, Szalai *et al.* 2000

Studies with isolated mitochondria incubated in the presence of physiological concentrations of phosphate indicate that Ca^{2+} forms a complex with P_i in the matrix when the matrix load exceeds 5–10 nmol per mg protein. Below this value, matrix free Ca^{2+}, $[Ca^{2+}]_m$, varies with Ca^{2+} load. As discussed in Chapter 8, these conditions of varying $[Ca^{2+}]_m$ allow for the control of matrix Ca^{2+}-activated enzymes such as pyruvate dehydrogenase, isocitrate dehydrogenase and 2-oxoglutarate dehydrogenase. Estimated levels of $[Ca^{2+}]_m$ for this regulatory role are in the region of 1 μM. Once the threshold for the formation of the Ca–P_i complex is exceeded, the law of mass action should imply that $[Ca^{2+}]_m$ should be invariant with total Ca^{2+} load, and inversely related to the free matrix P_i concentration. This is reflected in the virtually constant 'set-point' observed with isolated mitochondria loading with increasing concentrations of Ca^{2+} in the presence of P_i. The Ca^{2+} efflux pathways from liver and brain or heart mitochondria respond to changes in $[Ca^{2+}]_m$ and so the set-point at which uptake and efflux balance will increase with $[Ca^{2+}]_m$. This rising set-point is found when mitochondria accumulate Ca-acetate *in vitro*, or when P_i concentrations are decreased. However, in the presence of excess P_i, the set-point remains virtually constant until the onset of the permeability transition.

The increase in $[Ca^{2+}]_m$ can be detected by the Ca^{2+}-dependent photo-protein aequorin targeted to the mitochondrial matrix. Aequorins can be engineered to produce suitable Ca^{2+} binding affinities and in the presence of the cofactor coelenterazine release a single photon upon Ca^{2+} binding, after which they inactivate. The very limited light output thus available precludes the use of conventional imaging and most studies are performed with a photomultiplier detecting the total light emission from a cell population. The peak intensity attained is a function of the third power of the free Ca^{2+}, which makes the technique very sensitive for the detection of high $[Ca^{2+}]_m$ levels. Indeed, peak values in excess of 50 μM have been detected in mitochondria in cells in response to Ca^{2+} release from adjacent endoplasmic reticulum, emphasizing the importance of local gradients of free Ca^{2+} in the cytoplasm (Fig. 9.4).

Fluorescent Ca^{2+} indicators bearing a delocalized positive charge and thus still membrane permeant in the acetoxymethyl (AM)-esterified form will in theory accumulate in the mitochondrial matrix in response to the mitochondrial membrane potential, $\Delta\psi_m$. The major technical problem is whether there is sufficient esterase activity within the matrix to regenerate the Ca^{2+} chelator, thus enabling $[Ca^{2+}]_m$ to be monitored. The most commonly employed probe is rhod-2, although its high affinity for Ca^{2+} means that it is less sensitive to high elevations in $[Ca^{2+}]_m$.

9.4 MITOCHONDRIA AND PROGRAMMED CELL DEATH

Reviews   Liu *et al.* 1996, Mannella 1998, Heiskanen *et al.* 1999, Krajewski *et al.* 1999, Chai *et al.* 2000, Crompton 2000, Desagher and Martinou 2000, Green 2000, Bernardi *et al.* 2001

Cells die by two main pathways: necrosis (Section 9.5), culminating in plasma membrane lysis; and apoptosis or programmed cell death, where an orderly shut-down of metabolism

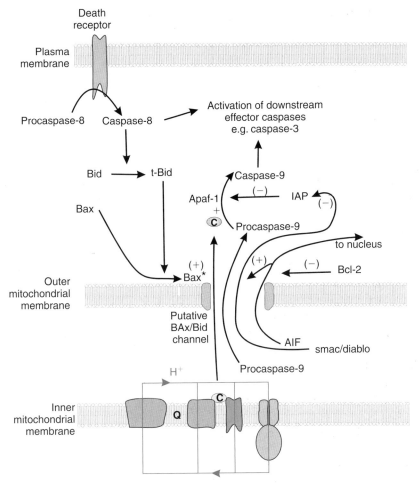

Figure 9.5   Two pathways of programmed cell death.
For details see text.

and digestion of cell contents occurs with minimal leakage of cell contents. Mitochondrial dysfunction is implicated in both pathways. This is very much an ongoing area of research, but we shall review the current status.

A family of proteases, selectively cleaving proteins on the C-side of asp residues, and hence termed caspases, play a key role in the proteolytic digestion of the cell contents during apoptosis. Caspases exist as inactive precursors in non-apoptotic cells and are activated by proteolytic cleavage. A key event in the study of programmed cell death took place in 1996, when it was found that cyt c was required in a cell-free cytosolic extract in order to observe the characteristic activation of caspases and DNA fragmentation. Together with a cytoplasmic factor, Apaf-1, and driven by ATP hydrolysis, cyt c forms an 'apoptosome' capable of aggregating procaspase-9, allowing the proteases monomers to cross-activate each other by proteolysis (Fig. 9.5). The active 'initiator' caspases then proteolytically cleave the procaspases of the 'executioner' caspases, which are responsible for the specific cleavages characteristic of apoptosis (Fig. 9.5). Caspase 3 is the dominant downstream caspase in most cells.

A second pathway for caspase activation is initiated by the binding of ligands to the tumour necrosis factor (TNF) family of so-called 'death receptors'. TNF receptors such as FAS/CD95 aggregate on the plasma membrane after ligand binding, recruiting cytoplasmic proteins and then procaspase-8. As in the case of the apoptosome, this complex allows oligomerization and autoproteolysis of the procaspase, which then proceeds to activate the downstream executioner caspases (Fig. 9.5). In some cells ('type 1'), this pathway is sufficient for apoptosis without an involvement of mitochondria. In 'type 2' cells, however, release of cyt c and other factors from mitochondria is involved in an amplification pathway necessary for efficient cell death.

Bid and Bax are inert in the cytoplasm of healthy cells; however, cleavage of Bid by caspase-8 (Fig. 9.5) generates a 15 kDa C-terminal fragment, t-Bid, that translocates to, and associates with, the outer mitochondrial membrane (Fig. 9.5). t-Bid may facilitate a conformational change in Bax, exposing a hydrophobic C-terminal facilitating its translocation and insertion into the outer membrane. The nature of the outer mitochondrial membrane receptor for these peptides is unknown, although there is evidence implicating the outer membrane porin VDAC, which co-immunoprecipitates with Bax from mitochondrial extracts. VDAC is also believed by some groups to be a component of the permeability transition pore together with ANT and the matrix-located cyclophilin-D (Section 8.2.5).

The least understood step in apoptosis is the release mechanism for cyt c (Fig. 9.5). There are two hypotheses: either Bax, in concert with other recruited proteins, forms or activates a specific pore in the outer membrane allowing for the selective efflux of cyt c from the intermembrane space, or there is a non-selective rupture of the outer membrane as a consequence of osmotic expansion of the matrix and unfolding of the inner membrane. Activation of the permeability transition is usually considered causal in the latter mechanism. Cyt c is loosely associated with the outer face of the inner mitochondrial membrane, where it catalyses the transfer of electrons between complex III and complex IV (Chapter 5). Cyt c appears to be freely mobile within the intermembrane space and so can be released *in vitro* upon rupture of the outer membrane. It must be emphasized that neither hypothesis in their present form appears to satisfactorily explain all observations.

The first protein to be described capable of controlling apoptosis was the 26 kDa anti-apoptotic protein Bcl-2. Together with Bcl-X_L, Bcl-2 stabilizes the mitochondrion against the release of cyt c and antagonizes the action of Bax and its homologues, perhaps by dimerizing with and inactivating the latter.

An early emphasis upon the role of the permeability transition in apoptosis was based in part upon incorrect interpretation of fluorescent membrane potential probe experiments in flow cytometry (Section 9.2.1). Since opening of the permeability transition pore, PTP, is associated with a collapse of $\Delta\psi_m$, it is an effective means of depleting the cell of ATP by ATP synthase reversal, which would be in conflict with the requirement for ATP in the apoptotic sequence. It seems clear that the PTP *can* be activated in cells under specialized conditions and that cyt c release follows as a direct consequence. One of the best investigated examples is the response of a hepatocyte suspension to depletion of mitochondrial reduced glutathione by the pro-oxidant *tert*-butylhydroperoxide, tBuOOH, which is a substrate for glutathione peroxidase and thus oxidized GSH to GSSG (Chapter 5). Hepatocyte mitochondria depolarize synchronously with the release of cyt c, visualized continuously as a cyt c–green fluorescent protein construct. This conjugated cyt c appears to localize like the native cytochrome and the punctate fluorescence of individual mitochondria disappears synchronously with mitochondrial depolarization, monitored in turn by the cationic probe TMRM$^+$.

Considerable caution must be taken when applying agents which are effective inhibitors of the permeability transition in isolated mitochondria to intact cells. Cyclosporin A effectively inhibits the PTP of isolated mitochondria, at least from liver and heart, but its use in intact cells is complicated by the fact that it is a potent inhibitor of the Ca^{2+}-dependent phosphatase, calcineurin, and thus induces a generalized hyperphosphorylated state in the cell. The usual controls are to use FK-506, which is a calcineurin inhibitor without acting on the mitochondrial cyclophilin D.

Accurate monitoring of $\Delta\psi_m$ during apoptosis is one means of determining whether the PTP is causal in the release of cyt c. Unless the PTP is deliberately induced by imposed oxidative stress, as above, most studies report a maintained $\Delta\psi_m$ until late in the apoptotic sequence. If the PTP and outer membrane rupture is not obligatory, how is the 15 kDa cytochrome c transported across the outer membrane? The alternative suggestion is that Bax and related proteins form a selective pore in the outer membrane allowing the efflux of cyt c, but not other proteins. Cyt c has a diameter of about 34 Å and so a putative pore would need to have an appropriate size. VDAC itself has a pore diameter of about 25 Å and so is too small (otherwise cyt c could spontaneously leak from the intermembrane space).

A further complication for the 'pore' mechanism is the increasing number of intermembrane space proteins that have been proposed to be released during apoptosis. In addition to cyt c, a 57 kDa flavoprotein termed apoptosis-inducing factor, AIF (Fig. 9.5), has been implicated in signalling to the nucleus and facilitating cyt c release itself. Furthermore, procaspase-9 has been reported to be located in the intermembrane space in many cells and to be coreleased with the cytochrome. Finally a regulatory protein, smac/diablo has been shown to be released during apoptosis and to neutralize the inhibitory effect of a cytoplasmic inhibitor of apoptosis (IAP) that prevents apoptosome activation of procaspase-9.

No current mechanism satisfactorily explains how these four proteins can be released from mitochondria without an extensive disruption of outer membrane integrity.

9.5 MITOCHONDRIA AND NECROTIC CELL DEATH

Review   Nicholls and Budd 2000

Disruption of the blood supply to the coronary arteries of the heart (heart attack) or to arteries supplying the brain (stroke) for more than 3–4 min results in irreversible damage to the tissue. In the case of the heart, damage occurs following the resumption of the blood supply (reperfusion injury), while in the brain there is rapid death of neurons at the central ischaemic core, while neurons in the 'penumbra' surrounding the core may undergo a delayed death again occurring after termination of the stroke. In both cases there is clear involvement of mitochondrial dysfunction, although the two fields have largely developed independently and will be considered separately here.

9.5.1 Reperfusion injury and the heart

Review   Crompton 1999

Reperfusion of the heart following an ischaemic episode or heart surgery creates conditions similar to those that, in isolated mitochondrial preparations, facilitate the permeability transition, namely matrix Ca^{2+} accumulation, low ATP and ADP, high P_i and oxidative stress. The cellular bioenergetic crisis created by the PTP would lead to rapid death of the cardiomyocyte. The PTP inhibitor cyclosporin-A protects isolated cardiomyocytes from losses of $\Delta\psi_m$ and cell viability caused by *in vitro* anoxia-reoxygenation. Cyclosporin A is a non-selective inhibitor that not only binds to cyclophilin D in the matrix, but binds also to the variety of other cyclophilins in the cell and to calcineurin; thus interpretation of the protection is difficult.

Independent evidence for the operation of the PTP in ischaemia–reperfusion comes from the appropriately named 'hot-dog' experiment, where hearts are perfused with radiolabelled 2-deoxyglucose (DOG), which accumulates in the cytoplasm as DOG-6P but cannot enter the matrix. After ischaemia–reperfusion, however, subsequently isolated mitochondria were found to contain DOG-6P, indicating permeabilization of the inner membrane during the perfusion.

9.5.2 Glutamate excitotoxicity and the brain

Review   Nicholls and Budd 2000

The amino acid glutamate is the major neurotransmitter in the mammalian brain. That a ubiquitous metabolite can function as a specific neurotransmitter is dependent upon precise compartmentation within the cell and maintenance of low extracellular levels preventing chronic receptor activation. This compartmentation is ATP dependent and is disrupted by even brief periods of ischaemia, which may be focal (as in stroke) or global during a heart attack. Glutamate concentrations in the cerebrospinal fluid (CSF) increase from about 1 μM to 100 μM within a few minutes of ischaemia. Neurons in the 'penumbra' surrounding a focal infarct are still oxygenated but their receptors are activated by glutamate leaking from the ischaemic 'core'. Similarly upon reperfusion following global ischaemia, the neurons are bathed in a high glutamate concentration sufficient to activate their glutamate receptors. N-Methyl-D-aspartate (NMDA) receptors are a subclass of glutamate receptor that play an essential role in the acquisition of memory and allow both Na^+ and Ca^{2+} to enter the neuron. Since a Na^+-gradient across the plasma membrane is required to drive the reaccumulation of glutamate into the cells, and since NMDA receptor activation dissipates this gradient, extracellular glutamate can persist resulting in chronic NMDA receptor activation and extensive Ca^{2+} loading of the neuron.

These events can be modelled with isolated neurons in culture. Cytoplasmic free Ca^{2+}, $[Ca^{2+}]_c$, in the cells is typically 50–100 nM, well below the 500 nM 'set-point' at which brain mitochondria become net accumulators of the cation (Section 8.2.3). Thus the neuronal mitochondrial matrix is largely depleted of Ca^{2+}. However, chronic NMDA receptor activation raises $[Ca^{2+}]_c$ above the set-point and net accumulation into the matrix occurs. Brain mitochondria are highly resistant to the permeability transition and this Ca^{2+} loading only lowers $\Delta\psi_m$ by a few millivolts as the cation is accumulated. The mitochondria continue to generate ATP as evidenced by a hyperpolarization upon addition of oligomycin to induce a state 3–state 4 transition. However, after a variable period of glutamate exposure, cytoplasmic Ca^{2+} homeostasis fails and the neurons undergo necrosis. PTP inhibitors have no effect upon this delayed Ca^{2+} deregulation (DCD) and the observed mitochondrial depolarization appears to be an effect, rather than a cause, of the failed cytoplasmic Ca^{2+} homeostasis.

Even a brief exposure of cultured neurons to glutamate can lead to subsequent DCD. The initial matrix Ca^{2+} loading is critical, and neurons whose mitochondria are depolarized by the combination of oligomycin plus a respiratory chain inhibitor prior to glutamate exposure, and are thus maintained by glycolytic ATP are much more resistant to glutamate excitotoxicity. The consensus hypothesis is that matrix Ca^{2+} loading initiates an increased production of reactive oxygen species that leads to cell death, although it has not been possible to reproduce this finding with isolated brain mitochondria.

9.6 THE MITOCHONDRIAL GENOME

Reviews   Scheffler 1999, Bauer *et al.* 2000, Schon 2000

Human mitochondrial DNA contains 37 genes within a 16.6 kilobase circular double-stranded DNA. Twenty-two of these encode transfer RNAs, another two encode ribosomal RNAs, leaving only 13 genes encoding polypeptides. These last are all components of the respiratory chain or ATP synthase (Table 9.1). The respiratory chain complexes contain at least 70 nuclear-encoded peptides synthesized in the cytoplasm and imported into the mitochondrion by specific import pathways (Section 9.7) and these must be assembled together with the mitochondrially encoded subunits to produce the functional complexes. The mitochondrial genome is extraordinarily compact (Fig. 9.6), indeed two pairs of genes overlap by a few bases, while the complementary light chain is also transcribed, coding for one complex I subunit and several tRNAs.

Table 9.1

	Location	
Pearson's syndrome, Kearns–Sayre syndrome, chronic progressive external ophthalmoplegia (CPEO)	'Common deletion' of 4977 base pairs between A8 and ND5	All mt protein synthesis abolished due to lack of tRNAs
MELAS (mitochondrial encephalomyopathy, lactic acidosis and stroke-like episodes)	$tRNA^{leu(UUR)}$	All mt protein synthesis abolished due to lack of tRNAs
MERRF (myoclonus, epilepsy, with ragged-red fibres)	$tRNA^{lys}$	All mt protein synthesis abolished due to lack of tRNAs
Leber's hereditary optic neuropathy (LHON)	6 point mutations in complex I ND genes	Loss of complex I activity
NARP (neuropathy, ataxia and retinitis pigmentosa)	A6	Inhibition of ATP synthase

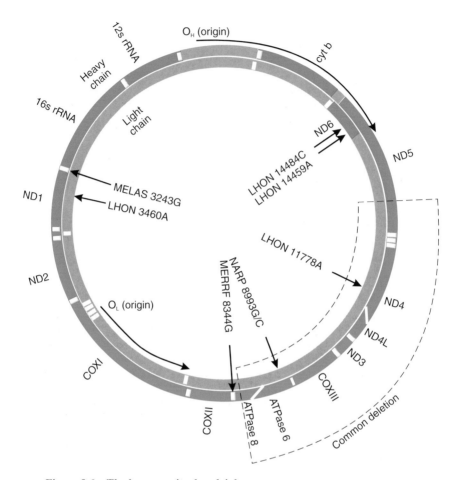

Figure 9.6   The human mitochondrial genome.
The 16 kb human mitochondrial genome encodes six complex I peptides, cyt *b*, three complex IV peptides and two peptides of the ATP synthase F_o. The two strands may be separated on a density gradient into heavy and light chains. Both chains are coding: the light chain for 8 tRNAs (white boxes) and the ND6 subunit of complex I, while the heavy chain encodes the other 14 tRNAs and remaining complex subunits. The genes for ND5 and ND6 and for ATPase 6 and ATPase 8 overlap by a few bases. Sites of pathological point mutations are shown by arrows and the region of the common deletion is outlined.

9.7   IMPORT AND ASSEMBLY OF MITOCHONDRIAL PROTEINS

Reviews   Neupert 1997, Koehler *et al.* 1999, Tokatlidis and Schatz 1999, Bauer *et al.* 2000

Despite the 13 mtDNA-encoded inner membrane proteins, the large majority of mitochondrial proteins are encoded by nDNA and synthesized in the cytoplasm. Imported proteins must contain information allowing them to be targeted to one of four locations: outer membrane, intermembrane space, inner membrane or matrix. Protein targeting and translocation into mitochondria has mostly been investigated with fungal mitochondria from yeast

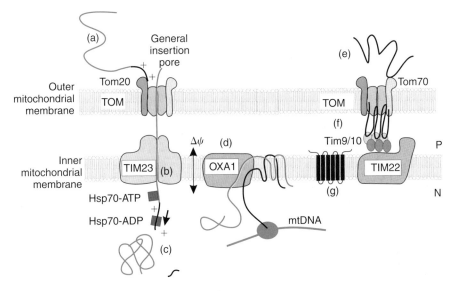

Figure 9.7   Mitochondrial protein import.
Protein precursors destined for the matrix possess positively charged leader sequences that are recognized by the Tom20 receptor of the TOM complex (a). (b) The peptide threads through the inner membrane TIM23 complex driven by $\Delta\psi$ and pulled by a ratchet mechanism involving reversible attachment of Hsp70. (c) The imported protein folds into its functional conformation in the matrix. (d) Mitochondrially synthesized peptides and nuclear peptides that co-assemble with them are inserted by the OXA1 complex. (e) Mitochondrial carrier proteins destined for the inner membrane lack a leader sequence and are recognized by signature sequences located at the future matrix loops. (f) A tripartite structure is stabilized by interaction with the intermembrane-space proteins Tim9 and 10, when the TIM22 complex catalyses insertion of the carrier (g).

or *Neurospora*. The following is a necessarily simplified description; for a full discussion of the multiple peptides involved in protein import, reference should be made to the above reviews.

All proteins destined for insertion in mitochondria must recognize the mitochondrion and then traverse the outer membrane. The outer membrane possesses a set of integral membrane proteins that comprise the *TOM* complex (Fig. 9.7). The TOM complex, which is used by all mitochondrial proteins regardless of final destination, comprises three exposed preprotein receptors and a large protein-conducting channel, termed the general insertion pore. Preproteins held in the correct conformation by cytoplasmic chaperonins are delivered to one of the TOM receptors and thread through the general insertion pore. Preproteins destined for the matrix carry a positively charged N-terminal targeting sequence; for their import, TOM interacts with an inner membrane protein translocase complex *TIM23*. A different inner membrane complex, *TIM22*, is required by many integral inner membrane proteins, including carriers such as ANT and the phosphate transporter. These proteins are not synthesized with cleavable N-terminal presequences, but instead have internal sequences that are recognized as targeting motifs. The nuclear-encoded subunits of complexes that contain additional mtDNA-encoded subunits (complexes I, II and IV and the ATP synthase) are imported initially into the matrix and then inserted from the matrix into the inner membrane together with the mitochondrially encoded subunits via a fourth protein translocase complex *OXA1*.

Matrix proteins are threaded through TIM23 as unfolded polypeptides. The driving force for the translocation of the positively charged N-terminal presequence across the inner membrane is $\Delta\psi_m$. Once the presequence emerges into the matrix, it is recognized and bound by a mitochondrial heat-shock protein mt-Hsp70, associated with the TIM23 complex. Hsp-70 is an ATPase possessing a peptide binding pocket that is open in the ATP conformation and tightly binding after the ATP is hydrolysed to ADP. An additional matrix protein, Mge1p, releases Hsp-70 from the TIM23 complex, allowing Hsp-70 to diffuse away from the membrane, pulling more of the peptide sequence through the translocase. Further, Hsp-70 binds to TIM23 and clamps on to the emerging peptide and the 'ratchet' process is repeated until the protein is fully imported. It is clearly important that TIM23 does not allow protons to leak across the membrane, thus the channel formed by a TIM23 dimer is closed until opened by interaction with the imported protein presequence.

The import of inner membrane carriers such as ANT poses the problem of how to insert a 6-transmembrane protein into the inner membrane. The carrier precursors are synthesized in the cytoplasm without cleavable presequences. However, each of the three matrix loops of the incorporated protein contains a conserved 'signature' sequence, which is recognized by a component (Tom70) of the outer membrane TOM complex and is distinct from the Tom20 receptor recognizing matrix-targeted protein presequences. The carrier protein is partially translocated through the outer membrane general insertion pore, but remains attached to the TOM complex at its C-terminal. Translocation occurs at contact sites between outer and inner membranes, and involves an association between TOM, the TIM22 complex in the inner membrane and soluble intermembrane space proteins termed Tim9 and Tim10. These two latter are Zn-finger proteins that recognize and bind the 'carrier signature' sequences and assist in the folding of the carrier into its three-module tertiary structure. The TIM22 complex then catalyses the insertion of the folded carrier into the inner membrane by a $\Delta\psi_m$-dependent process.

The import of outer membrane and intermembrane space proteins is arrested at appropriate stages in the above pathways. The components of the OXA1 complex for the insertion into the inner membrane of nuclear-encoded and mitochondrially encoded peptides from the matrix are currently incompletely characterized. It is unclear whether the latter are inserted cotranslationally or after synthesis and binding to matrix chaperones.

9.8 MITOCHONDRIAL GENETIC DISEASES

Reviews   Inoue *et al.* 2000, Schon 2000

A single cell can contain hundreds of mitochondria and each mitochondria contains multiple copies of mtDNA. It follows that a mtDNA mutation need not be uniform, but can occur in a subset of genomes within a cell or organelle ('heteroplasmy'). Thus the severity of a bioenergetic defect resulting from a pathogenic mutation can differ between organs and between individuals, and indeed will only produce a phenotype when a critical threshold (which can differ between cell types) is exceeded. The proportion of mutant mtDNA can also vary with time, depending on the effect a mutant subpopulation of mtDNA has upon the replication kinetics of their mitochondria. Finally, mitochondria are maternally inherited.

Table 9.1 lists some of the major mitochondria genetic diseases. The 4977-bp 'common deletion' excises the mtDNA between genes A8 and ND5 (Fig. 9.6) and thus affects genes encoding seven respiratory chain polypeptides and five of the 22 tRNAs necessary for mitochondrial protein synthesis. MELAS, MERRF and related mutations causes a generalized loss of mitochondrial function by mutating essential tRNAs, while NARP and LHON are caused by specific point mutations in, respectively, complex I and F_0.

The distinctive pathology associated with these diseases relate to the segregation of the mutant mitochondria during development and ageing, and the threshold beyond which a genetic deficit produces a bioenergetic phenotype. Many of these diseases manifest themselves as encephalomyopathies, i.e. in brain and muscle, perhaps reflecting the high peak bioenergetic demands placed upon these organs. A recent advance has been the generation of a mouse model bearing mutant mtDNA that had been introduced into fertilized eggs from cybrids (Section 9.8.1). The mutant mtDNA was transmitted maternally and induced mitochondrial dysfunction in different tissues.

9.8.1 Rho-0 cells and cybrids

Reviews   Inoue *et al*. 1997, Buchet and Godinot 1998, Porteous *et al*. 1998

Ethidium bromide intercalates into DNA and prevents replication. Since ethidium is a membrane-permeant cation, it accumulates to much higher concentrations within the mitochondrial matrix than the nucleus and can thus be used selectively to inhibit mtDNA replication. Over a period of time the turnover of proteins creates a population of 'ghost' mitochondria, lacking the 13 mitochondrially encoded subunits but still maintaining the overall mitochondrial morphology. The mitochondria within these rho-0 (ρ-0) cells lack a functional respiratory chain or ATP synthase. Interestingly, they still maintain a (suboptimal) $\Delta\psi_m$ owing to the electrogenicity of the adenine nucleotide translocator, ANT (Section 8.5). Glycolytic ATP enters through the ANT and is hydrolysed by F_1 (which is still functional in the direction of ATP hydrolysis, but no longer correctly associated with F_0, owing to the absence of mitochondrially encoded subunits A6 and A8 of F_0). The efflux of ADP together with P_i is associated with the efflux of one H^+ by the reversal of the process shown in Fig. 8.5.

Rho-0 cells are the starting point for the construction of hybrid cells containing 'foreign' mitochondria. These '*cybrids*' are created by fusing the ρ-0 cells with enucleated fibroblasts or platelets or isolated nerve terminals (synaptosomes), which contain mitochondria but lack nDNA. Since mitochondrial mutations are (fortunately) rare, cybrid technology provides a powerful tool to examine the bioenergetic consequences of mitochondrial mutations, as well as less well-defined changes in mitochondrial function in neurodegenerative diseases such as Alzheimer's and Parkinson's, in robust cell lines rather than rare biopsy samples. For example, incorporation of mitochondria containing the common deletion (Section 9.6) into a cell line created cybrids with varying percentages of the truncated mtDNA. Cybrids containing 50% or more normal mtDNA showed no bioenergetic deficit, but increasing the proportion of truncated mtDNA led to a progressively severe decrease in $\Delta\psi_m$ and ATP synthesis.

9.9 MITOCHONDRIAL INVOLVEMENT IN NEURODEGENERATIVE DISEASES

Review   Schapira 1999

In addition to the mitochondrial mutations discussed in Section 9.8, there is increasing evidence that some of the most common and debilitating neurodegenerative disorders are associated in some way with mitochondrial dysfunction, although cause and effect remains to be established in most cases. The general principle is that even a slight restriction in respiratory chain activity may substantially increase the chances of neuronal cell death; however, much remains to be discovered before the link between the bioenergetic deficiency and the pattern of neuronal susceptibility can be established.

9.9.1 Parkinson's disease

Reviews and further reading   Davis et al. 1979, Langston et al. 1983, Nicklas et al. 1985, Schapira 1999, Betarbet et al. 2000

The lifetime risk of developing Parkinson's disease (PD) is about 1 in 40. The main clinical characteristics, slowing of movement (bradykinesia), rigidity and tremor, can be ascribed to the loss of dopaminergic neurons with their cell bodies in the substantia nigra and projecting into the striatum. The causes of PD are unknown; in some instances there is evidence for a genetic component, while chronic exposure to environmental toxins has also been proposed as a contributing factor (see below).

   The first indication of a mitochondrial component to the disease came with the fortuitous discovery (for the scientists but not the victims) that an illegal pethidine 'designer drug' analogue induced Parkinsonism in its users. An impurity in the drug, MPTP, was oxidized by monoamine oxidase-B in glial cells to MPP$^+$ (1-methyl-4-phenyl-pyridinium). MPP$^+$ is transported by the dopamine transporter in the plasma membranes of surrounding dopaminergic neurons and behaves in the cells as a lipophilic cation, being accumulated into the mitochondrial matrix in response to $\Delta\psi_m$. In inverted submitochondrial particles, MPP$^+$ is a weak inhibitor of complex I with a K_i of about 10^{-4}M; however, the combined concentrative capacity of the plasma membrane transporter and the mitochondrial membrane potential implies that nanomolar concentrations of extracellular MPP$^+$ would be toxic in vivo. The specificity of MPP$^+$ for dopaminergic neurons can be ascribed to the selective uptake into those neurons via the dopamine transporter; however, careful titration of the systemic complex I inhibitor rotenone, administered in vivo chronically to rats, also leads to a selective degeneration of dopaminergic neurons in the substantia nigra and their terminals in the striatum. It thus appears that these neurons are the most sensitive to complex I inhibition. It should be emphasized that inhibition in these disease models need not be total for neuronal death to occur; indeed, maximal complex I inhibition results in rapid cell death. In contrast, neurons may survive for prolonged periods with a partial block of complex I (20–30%) until faced with a maximal energy demand, when neurons with mitochondria that are unable to meet this demand would succumb. Excitotoxicity resulting from pathological release of glutamate (Section 9.5.2) is potentiated in energy-compromised neurons, as is the generation of reactive oxygen species (Chapter 5). The selective susceptibility of dopaminergic neurons to systemic complex I inhibition may be due to the oxidative stress associated with dopamine synthesis and metabolism.

There is good evidence for a selective partial deficiency of complex I activity in the substantia nigra, but not in other brain areas, in some PD patients. Paradoxically, complex I activity is also reduced in blood platelets from these patients and this finding has been used for the construction of cybrids (Section 9.8.1) in order to determine whether this is caused by a mtDNA defect. If the complex I deficiency were due to nuclear gene expression changes or post-translational modification, it would not survive in the cybrids. An average 25% reduction in complex I activity in the cybrids was found. Individual cells showed variable staining with a complex I monoclonal antibody suggesting that the mtDNA defect is heteroplasmic. This study does not establish a genetic basis for the deficiency, since it could also have arisen somatically.

9.9.2 Huntington's disease

Reviews and further reading   Gu *et al.* 1996, Ferrante *et al.* 1997, Onodera *et al.* 1997, Beal 1999, Perutz 1999, Schapira 1999, Beal 2000

Huntington's disease (HD) is an autosomal dominant disease caused by an abnormal CAG expansion in the gene encoding huntingtin, a ubiquitously expressed nuclear-encoded 350 kDa protein of unknown function. A chain of less than 38 glutamines (encoded by CAG) gives no pathology, while above 40 amino acids the huntingtin begins to aggregate, and this is associated with the symptoms of HD (ataxia, uncontrolled movement (chorea) and dementia), the severity of which increases with the length of the extension. Gamma-aminobutyrate (GABA)ergic neurons of the caudate nucleus of the neostriatum are most affected in HD. Although huntingtin is not a mitochondrially targeted protein (indeed, the abnormal aggregated form appears to form aggregates within the nucleus and cytoplasm), HD is associated with a major deficiency of complex II/III activity in the caudate nucleus and a severe reduction in aconitase activity. This last is diagnostic of oxidative stress, since the citric acid cycle Fe/S protein is highly susceptible to inhibition by $O_2^{\bullet-}$ and other reactive oxygen species.

The importance of the complex II deficiency has been demonstrated in *in vivo* studies where administration of 3-nitropropionic acid, an irreversible inhibitor of succinate dehydrogenase, or chronic administration of the reversible inhibitor malonate or its more permeant ester methylmalonate, produced behavioural changes and pathology in primates similar to HD.

9.9.3 Friedreich's ataxia

Reviews and further reading   Delatycki *et al.* 2000, Gordon 2000, Puccio *et al.* 2001

Friedreich's ataxia (FRDA) is an autosomal recessive mitochondrial disorder caused by decreased expression of frataxin, a mitochondrial protein whose function is unknown, as a result of an abnormal GAA expansion in intron 1 of the gene. It is characterized by degeneration of large sensory neurons, cardiomyopathy and an increased incidence in diabetes. Yeast knockout models accumulate Fe within their mitochondria and yet show decreased synthesis of Fe/S proteins, producing deficiencies in complexes I–III and aconitase. The overload of the mitochondria with free Fe leads to an increased sensitivity to oxidative stress. A neuronal frataxin-deficient mouse model indicates that the Fe accumulation occurs after the onset of pathology and inactivation of the Fe/S enzymes and so is unlikely to be the primary event.

9.9.4 Alzheimer's disease

Reviews and further reading   Huang *et al.* 1999, Kish *et al.* 1999, Beal 2000, Blass *et al.* 2000, Hirai *et al.* 2001, Valla *et al.* 2001

Alzheimer's disease (AD) is the most prevalent neurodegenerative disorder and is the subject of a vast body of current research. It must be emphasized that no generally accepted mechanism has emerged to date and all conclusions are tentative. Here we shall restrict discussion to the circumstantial evidence for the involvement of mitochondrial dysfunction, either as a contributing factor or a consequence of the pathology of the disease.

Decreased cerebral metabolism has been detected by positron emission tomography (PET) in patients at risk for AD before the development of clinical symptoms. The possibility of a decreased complex IV activity in AD has been intensively investigated since a report, later disproved, that complex IV mutations occurred with higher frequencies in AD patients than the population at large. Despite this, there is general agreement that complex IV activity may be selectively reduced; thus, in one recent study, up to a 39% reduction in complex IV activity was detected in a specific region of the cerebral cortex, layer 1 of the posterior cingulate, that showed a large *in vivo* decrease by PET. No change in succinate dehydrogenase was detected, ruling out a generalized decrease in mitochondrial function. However, it must be emphasized that such a study does not establish that the complex IV deficiency is causative in the progression of the disease. Thus both mtDNA- and nDNA-encoded complex IV subunits are reduced in AD, as well as some other neurodegenerative disorders, and an alternative view is that this may simply reflect a non-specific decrease in the degenerating brain.

The amyloid precursor protein (APP) family of integral membrane glycoproteins are cell adhesion and/or neurite growth proteins whose extracellular domains may be cleaved by secretase enzymes, liberating β-amyloid (Aβ) peptides. Aggregated Aβ is the major component of the plaques that are the most characteristic features of AD. In the presence of Fe(III) or Cu(II), Aβ generates H_2O_2 and thus contributes to oxidative stress. Addition of aggregated Aβ to isolated nerve terminals (synaptosomes) or neurons results in an increase in Ca^{2+} entry, oxidative stress and mitochondrial dysfunction. It is unclear whether this occurs *in vivo*.

REFERENCES

Abrahams, J.P., Leslie, A.G., Lutter, R. and Walker, J.E. (1994) Structure at 2.8 Å resolution of F_1-ATPase from bovine heart mitochondria. *Nature* **370**, 621–628.

Abramson, J., Riistama, S., Larsson, G., Jasaitis, A., Svensson, E., Laakkonen, L., Puustinen, A., Iwata, S. and Wikstrom, M. (2000) The structure of the ubiquinol oxidase from *Escherichia coli* and its ubiquinone binding site. *Nature Struct. Biol.* **7**, 910–917.

Abramson, J., Svensson, E., Byrne, B. and Iwata, S. (2001) Structure of cytochrome *c* oxidase: a comparison of the bacterial and mitochondrial enzymes. *Biochim. Biophys. Acta* **1544**, 1–9.

Affourtit, C., Krab, K. and Moore, A.L. (2001) Control of plant mitochondrial respiration. *Biochim. Biophys. Acta* **1504**, 58–69.

Albracht, S.P. and Hedderich, R. (2000) Learning from hydrogenases: location of a proton pump and of a second FMN in bovine NADH–ubiquinone oxidoreductase (Complex I). *FEBS Lett.* **485**, 1–6.

Allen, J.F. and Forsberg, J. (2001) Molecular recognition in thylakoid structure and function. *Trends Plant Sci.* **6**, 317–326.

Anthony, C. (2000) Methanol dehydrogenase, a PQQ-containing quinoprotein dehydrogenase. *Subcell. Biochem.* **35**, 73–117.

Arechaga, I. and Jones, P.C. (2001) The rotor in the membrane of the ATP synthase and relatives. *FEBS Lett.* **494**, 1–5.

Arsenijevic, D., Onuma, H., Pecqueur, C., Raimbault, S., Manning, B.S., Miroux, B., Couplan, E., Alves-Guerra, M.C., Goubern, M., Surwit, R., Bouillaud, F., Richard, D., Collins, S. and Ricquier, D. (2000) Disruption of the uncoupling protein-2 gene in mice reveals a role in immunity and reactive oxygen species production. *Nature Genet.* **26**, 435–439.

Babcock, G.T. (1999) How oxygen is activated and reduced in respiration. *Proc. Natl Acad. Sci. USA* **96**, 12971–12973.

Bader, M., Muse, W., Ballou, D.P., Gassner, C. and Bardwell, J.C. (1999) Oxidative protein folding is driven by the electron transport system. *Cell* **98**, 217–227.

Baker, S.C., Ferguson, S.J., Ludwig, B., Pages, M.D., Richter, O.H. and van Spanning, R.J.M. (1998) Molecular genetics of the genus Paracoccus: metabolically versatile bacteria with bioenergetic flexibility. *Microbiol. Mol. Biol. Rev.* **62**, 1046–1078.

Barja, G. (1999) Mitochondrial oxygen radical generation and leak: sites of production in state 4 and 3, organ specificity, and relation to aging and longevity. *J. Bioenerg. Biomembr.* **31**, 347–366.

Barker, P.D. and Ferguson, S.J. (1999) Still a puzzle: why is haem covalently attached in c-type cytochromes? *Struct. Fold. Des.* **7**, R281–R290.

Bauer, M.F., Hofmann, S., Neupert, W. and Brunner, M. (2000) Protein translocation into mitochondria: the role of TIM complexes. *Trends Cell Biol.* **10**, 25–31.

Beal, M.F. (1999) Mitochondria, NO and neurodegeneration. In: *Mitochondria in the Life and Death of the Cell* (Brown, G.C., Nicholls, D.G. and Cooper, C., eds), Portland Press, London, pp. 43–54.

Beal, M.F. (2000) Energetics in the pathogenesis of neurodegenerative diseases. *Trends Neurosci.* **23**, 298–304.

Behr, J., Michel, H., Mantele, W. and Hellwig, P. (2000) Functional properties of the heme propionates in cytochrome *c* oxidase from *Paracoccus denitrificans*. Evidence from FTIR difference spectroscopy and site-directed mutagenesis. *Biochemistry* **39**, 1356–1363.

Beja, O., Aravind, L., Koonin, E.V., Suzuki, M.T., Hadd, A., Nguyen, L.P., Jovanovich, S.B., Gates, C.M., Feldman, R.A., Spudich, J.L., Spudich, E.N. and DeLong, E.F. (2000) Bacterial rhodopsin: evidence for a new type of phototrophy in the sea. *Science* **289**, 1902–1906.

Berdall, D.S. and Marasse R.S. (1995) Cyclic photophosphorylation and electron transport. *Biochim. Biophys. Acta* **1229**, 22–38.

Berks, B.C., Ferguson, S.J., Moir, J.W. and Richardson, D.J. (1995) Enzymes and associated electron transport systems that catalyse the respiratory reduction of nitrogen oxides and oxyanions. *Biochim. Biophys. Acta* **1232**, 97–173.

Berks, B.C., Sargent, F. and Palmer, T. (2000) The Tat protein export pathway. *Mol. Microbiol.* **35**, 260–274.

Bernardi, P., Basso, E., Colonna, R., Costantini, P., Di Lisa, F., Eriksson, O., Fontaine, E., Forte, M., Ichas, F., Massari, S., Nicolli, A., Petronilli, V. and Scorrano, L. (1998) Perspectives on the mitochondrial permeability transition. *Biochim. Biophys. Acta* **1365**, 200–206.

Bernardi, P., Petronilli, V., Di Lisa, F. and Forte, M. (2001) A mitochondrial perspective on cell death. *Trends Biochem. Sci.* **26**, 112–117.

Berry, E.A., Huang, L.S., Zhang, Z. and Kim, S.H. (1999) Structure of the avian mitochondrial cytochrome bc1 complex. *J. Bioenerg. Biomembr.* **31**, 177–190.

Berry, E.A., Guergova, K., Huang, L.S. and Crofts, A.R. (2000) Structure and function of cytochrome bc complexes. *Annu. Rev. Biochem.* **69**, 1005–1075.

Berry, R.M. and Armitage, J.P. (1999) The bacterial flagella motor. *Adv. Microb. Physiol.* **41**, 291–337.

Betarbet, R., Sherer, T.B., MacKenzie, G., Garcia-Osuna, M., Panov, A.V. and Greenamyre, J.T. (2000) Chronic systemic pesticide exposure reproduces features of Parkinson's disease. *Nature Neurosci.* **3**, 1301–1306.

Bibby, T.S., Nield, J. and Barber, J. (2001) Iron deficiency induces the formation of an antenna ring around trimeric photosystem I in cyanobacteri. *Nature* **412**, 743–745.

Blake, R.C., Shute, E.A., Greenwood, M.M., Spencer, G.H. and Ingledew, W.J. (1993) Enzymes of aerobic respiration on iron. *FEMS Microbiol. Rev.* **11**, 9–18.

Blass, J.P., Sheu, R.K. and Gibson, G.E. (2000) Inherent abnormalities in energy metabolism in Alzheimer disease. Interaction with cerebrovascular compromise. *Ann. NY Acad. Sci.* **903**, 204–221.

Boekema, E.J., Hifney, A., Yakushevska, A.E., Piotrowski, M., Keegstra, W., Berry, S., Michael, K-P., Pistorius, E.K. and Krulp, J. (2001) A giant chlorophyll–protein complex is induced by iron-deficiency in cyanobacteria. *Nature* **412**, 745–747.

Bouillaud, F., Couplan, E., Pecqueur, C. and Ricquier, D. (2001) Homologues of the uncoupling protein from brown adipose tissue (UCP1): UCP2, UCP3, BMCP1 and UCP4. *Biochim. Biophys. Acta* **1504**, 107–119.

Boveris, A., Costa, L.E., Poderoso, J.J., Carreras, M.C. and Cadenas, E. (2000) Regulation of mitochondrial respiration by oxygen and nitric oxide. *Ann. NY Acad. Sci.* **899**, 121–135.

Boyer, P.D. (1997) The ATP synthase – a splendid molecular machine. *Annu. Rev. Biochem.* **66**, 717–749.

Brand, M.D. (1985) The stoichiometry of the exchange catalysed by the mitochondrial calcium/sodium antiporter. *Biochem. J.* **229**, 161–166.

Brand, M.D., Reynafarje, B. and Lehninger, A.L. (1976) Re-evaluation of the H^+/site ratio of mitochondrial electron transport with the oxygen pulse technique. *J. Biol. Chem.* **251**, 5670–5679.

oxidase subunits in Alzheimer's disease and in hereditary spinocerebellar ataxia disorders: a nonspecific change? *J. Neurochem.* **72**, 700–707.

Kobayashi, T. and Ito, K. (1999) Respiratory chain strongly oxidizes the CXXC motif of DsbB in the *Escherichia coli* disulfide bond formation pathway. *EMBO J.* **18**, 1192–1198.

Koehler, C.M., Merchant, S. and Schatz, G. (1999) How membrane proteins travel across the mitochondrial intermembrane space. *Trends Biochem. Sci.* **24**, 428–432.

Kojima, S. and Blair, D.F. (2001) Conformational changes in the stator of the bacterial flagellar motor. *Biochemistry* **40**, 13041–13050.

Kolbe, M., Besir, H., Essen, L.O. and Oesterhelt, D. (2000) Structure of the light-driven chloride pump halorhodopsin at 1.8 Å resolution. *Science* **288**, 1390–1396.

Konings, W.N., Lolkema, J.S. and Poolman, B. (1995) The generation of metabolic energy by solute transport. *Arch. Microbiol.* **164**, 235–242.

Koronakis, V., Sharff, A., Koronakis, E., Luisi, B. and Hughes, C. (2000) Crystal structure of the bacterial membrane protein TolC central to multidrug efflux and protein export. *Nature* **405**, 914–919.

Kowaltowski, A.J., Seetharaman, S., Paucek, P. and Garlid, K.D. (2001) Bioenergetic consequences of opening the ATP-sensitive K^+ channel of heart mitochondria. *Am. J. Physiol. Heart Circ. Physiol.* **280**, H649–H657.

Krajewski, S., Krajewska, M., Ellerby, L.M., Welsh, K., Xie, Z.H., Deveraux, Q.L., Salvesen, G.S., Bredesen, D.E., Rosenthal, R.E., Fiskum, G. and Reed, J.C. (1999) Release of caspase-9 from mitochondria during neuronal apoptosis and cerebral ischemia. *Proc. Natl Acad. Sci. USA* **96**, 5752–5757.

Krulwich, T.A. (1995) Alkaliphiles: 'basic' molecular problems of pH tolerance and bioenergetics. *Mol. Microbiol.* **15**, 403–410.

Kuhlbrandt, W., Wang, D.N. and Fujiyoshi, Y. (1994) Atomic model of plant light-harvesting complex by electron crystallography. *Nature* **367**, 614–621.

Lancaster, C.R., Kroger, A., Auer, M. and Michel, H. (1999) Structure of fumarate reductase from *Wolinella succinogenes* at 2.2 Å resolution. *Nature* **402**, 377–385.

Langston, J.W., Ballard, P., Tetrud, J.W. and Irwin, I. (1983) Chronic Parkinsonism in humans due to a product of meperidine-analog synthesis. *Science* **219**, 979–980.

Lanyi, J.K. (2000) Special issue: Bacteriorhodopsin. *Biochim. Biophys. Acta* **1460**, 1–239.

Lanyi, J.K. and Luecke, H. (2001) Bacteriorhodopsin. *Curr. Opin. Struct. Biol.* **11**, 415–419.

Li, W., Shariat-Madar, Z., Powers, M., Sun, X., Lane, R.D. and Garlid, K.D. (1992) Reconstitution, identification, purification, and immunological characterization of the 110-kDa Na^+/Ca^{2+} antiporter from beef heart mitochondria. *J. Biol. Chem.* **267**, 17983–17989.

Liu, X., Kim, C.N., Yang, J., Jemmerson, R. and Wang, X. (1996) Induction of apoptotic program in cell-free extracts: requirement for dATP and cytochrome *c. Cell* **86**, 147–157.

Lombardi, A., Lanni, A., Moreno, M., Brand, M.D. and Goglia, F. (1998) Effect of 3,5-di-iodo-L-thyronine on the mitochondrial energy-transduction apparatus. *Biochem. J.* **330**, 521–526.

Mannella, C.A. (1998) Conformational changes in the mitochondrial channel protein, VDAC, and their functional implications. *J. Struct. Biol.* **121**, 207–218.

Manting, E.H. and Driessen, A.J.M. (2000) *Escherichia coli* translocase: the unravelling of a molecular machine. *Mol. Microbiol.* **37**, 226–238.

Marger, M.D. and Saier, M.H. (1993) A major superfamily of transmembrane facilitators that catalyse uniport, symport and antiport. *Trends Biochem. Sci.* **18**, 13–20.

Mather, M. and Rottenberg, H. (2000) Aging enhances the activation of the permeability transition pore in mitochondria. *Biochem. Biophys. Res. Commun.* **273**, 603–608.

Menz, R.I., Walker, J.E. and Leslie, A.G. (2001) Structure of bovine mitochondrial F1-ATPase with nucleotide bound to all three catalytic sites: implications for the mechanism of rotary catalysis. *Cell* **106**, 331–341.

Mitchell, P. and Moyle, J. (1967) Respiration-driven proton translocation in rat liver mitochondria. *Biochem. J.* **105**, 1147–1162.

Mitchell, P. and Moyle, J. (1969a) Estimation of membrane potential and pH difference across the cristae membrane of rat liver mitochondria. *Eur. J. Biochem.* **7**, 471–484.

Mitchell, P. and Moyle, J. (1969b) Translocation of some anions cations and acids in rat liver mitochondria. *Eur. J. Biochem.* **9**, 149–155.

Moncada, S. and Erusalimsky, J.D. (2002) Does nitric oxide modulate mitochondrial energy generation and apoptosis. *Nature Mol. Cell. Biol. Rev.* **3**, 214–220.

Mootha, V.K., French, S. and Balaban, R.S. (1996) Neutral carrier-based 'Ca^{2+}-selective' microelectrodes for the measurement of tetraphenylphosphonium. *Analyt. Biochem.* **236**, 327–330.

Murphy, M.P. (2001) How understanding the control of energy metabolism can help investigation of mitochondrial dysfunction, regulation and pharmacology. *Biochim. Biophys. Acta* **1504**, 1–11.

Nakamoto, R.K., Ketchum, C.J. and al Shawi, M.K. (1999) Rotational coupling in the F0F1 ATP synthase. *Annu. Rev. Biophys. Biomol. Struct.* **28**, 205–234.

Neupert, W. (1997) Protein import into mitochondria. *Annu. Rev. Biochem.* **66**, 863–917.

Nicholls, D.G. (1974) The influence of respiration and ATP hydrolysis on the proton electrochemical potential gradient across the inner membrane of rat liver mitochondria as determined by ion distribution. *Eur. J. Biochem.* **50**, 305–315.

Nicholls, D.G. and Åkerman, K.E.O. (1982) Mitochondrial calcium transport. *Biochim. Biophys. Acta* **683**, 57–88.

Nicholls, D.G. and Budd, S.L. (2000) Mitochondria and neuronal survival. *Physiol. Rev.* **80**, 315–360.

Nicholls, D.G. and Locke, R.M. (1984) Thermogenic mechanisms in brown fat. *Physiol. Rev.* **64**, 1–64.

Nicholls, D.G. and Rial, E. (1999) A history of UCP1: the first uncoupling protein. *J. Bioenerg. Biomembr.* **31**, 399–406.

Nicholls, D.G. and Ward, M.W. (2000) Mitochondrial membrane potential and cell death: mortality and millivolts. *Trends Neurosci.* **23**, 166–174.

Nicholls, D.G., Grav, H.J. and Lindberg, O. (1972) Mitochondria from brown adipose tissue: regulation of respiration in vitro by variations in volume of the matrix compartment. *Eur. J. Biochem.* **31**, 526–533.

Nicklas, W.J., Vyas, I. and Heikkila, R.E. (1985) Inhibition of NADH-linked oxidation in brain mitochondria by 1-methyl-4-phenyl-pyridine, a metabolite of the neurotoxin, 1-methyl-4-phenyl-1,2,5,6-tetrahydropyridine. *Life Sci.* **36**, 2503–2508.

Noji, H., Yasuda, R., Yoshida, M. and Kinosita, K. (1997) Direct observation of the rotation of F$_1$-ATPase. *Nature* **386**, 299–302.

Ohnishi, T., Moser, C.C., Page, C.C., Dutton, P.L. and Yano, T. (2000) Simple redox-linked proton-transfer design: new insights from structures of quinol–fumarate reductase. *Struct. Fold. Des.* **8**, R23–R32.

Onodera, O., Burke, J.R., Miller, S.E., Hester, S., Tsuji, S., Roses, A.D. and Strittmatter, W.J. (1997) Oligomerization of expanded-polyglutamine domain fluorescent fusion proteins in cultured mammalian cells. *Biochem. Biophys. Res. Commun.* **238**, 599–605.

Page, C.C., Moser, C.C., Chen, X.X. and Dutton, P.L. (1999) Natural engineering principles of electron tunnelling in biological oxidation–reduction. *Nature* **402**, 47–52.

Palmieri, F., Bisaccia, F., Capobianco, L., Dolce, V., Fiermonte, G., Iacobazzi, V., Indiveri, C. and Palmieri, L. (1996) Mitochondrial metabolite transporters. *Biochim. Biophys. Acta* **1275**, 127–132.

Palmieri, F.M., Lasorsa, F.M., Vozza, A., Agrimi, G., Fiermonte, G., Runswick, M.J., Walker, J.E. and Palmieri, F. (2000) Identification and functions of new transporters in yeast mitochondria. *Biochim. Biophys. Acta Bio-Energet.* **1459**, 363–369.

Pecqueur, C., Alves-Guerra, M.C., Gelly, C., Lévi-Meyrueis, C., Couplan, E., Collins, S., Ricquier, D., Bouillaud, F. and Miroux, B. (2001) Uncoupling protein 2, *in vivo* distribution, induction upon oxidative stress, and evidence for translational regulation. *J. Biol. Chem.* **276**, 8705–8712.

Perez, J.A. and Ferguson, S.J. (1990) Kinetics of oxidative phosphorylation in *Paracoccus denitrificans*. 1. Mechanism of ATP synthesis at the active sites of $F_0 F_1$ ATPase. *Biochemistry* **29**, 10503–10518.

Perutz, M.F. (1999) Glutamine repeats and neurodegenerative diseases: molecular aspects. *Trends Biochem. Sci.* **24**, 58–63.

Poole, R.K. and Cook, G.M. (2000) Redundancy of aerobic respiratory chains in bacteria? Routes, reasons and regulation. *Adv. Microb. Physiol.* **43**, 165–224.

Porteous, W.K., James, A.M., Sheard, P.W., Porteous, C.M., Packer, M.A., Hyslop, S.J., Melton, J.V., Pang, C.Y., Wei, Y.H. and Murphy, M.P. (1998) Bioenergetic consequences of accumulating the common 4977-bp mitochondrial DNA deletion. *Eur. J. Biochem.* **257**, 192–201.

Postma, P.W., Lengeler, J.W. and Jacobson, G.R. (1993) Phosphoenolpyruvate : carbohydrate phosphotransferase systems of bacteria. *Microbiol. Rev.* **57**, 543–594.

Proshlyakov, D.A., Pressler, M.A., DeMaso, C., Leykam, J.F., DeWitt, D.L., and Babcock, G.T. (2000) Oxygen activation and reduction in respiration: involvement of redox-active tyrosine 244. *Science* **290**, 1588–1591.

Puccio, H., Simon, D., Cossée, M., Criqui-Filipe, P., Tiziano, F., Melki, J., Hindelang, C., Matyas, R., Rustin, P. and Koenig, M. (2001) Mouse models for Friedreich ataxia exhibit cardiomyopathy, sensory nerve defect and Fe-S enzyme deficiency followed by intramitochondrial iron deposits. *Nature Genet.* **27**, 181–186.

Putman, M., van Veen, H.W. and Konings, W.N. (2000) Molecular properties of bacterial multidrug transporters. *Microbiol. Mol. Biol. Rev.* **64**, 672–693.

Reynafarje, B., Brand, M.D. and Lehninger, A.L. (1976) Evaluation of the H^+/site ratio of mitochondrial electron transport from rate measurements. *J. Biol. Chem.* **251**, 7442–7451.

Rial, E., Poustie, E.A. and Nicholls, D.G. (1983) Brown adipose tissue mitochondria: the regulation of the 32,000 Mr uncoupling protein by fatty acids and purine nucleotides. *Eur. J. Biochem.* **137**, 197–203.

Richardson, D.J. (2000) Bacterial respiration: a flexible process for a changing environment. *Microbiology* **146**, 551–571.

Rizzuto, R., Bernardi, P. and Pozzan, T. (2000) Mitochondria as all-round players of the calcium game. *J. Physiol. (Lond.)* **529**, 37–47.

Rizzuto, R., Simpson, A.W.M., Brini, M. and Pozzan, T. (1992) Rapid changes of mitochondrial Ca^{2+} revealed by specifically targeted recombinant aequorin. *Nature* **358**, 325–327.

Rohwer, J.M., Meadow, N.D., Roseman, S., Westerhoff, H.V. and Postma, P.W. (2000) Understanding glucose transport by the bacterial phosphoenolpyruvate : glycose phosphotransferase system on the basis of kinetic measurements *in vitro*. *J. Biol. Chem.* **275**, 34909–34921.

Rottenberg, H. and Marbach, M. (1990) The Na^+-independent Ca^{2+} efflux system in mitochondria is a $Ca^{2+}/2H^+$ exchange system. *FEBS Lett.* **274**, 65–68.

Rottenberg, H. and Wu, S.L. (1998) Quantitative assay by flow cytometry of the mitochondrial membrane potential in intact cells. *Biochim. Biophys. Acta* **1404**, 393–404.

Rouhani, S., Cartailler, J-P., Facciotti, M.T., Walain, P., Needelman, R., Lanyi, J.K., Gaeser, R.M. and Luecke, H. (2001) Crystal structure of the D85S mutant of bacteriorhodopsin: model of an O-like intermediate. *J. Mol. Biol.* **313**, 615–628.

Rutherford, A.W. and Faller, P. (2001) The heart of photosynthesis in glorious 3D. *Trends Biochem. Sci.* **26**, 341–344.

Rutter, G.A., Burnett, P., Rizzuto, R., Brini, M., Murgia, M., Pozzan, T., Tavare, J.M. and Denton, R.M. (1996) Subcellular imaging of intramitochondrial Ca^{2+} with recombinant targeted aequorin: significance for the regulation of pyruvate dehydrogenase activity. *Proc. Natl Acad. Sci. USA* **93**, 5489–5494.

Sahin, T. and Kaback, H.R. (2001) Arg-302 facilitates deprotonation of Glu-325 in the transport mechanism of the lactose permease from *Escherichia coli*. *Proc. Natl Acad. Sci. USA* **98**, 6068–6073.

Sahin, T., Karlin, A. and Kaback, H.R. (2000) Unraveling the mechanism of the lactose permease of *Escherichia coli*. *Proc. Natl Acad. Sci. USA* **97**, 10729–10732.

Shäfer, G., Engelhard, M. and Müller, V. (1999) Bioenergetics of the Archaea. *Microbiol. Mol. Biol. Rev.* **63**, 570–620.

Saraste, M. (1999) Oxidative phosphorylation at the *fin de siècle*. *Science* **283**, 1488–1493.

Sazanov, L.A. and Walker, J.E. (2000) Cryo-electron crystallography of two sub-complexes of bovine complex I reveals the relationship between the membrane and peripheral arms. *J. Mol. Biol.* **302**, 455–464.

Scaduto, R.C. and Grotyohann, L.W. (1999) Measurement of mitochondrial membrane potential using fluorescent rhodamine derivatives. *Biophys. J.* **76**, 469–477.

Schafer, F.Q. and Buettner, G.R. (2001) Redox environment of the cell as viewed through the redox state of the glutathione disulfide/glutathione couple. *Free Radic. Biol. Med.* **30**, 1191–1212.

Schapira, A.H.V. (1999) Mitochondrial involvement in Parkinson's disease, Huntington's disease, hereditary spastic paraplegia and Friedreich's ataxia. *Biochim. Biophys. Acta* **1410**, 159–170.

Scheffler, I.E. (1999) *Mitochondria*. Wiley-Liss, New York.

Schnable, P.S. and Wise, R.P. (1998) The molecular basis of cytoplasmic male sterility and fertility restoration. *Trends Plant Sci.* **3**, 175–180.

Schon, E.A. (2000) Mitochondrial genetics and disease. *Trends Biochem. Sci.* **25**, 555–560.

Schultz, B.E. and Chan, S.I. (2001) Structures and proton-pumping strategies of mitochondrial respiratory enzymes. *Annu. Rev. Biophys. Biomol. Struct.* **30**, 23–65.

Sharff, A., Fanutti, C., Shi, J., Calladine, C. and Luisi, B. (2001) The role of the TolC family in protein transport and multidrug efflux. *Eur. J. Biochem.* **268**, 5011–5026.

Shäfer, G., Engelhard, M. and Müller, V. (1999) Bioenergetics of the Archaea. *Microbiol. Mol. Biol. Rev.* **63**, 570–620.

Smith, D.G., Russell, W.C., Ingledew, W.J. and Thirkell, D. (1993) Hydrolysis of urea by *Ureaplasma urealyticum* generates a transmembrane potential with resultant ATP synthesis. *J. Bacteriol.* **175**, 3253–3258.

Soriano, G.M., Ponamarev, M.V., Carrell, C.J., Xia, D., Smith, J.L. and Cramer, W.A. (1999) Comparison of the cytochrome bc1 complex with the anticipated structure of the cytochrome b6f complex. *J. Bioenerg. Biomembr.* **31**, 201–213.

Sparagna, G.C., Gunter, K.K., Sheu, S.S. and Gunter, T.E. (1995) Mitochondrial calcium uptake from physiological-type pulses of calcium. A description of the rapid uptake mode. *J. Biol. Chem.* **270**, 27510–27515.

Stock, D., Leslie, A.G. and Walker, J.E. (1999) Molecular architecture of the rotary motor in ATP synthase. *Science* **286**, 1700–1705.

Stock, D., Gibbons, C., Arechaga, I., Leslie, A.G. and Walker, J.E. (2000) The rotary mechanism of ATP synthase. *Curr. Opin. Struct. Biol.* **10**, 672–679.

Szalai, G., Csordás, G., Hantash, B.M., Thomas, A.P. and Hajnóczky, G. (2000) Calcium signal transmission between ryanodine receptors and mitochondria. *J. Biol. Chem.* **275**, 15305–15313.

Takahashi, A., Camacho, P., Lechleiter, J.D. and Herman, B. (1999) Measurement of intracellular calcium. *Physiol. Rev.* **79**, 1089–1125.

Tanaka, T., Nakamura, H., Nishiyama, A., Hosoi, F., Masutani, H., Wada, H. and Yodoi, J. (2000) Redox regulation by thioredoxin superfamily; protection against oxidative stress and aging. *Free Radic. Res.* **33**, 851–855.

Thauer, R.K. (1998) Biochemistry of methanogenesis: a tribute to Marjory Stephenson. 1998 Marjory Stephenson Prize Lecture. *Microbiology* **144**, 2377–2406.

Tokatlidis, K. and Schatz, G. (1999) Biogenesis of mitochondrial inner membrane proteins. *J. Biol. Chem.* **274**, 35285–35288.

Tsunoda, S.P., Aggeler, R., Yoshida, M. and Capaldi, R.A. (2001a) Rotation of the c subunit oligomer in fully functional F_1F_0 ATP synthase. *Proc. Natl Acad. Sci. USA* **98**, 898–902.

Tsunoda, S.P., Rodgers, A.J., Aggeler, R., Wilce, M.C., Yoshida, M. and Capaldi, R.A. (2001b) Large conformational changes of the epsilon subunit in the bacterial F_1F_0 ATP synthase provide a ratchet action to regulate this rotary motor enzyme. *Proc. Natl Acad. Sci. USA* **98**, 6560–6564.

Turrens, J.F. (1997) Superoxide production by the mitochondrial respiratory chain. *Biosci. Rep.* **17**, 3–8.

Unden, G. and Bongaerts, J. (1997) Alternative respiratory pathways of *Escherichia coli*: energetics and transcriptional regulation in response to electron acceptors. *Biochim. Biophys. Acta* **1320**, 217–234.

Valla, J., Berndt, J.D. and Gonzalez-Lima, F. (2001) Energy hypometabolism in posterior cingulate cortex of Alzheimer's patients: superficial laminar cytochrome oxidase associated with disease duration. *J. Neurosci.* **21**, 4923–4930.

van Brederode, M.E. and Jones, M.R. (2000) Reaction centres of purple bacteria. *Subcell. Biochem.* **35**, 621–676.

Vogel, R., Wiesinger, H., Hamprecht, B. and Dringen, R. (1999) The regeneration of reduced glutathione in rat forebrain mitochondria identifies metabolic pathways providing the NADPH required. *Neurosci. Lett.* **275**, 97–100.

Walker, J.E. (ed.) (2000a) Protonmotive ATPases. *J. Exp. Biol.* **203**, 1–170.

Walker, J.E. (2000b) Special issue: The mechanism of F_1F_0-ATPase. *Biochim. Biophys. Acta* **1458**, 221–510.

Ward, M.W., Rego, A.C., Frenguelli, B.G. and Nicholls, D.G. (2000) Mitochondrial membrane potential and glutamate excitotoxicity in cultured cerebellar granule cells. *J. Neurosci.* **20**, 7208–7219.

White, D. (2001) *The Physiology and Biochemistry of Prokaryotes*. Oxford University Press, New York.

Williams, K.A. (2000) Three-dimensional structure of the ion-coupled transport protein NhaA. *Nature* **403**, 112–115.

Williams, P.A., Fulop, V., Leung, Y-C., Chan, C., Moir, J.W.B., Ferguson, S.J., Radford, S.E. and Hajdu, J. (1995) Pseudospecific docking surfaces on electron transfer proteins as illustrated by pseudoazurin, cytochrome c_{550} and cytochrome cd_1 nitrite reductase. *Nat. Struct. Biol.* **2**, 975–982.

Wingrove, D.E. and Gunter, T.E. (1986) Kinetics of mitochondrial calcium transport II. A kinetic description of the sodium-dependent calcium efflux mechanism of liver mitochondria and inhibition by ruthenium red and by tetraphenylphosphonium. *J. Biol. Chem.* **261**, 15166–15171.

Yagi, T., Yano, T., Di, B. and Matsuno, Y. (1998) Procaryotic complex I (NDH-1), an overview. *Biochim. Biophys. Acta* **1364**, 125–133.

Yasuda, R., Noji, H., Yoshida, M., Kinosita, K. and Itoh, H. (2001) Resolution of distinct rotational substeps by submillisecond kinetic analysis of F_1-ATPase. *Nature* **410**, 898–904.

Yoshida, M., Muneyuki, E. and Hisabori, T. (2001) ATP synthase – a marvellous rotary engine of the cell. *Nature Mol. Cell Biol. Rev.* **2**, 669–677.

Yu, C.A., Xia, D., Kim, H., Deisenhofer, J., Zhang, L., Kachurin, A.M. and Yu, L. (1998) Structural basis of functions of the mitochondrial cytochrome $bc1$ complex. *Biochim. Biophys. Acta* **1365**, 151–158.

Zaslavsky, D. and Gennis, R.B. (2000) Proton pumping by cytochrome oxidase: progress, problems and postulates. *Biochim. Biophys. Acta* **1458**, 164–179.

Zeuthen, T. (2001) How water molecules pass through aquaporins. *Trends Biochem. Sci.* **26**, 77–79.

Zhang, Z., Huang, L., Shulmeister, V.M., Chi, Y.I., Kim, K.K., Hung, L.W., Crofts, A.R., Berry, E.A. and Kim, S.H. (1998) Electron transfer by domain movement in cytochrome bc_1. *Nature* **392**, 677–684.

Zhang, Z., Berry, E.A., Huang, L.S. and Kim, S.H. (2000) Mitochondrial cytochrome bc_1 complex. *Subcell. Biochem.* **35**, 541–580.

Zoccarato, F. and Nicholls, D.G. (1982) The role of phosphate in the regulation of the Ca efflux pathway of liver mitochondria. *Eur. J. Biochem.* **127**, 333–338.

Zouni, A., Witt, H.T., Kern, J., Fromme, P., Krauss, N., Saenger, W. and Orth, P. (2001) Crystal structure of photosystem II from *Synechococcus elongatus* at 3.8 Å resolution. *Nature* **409**, 739–743.

APPENDIX

PROTEIN STRUCTURES

Understanding of the molecular mechanisms that underpin bioenergetics continues to advance as more 3-D structures of proteins are determined. Some of these structures are illustrated in this book (especially in Plates A to K) but many more are now available. These can be accessed on the *Protein Data Bank* (PDB) http://www.rcsb.org. Each structure is given a code. A selection of such codes, together with the proteins they refer to, are given below for proteins that are relevant to the subject matter of this book. Where the protein is not described in the text, brief details of interesting features are noted.

A very valuable compilation of protein structure/function information for many of the metalloproteins discussed in this book can be found in *Handbook of Metalloproteins*, Volumes 1 and 2, edited by A. Messerschmidt, R. Huber, T. Poulos and K. Wieghardt and published by J. Wiley in 2001.

*Indicates periplasmic location in bacteria

Protein	Code	Comments
A. Respiratory electron transfer proteins: (Chapter 5)		
Amine dehydrogenase	1JMX	*A quinohemoprotein with a novel cross-linked active site cysteine tryptophylquinone and three novel thioether cross links. From *Pseudomonas putida*; passes electrons to *c*-type cytochromes or cupredoxins
Azurin	4AZU	*Type 1 copper protein – mainly β sheet structure
Carbon monoxide dehydrogenase (aerobic)	1QJ2	Has active site Mo co-ordinated by 1MCD (molybdopterin-cytosine dinucleotide) and an S-selenylcysteine. Associated with proteins carrying Fe/S centres and FAD. Feeds electrons into a CO-insensitive respiratory chain

Protein	Code	Comments
Cytochrome aa_3 oxidase	1AR1	Two subunits of *Paracoccus denitrificans* enzyme
Cytochrome aa_3 oxidase	1OCC	Bovine heart enzyme oxidized
Cytochrome ba_3 oxidase	1EHK	A distinctive oxidase from the thermophile *Thermus thermophilus*
Cytochrome bc_1 complex	1BCC	Chicken heart
Cytochrome bc_1 complex	1BE3	Bovine heart
Cytochrome bc_1 complex	1BGY	Bovine heart
Cytochrome bc_1 complex	1E2V	Yeast
Cytochrome c	1HRC	Mitochondrial from horse heart
Cytochrome c_2	2CXB	*Rhodobacter sphaeroides*
Cytochrome c_3	2CYR	*A tetra heme c-type cytochrome found in sulfate reducing bacteria
Cytochrome c peroxidase	1EB7	*From *Pseudomonas aeruginosa*; is a diheme c-type cytochrome
	2CYP	A b-type cytochrome from the inter-membrane space of yeast mitochondria
Cytochrome c_{550}	1COT	*Paracoccus denitrificans*
D. lactate dehydrogenase	1FOX	Peripheral membrane protein from *E. coli*
DMSO (dimethylsulphoxide) reductase	1CXS 1DMR 4DMR 1DMS 1EU1	*Has active site Mo co-ordinated by two MGD (molybdopterin-guanosine dinucleotide) molecules, each of which provides two sulfur ligands; other ligands to Mo vary
Formate dehydrogenase	1KQF	See Plate E
Fumarate reductase	1FUM 1QLA 1D4E	See Plate E *A flavo c-type cytochrome enzyme from *Shewanella putrefaciens*
Hydrogenase	2FRV	*From *Desulfovibrio gigas* has a Ni-Fe active site. Donates electrons to cytochrome c_3
Hydroxylamine oxidoreductase	1FGJ	*Trimeric enzyme that catalyses oxidation of hydroxylamine to nitrite. Has eight c-type hemes per polypeptide one of which, with a cross-link to a tyrosine, forms the active site. Found in *Nitrosomonas* bacteria

Protein	Code	Comments
Methanol dehydrogenase	1H4I	*Active site PQQ within an eight bladed β propeller
Methylamine dehydrogenase	2BBK	*Contains two cross-linked and modified tryptophans at active site within a seven bladed β propeller
Methyl-coenzyme M reductase	1MRO	Contains Ni at active site as part of F_{430}
Nitrate reductase (periplasmic-Nap)	2NAP	*Contains active site Mo bound by two MGD molecules
Nitrite reductase with five c-type hemes	1QDB	*Ammonia is reaction product. Unusual CXXCK heme binding motif at active site
Nitrite reductase cytochrome cd_1		*Nitric oxide is reaction product; c-type cytochrome domain accepts electrons from c-type cytochrome or cupredoxins and is α helical; the specialized active site d_1 heme is located within an eight bladed β propeller structure. Differences in structure between the enzymes from *Pseudomonas aeruginosa* and *Paracoccus pantotrophus*; the latter undergoes a remarkable conformational change upon reduction
	1AOF	From *P. pantotrophus* reduced
	1QKS	From *P. pantotrophus* oxidized
	1NIR	From *P. aeruginosa* oxidized
Nitrite-reductase Cu-type	1AFN	*Nitric oxide is reaction product. A trimer. Three type 1 Cu ions act as the electron acceptors from other proteins; three type 2 Cu ions are at the active sites
Nitrous oxide reductase	1QNI 1FWX	*Contains both Cu_A site seen in cytochrome aa_3 oxidase plus a novel active site Cu_Z with 4 Cu bridged by an S atom. The active site is enclosed in a seven-bladed β-propeller structure. From *Pseudomonas nautica* and *P. denitrificans*
Pseudoazurin	3PAZ 8PAZ	*Type 1 copper protein – mainly β sheet structure. Oxidized and reduced forms from *Alcaligenes faecalis*

Protein	Code	Comments
Rusticyanin	1A8Z	*Type 1 Cu protein from *T. ferrooxidans*
Sulfite oxidase	1SOX	Chicken liver from inter-mitochondrial space. Active site Mo co-ordinated by a pterin
TMAO (Trimethylamine-N-oxide) reductase	1TMO	*From *Shewanella massilia*; has active site Mo co-ordinated by two MGD groups
B. Photosynthetic proteins: (Chapter 6) Bacteriorhodopsin	1C3W	
Bacteria photosynthetic reaction centre	1PCR 2PRC	From *R. sphaeroides* From *R. viridis*
Cytochrome-f	1HCZ	Unusual *c*-type cytochrome with β-sheet structure
Halorhodopsin	1E12	
LHCII	Not deposited	Spinach
LH2	1K2U	Circular $\alpha_9\beta_9$ structure *R. acidophilia*
Photosystem 1	1JBO	From *S. elongatus*
Photosystem 2	1FE1	From *S. elongatus*
Plastocyanin	1AG6 7PCY	From spinach
C. ATP synthase: (Chapter 7) Bovine heart F_1	1BMF	The original 2.8 Å structure of the F_1 part
Bovine heart F_1	1E79	Contains the central stalk region originally not seen (2.4 Å)
Bovine heart F_1	1H8E	A form with all catalytic sites occupied
Bovine heart F_1 inhibitor	1GMU	Coiled coil
E. coli: F_1 δ subunit	1ABV	N-terminal domain
Thylakoid F_1	1KMH	Has inhibitor tentoxin bound
Yeast F_1 + c ring	1QO1	3.9 Å resolution but shows ring of 10 c subunits
D. Transport proteins: (Chapter 8) *E. coli* glycerol facilitator	1FX8	A passive facilitation of the cytoplasmic membrane
E. coli Msb protein	1JSQ	First structural information at 4.5 Å resolution for an ABC-type ATPase. A homodimer with each subunit providing six trans-membrane α helices and a nucleotide-binding domain. The helices pack so as to provide a chamber that is exposed to the cytoplasmic half of the bilayer

INDEX

Note: Numbers in *italics* refer to figures and tables.